Abiotic Stress Signaling and Responses in Plants

Abiotic Stress Signaling and Responses in Plants

Editors

Małgorzata Nykiel
Mateusz Labudda
Beata Prabucka
Marta Gietler
Justyna Fidler

MDPI

Basel • Beijing • Wuhan • Barcelona • Belgrade • Novi Sad • Cluj • Manchester

Editors
Małgorzata Nykiel
Department of Biochemistry and Microbiology
Warsaw University of Life Sciences-SGGW
Warsaw
Poland

Mateusz Labudda
Department of Biochemistry and Microbiology
Warsaw University of Life Sciences-SGGW
Warsaw
Poland

Beata Prabucka
Department of Biochemistry and Microbiology
Warsaw University of Life Sciences-SGGW
Warsaw
Poland

Marta Gietler
Department of Biochemistry and Microbiology
Warsaw University of Life Sciences-SGGW
Warsaw
Poland

Justyna Fidler
Department of Biochemistry and Microbiology
Warsaw University of Life Sciences-SGGW
Warsaw
Poland

Editorial Office
MDPI
St. Alban-Anlage 66
4052 Basel, Switzerland

This is a reprint of articles from the Special Issue published online in the open access journal *Plants* (ISSN 2223-7747) (available at: www.mdpi.com/journal/plants/special_issues/Stress_Signaling).

For citation purposes, cite each article independently as indicated on the article page online and as indicated below:

Lastname, A.A.; Lastname, B.B. Article Title. *Journal Name* **Year**, *Volume Number*, Page Range.

ISBN 978-3-0365-9255-8 (Hbk)
ISBN 978-3-0365-9254-1 (PDF)
doi.org/10.3390/books978-3-0365-9254-1

Contents

About the Editors

Małgorzata Nykiel

Malgorzata Nykiel obtained her doctoral degree (PhD) in agricultural sciences (specialization in Agronomy) in 2007. From 1996 to 2008, she worked at the Institute of Plant Breeding and Acclimatization-PIB, where she studied the impact of drought on physiological and biochemical processes in plants. At present, she is a researcher and an associate professor in the Department of Biochemistry and Microbiology, Institute of Biology, Warsaw University of Life Sciences-SGGW (WULS-SGGW), Poland. In her scientific work, she focuses on intracellular signaling in cereal responses to various abiotic factors. She also leads investigation on chemical priming, which contributes to increasing plants' resistance to stress. She has been awarded by the Rector of WULS-SGGW for scientific achievements and received many team awards. She is also member of the Polish Botanical Society.

Mateusz Labudda

Mateusz Labudda received his post-doctoral habilitation degree (DSc) in biological sciences in 2021. He is currently a researcher in the Department of Biochemistry and Microbiology, Institute of Biology, Warsaw University of Life Sciences-SGGW (WULS-SGGW), Poland. His scientific interests focus on the molecular, biochemical, physiological, and structural changes in host plants infested with nematodes, mites, and fungi. He is also interested in the holistic defense responses of plants subjected to heavy metals, drought, and salinity. He has a special interest on the regulation of nitrogen metabolism, including proteolysis and proteome modifications, as well as the biochemistry of oxidative and nitrosative stresses induced by various stress conditions. He has been awarded by the Rector of WULS-SGGW for scientific achievements and received many team and individual awards. He is involved in the activities of national and foreign scientific societies. He is also an Associate Editor for *Frontiers in Plant Science*, *Scientific Reports*, *Acta Agrobotanica*, *Acta Physiologiae Plantarum*, *Plants*, and *International Journal of Molecular Sciences*.

Beata Prabucka

Beata Prabucka received her doctoral degree (PhD) in agricultural sciences (specialization in Plant Biochemistry) in 2003. She is currently a researcher and an associate professor in the Department of Biochemistry and Microbiology, Institute of Biology, Warsaw University of Life Sciences-SGGW (WULS-SGGW), Poland. Her scientific interests include the participation of peptidases in the hydrolysis of storage proteins during the germination of cereal grains, mechanisms of regulation of the expression and activity of cysteine endopeptidases from both the papain and legumain families, and serine endopeptidases in response to biotic stress. Her scientific interests in the area of cysteine endopeptidase activity regulation focus mainly on the participation of phytocystatins in this process. In addition, her scientific interests include the overexpression, purification, and structure of proteins and their post-translational modifications. Dr. Beata Prabucka is a co-author of many scientific publications, for which she has been honored with team awards by the Rector of WULS-SGGW. She is a member of Polish Biochemical Society.

Marta Gietler

Marta Gietler is a scientist and an educator specializing in biochemistry. She earned her PhD degree in Biological Sciences from the Faculty of Agriculture and Biology at the Warsaw University of Life Sciences (WULS-SGGW) in 2017, and she has continued her scientific work at the Department of Biochemistry and Microbiology, Institute of Biology of WULS-SGGW. Dr. Marta Gietler has also engaged in foreign internships to expand her knowledge and experience. She completed a month-long scientific internship at the University of Potsdam in Germany. Throughout her career, Dr. Marta Gietler has received several awards including individual and team awards funded by the Rector of WULS-SGGW for outstanding scientific and organizational achievements. She is a co-author of numerous scientific papers exploring the subject of plants responses to stresses at the protein and biochemical levels. She is also a board member of the Warsaw branch of the Polish Botanical Society.

Justyna Fidler

Justyna Fidler received her PhD in biological sciences in 2017. She is a researcher and lecturer at the Department of Biochemistry and Microbiology, Institute of Biology, Warsaw University of Life Sciences-SGGW (WULS-SGGW), Poland. Her scientific interests include plant molecular biology and biochemistry. In her scientific work, she focuses on the regulation of abscisic acid metabolism and signaling, both in physiological processes and under stress conditions in various plant species, mainly cereals. She also conducts research on chemical priming, aiming at increasing the resistance of plants subjected to abiotic environmental stresses. She is a co-author of many scientific publications, for which she has received awards from the Rector of WULS-SGGW. She is involved as a member of the Polish Botanical Society.

Preface

In the natural world, plants have long been known for their ability to withstand various environmental challenges. These challenges, often manifesting as abiotic stress factors, evoke responses at the molecular, biochemical, and physiological levels. Before a plant undertakes its comprehensive response to such stressors, it initiates an intricate and multi-component signaling system. This Special Issue, entitled "Abiotic Stress Signaling and Responses in Plants", is a compendium of 12 papers, both original research and review, exploring the vast realm of plant signaling in the face of diverse abiotic stressors.

The articles included in this Special Issue provide insights into the world of plants' resistance against environmental factors. They uncover the complex mechanisms of signal transduction cascades and offer potential strategies to enhance plant resilience, resource efficiency, and agricultural productivity. We, as Guest Editors, extend our heartfelt gratitude to the authors who have chosen to contribute their work to this Special Issue. Their efforts enrich our understanding of the remarkable adaptability and response mechanisms of plants to stresses.

Małgorzata Nykiel, Mateusz Labudda, Beata Prabucka, Marta Gietler, and Justyna Fidler
Editors

plants

Editorial

Abiotic Stress Signaling and Responses in Plants

Małgorzata Nykiel *, Marta Gietler , Justyna Fidler , Beata Prabucka and Mateusz Labudda

Department of Biochemistry and Microbiology, Institute of Biology, Warsaw University of Life Sciences-SGGW, Nowoursynowska 159, 02-776 Warsaw, Poland; marta_gietler@sggw.edu.pl (M.G.); justyna_fidler@sggw.edu.pl (J.F.); beata_prabucka@sggw.edu.pl (B.P.); mateusz_labudda@sggw.edu.pl (M.L.)

* Correspondence: malgorzata_nykiel@sggw.edu.pl; Tel.: +48-22-59326-74

1. Introduction

The responses of plants to stress factors are extremely elaborate. These responses occur at multiple levels, ranging from alterations in the molecular processes to structural changes in both the underground and aboveground parts of plants. The development of a comprehensive response to stress by plants is preceded by the activation of an effective system of signals. This Special Issue, under the title "Abiotic Stress Signaling and Responses in Plants", is a miscellany of twelve papers (both original research and reviews) related to plant signaling under various abiotic stress factors.

2. Salt Stress

In the study of Wang et al. [1], the authors delved into the effects of salt stress on plant reproductive structures. After subjecting plants to high salt concentrations, there was a sharp decrease in water potential within these structures, leading to a notable increase in seed abortion. The researchers suspected that this seed failure might be associated with the accumulation of reactive oxygen species (ROS) in the plant's ovules. Further investigation revealed genes responsible for ROS scavenging. Mutations in certain genes, such as iron-dependent superoxide dismutase (FSD2), ascorbate peroxidase (APX4), and three peroxidases (POD) (PER17, PER28, and PER29), resulted in a significant increase in seed failure under normal conditions. This study highlighted the complex relationship between ROS and seed development.

Similarly, in the study by Lu et al. [2], in which the effect of salinity on selected biochemical parameters in *Robinia pseudoacacia* seedlings was investigated, it was observed that NaCl treatment resulted in increased ROS accumulation. At a relatively low NaCl concentration (50 mM), an increase in the activity of antioxidant enzymes was observed, as well as an increase in the expression of genes related to the response to salinity, such as Na^+/H^+ exchanger 1 (*NHX1*) and salt overly sensitive 1 (*SOS1*). In turn, higher NaCl concentrations (100–200 mM) resulted in a decrease in antioxidant activity and the downregulation of the expression of the abovementioned genes, as well as a decrease in photosynthetic parameters and damage to the chloroplast structure. The authors indicated that *Robinia pseudoacacia* seedlings can tolerate low salt concentrations, while higher concentrations cause significant damage and metabolic disorders, resulting in a reduction in plant biomass.

This Special Issue also includes a paper on the Arabidopsis ABSCISIC ACID INTENSIVE 4 (ABI4) transcription factor (TF), which is involved in abscisic acid signaling and the related response to abiotic stresses. Maymon et al. (2022) [3] studied the influence of various factors on the stability of the ABI4 protein using transgenic Arabidopsis plants. It was observed that the level of ABI4 increased significantly in seedlings under NaCl stress. The obtained results indicate that ABAI4 is a highly unstable protein and is degraded by the 26S proteasome. In turn, the phosphorylation of serine 114 catalyzed by MAP kinases was responsible for the stabilization of the ABI4 protein. The authors suggested that the regulation of both gene expression and ABI4 protein levels is essential for the regulation of the activity of this key TF in abscisic acid signaling.

Citation: Nykiel, M.; Gietler, M.; Fidler, J.; Prabucka, B.; Labudda, M. Abiotic Stress Signaling and Responses in Plants. *Plants* **2023**, *12*, 3405. https://doi.org/10.3390/plants12193405

Academic Editor: Antonio Scopa

Received: 19 September 2023
Accepted: 22 September 2023
Published: 27 September 2023

3. Water Stress

The study of Kim and Sung (2023) [4] focused on rice, a crop known for its high water consumption. To address this concern and enhance resource efficiency, the researchers examined the benefits of alternative wetting and drying (AWD) compared to continuous flooding in rice cultivation. AWD proved effective in acquiring soil nitrate, and it led to an abundance of amino acids in shoots, suggesting a rearrangement of amino acid pools to support protein production during the shift from vegetative to reproductive growth. This research suggested the potential advantages of AWD in rice production, emphasizing the need for the further exploration of form-dependent nitrogen metabolism and root development.

In the study of Zhang et al. (2023) [5], the researchers investigated the response of peach seedlings to drought stress and the role of lauric acid (LA). Their findings indicated that LA pretreatment mitigated the negative effects of drought stress on peach seedlings. LA played a role in preserving photosynthetic pigments, preventing the closing of pores, and increasing the photosynthetic rate. Additionally, LA reduced the levels of superoxide anion, hydrogen peroxide and malondialdehyde (MDA) by enhancing the activity of antioxidant enzymes like catalase (CAT), POD, superoxide dismutase (SOD), and ascorbate peroxidase (APX). RNA-Seq analysis revealed that LA affected the expression of genes associated with plant–pathogen interactions, phenylpropanoid biosynthesis, and calcium signaling pathways. These findings provided insights into the molecular mechanisms through which LA enhances drought resistance in peach trees.

The study of Xiong et al. (2022) [6] in maize explored the role of a class-II LBD TF called ZmLBD5 in response to drought stress. ZmLBD5 was found to regulate plant growth and drought tolerance. The overexpression of ZmLBD5 in Arabidopsis increased susceptibility to drought, which was characterized by higher rates of water loss and altered stomatal behavior. Importantly, ZmLBD5 also influenced the level of ROS, with increased activities of antioxidant enzymes like SOD and POD. This research highlighted the significance of ZmLBD5 in the plant's response to drought stress, shedding light on its molecular mechanisms.

Waterlogging is an additional abiotic stress which inhibits aerobic respiration, resulting in the reduction of plant growth and significant yield losses in many plants. Waterlogging causes hypoxic conditions due to the slow diffusion of molecular oxygen in water, leading to various morphological and cellular acclimation responses. In the study of Li et al. (2023) [7] it was shown that waterlogging treatment reduced the levels of chlorophyll, soluble protein, soluble sugars, proline, and MDA in mulberry plants. Additionally, it decreased the activities of enzymes like APX, POD, and CAT. Waterlogging affected the rate of photosynthesis, stomatal conductance, and transpiration rate. It has been shown that the expression of 10,394 genes in total was changed under waterlogging, and photosynthesis-related genes such as cytochrome b6/f complex gene *petC* and photosynthetic electron transport gene *petE* were significantly downregulated. The research was conducted on mulberry cultivars with two propagation methods (cutting and grafting). Cutting groups were found to have a better ability to recover from waterlogging stress than grafted mulberries. These findings provide valuable insights into the underlying mechanisms of dual-method mulberry propagation in response to waterlogging stress and highlight the potential for developing waterlogging-tolerant mulberry cultivars.

4. Varia

Another important environmental stress is grazing, which significantly disturbs the growth and metabolism of plants, both through mechanical damage caused by foraging and trampling and the accumulation of feces. In the experiments conducted by Wang et al. (2022) [8], the influence of different grazing intensities on the transcriptome of *Taraxacum mongolicum* was examined. The greatest changes at the transcriptomic level of grazed plants compared to untreated plants concerned genes related to cell signaling and phytohormones, as well as those related to secondary metabolism and photosynthesis. Furthermore, it was observed that heavy grazing resulted in a more intense response at the transcriptomic level compared to light grazing. The authors indicated that the expression of genes related to secondary metabolism and photosynthesis may contribute to the increased stress tolerance of the studied species, and the obtained results provide important data on the molecular response to grazing.

Seed germination is a vigorous stage in the plant life cycle, and it begins with seed rehydration and imbibition. Due to the resumption of respiratory activity following imbibition, the oxygen content in the seed tissue is rapidly diminished. Therefore, the supply of the oxygen through the seed coat to the embryo becomes limited, which generates the hypoxic environment in the seed. The study of Jayawardhane et al. (2021) [9] on two species, hypoxia-tolerant rice and hypoxia-sensitive barley, showed differences in their metabolic activity, necessary for subsequent germination steps. It has been shown that the embryos of rice seeds have higher alcohol dehydrogenase activity, indicating more efficient anaerobic fermentation and an elevated nitric oxide (NO) level corresponding to a higher NO turnover rate via the phytoglobin–NO cycle. Both fermentation and NO turnover resulted in a higher ATP/ADP ratio in rice embryos prior to radicle protrusion, as compared to barley. Additionally, the activities of antioxidant enzymes (SOD, APX, monodehydroascorbate reductase, and dehydroascorbate reductase) in imbibed embryos were higher in rice than in barley, which corresponded to the reduced levels of ROS, MDA, and electrolyte leakage. In summary, the observation of metabolic changes in the embryos of two cereal species differing in tolerance to hypoxia can explain the adaptation of rice to low-oxygen environments.

This Special Issue also includes three review articles that analytically and critically summarize contemporary research in the area of plant signaling and response to abiotic stress. Each of them presents and analyzes research results from a different angle, the primary goal of which is to understand the induction processes and mechanisms that allow plants to survive unfavorable environmental and climatic conditions.

Thus, Jeyasri et al. [10] summarize the impact of important environmental factors causing abiotic stresses, such as atmosphere, soil, and water, on the growth and development of the world's most important cereal species: rice, wheat, sorghum, and maize. In their article, they focus on the use of systems biology and advanced sequencing approaches in genomics to explain the mechanism of the response to abiotic stresses in cereals. The authors present a holistic approach, enabling an understanding the mechanism of response to abiotic stresses (bioinformatics and functional omics, gene mining and agronomic traits, genome-wide association studies, and TF family). In the article, they also highlight how progress in omics studies facilitates the identification of genes responsive to abiotic stresses and influences the understanding of the interactions between signaling pathways, molecular insights, novel traits, and their importance in cereal crops. The authors emphasize the need for further research to obtain information from integrated omics databases to understand the mechanisms of abiotic stresses, which gives hope for the development of plant production with a large spectrum of stress tolerance.

The review article by Nykiel et al. [11] focuses on the discussion of signal transduction pathways in cereals under the influence of factors such as drought, salinity, heavy metals, and attack by pathogens and pests, but also, which is especially worth emphasizing, the crosstalk between reactions in response to double stress. This work summarizes the latest discoveries regarding signal transduction pathways and integrates information available in the contemporary literature, which is an important starting point not only for the precise formulation of future research tasks that will lead to a full explanation of the mechanism of plant response to stress factors, but also for further progress in the creative breeding of stress-tolerant cultivars of cereals.

In turn, Radani et al. [12] focus in their article on the Helix–Loop–Helix (bHLH) plant family TFs, involved in responses to abiotic stress, representing one of the most important families of eukaryotic genes containing the highly conserved bHLH motif. These factors activate or repress the transcription of specific response genes and thus influence the response to abiotic stresses such as drought, climate change, mineral deficiencies, excessive salinity, and water stress, and the regulation of these factors is crucial in achieving a better control of their activity. With the use of figures, the authors present the structural features, classification, functions, and regulatory mechanisms of the expression of bHLH TFs at the transcriptional level by other pre- and post-translational components (ubiquitination, phosphorylation, and glycosylation) during their response to various abiotic stress conditions.

5. Summary

Taken together, the contributions presented in this Special Issue contribute significantly to our understanding of plant responses to environmental stressors, including signal transduction cascades, and indicates potential strategies to enhance the resilience, resource efficiency, and overall agricultural productivity of plants. As Guest Editors, we are thankful to all of the authors for choosing to publish their articles in our Special Issue. We also recognize and appreciate the engagement of the reviewers who reviewed all of the submitted manuscripts.

Author Contributions: Conceptualization, M.N.; formal analysis, M.L.; writing—original draft preparation, M.N., M.G., J.F., B.P. and M.L.; writing—review and editing, M.L.; supervision, M.N. All authors have read and agreed to the published version of the manuscript.

Funding: This research received no external funding.

Data Availability Statement: Not applicable.

Conflicts of Interest: The authors declare no conflict of interest.

References

1. Wang, Y.-Y.; Head, D.J.; Hauser, B.A. During Water Stress, Fertility Modulated by ROS Scavengers Abundant in Arabidopsis Pistils. *Plants* **2023**, *12*, 2182. [CrossRef] [PubMed]
2. Lu, C.; Li, L.; Liu, X.; Chen, M.; Wan, S.; Li, G. Salt Stress Inhibits Photosynthesis and Destroys Chloroplast Structure by Downregulating Chloroplast Development–Related Genes in *Robinia pseudoacacia* Seedlings. *Plants* **2023**, *12*, 1283. [CrossRef] [PubMed]
3. Maymon, T.; Eisner, N.; Bar-Zvi, D. The ABCISIC ACID INSENSITIVE (ABI) 4 Transcription Factor Is Stabilized by Stress, ABA and Phosphorylation. *Plants* **2022**, *11*, 2179. [CrossRef] [PubMed]
4. Kim, G.; Sung, J. Transcriptional Expression of Nitrogen Metabolism Genes and Primary Metabolic Variations in Rice Affected by Different Water Status. *Plants* **2023**, *12*, 1649. [CrossRef] [PubMed]
5. Zhang, B.; Du, H.; Yang, S.; Wu, X.; Liu, W.; Guo, J.; Xiao, Y.; Peng, F. Physiological and Transcriptomic Analyses of the Effects of Exogenous Lauric Acid on Drought Resistance in Peach (*Prunus persica* (L.) Batsch). *Plants* **2023**, *12*, 1492. [CrossRef] [PubMed]
6. Xiong, J.; Zhang, W.; Zheng, D.; Xiong, H.; Feng, X.; Zhang, X.; Wang, Q.; Wu, F.; Xu, J.; Lu, Y. ZmLBD5 Increases Drought Sensitivity by Suppressing ROS Accumulation in Arabidopsis. *Plants* **2022**, *11*, 1382. [CrossRef] [PubMed]
7. Li, Y.; Huang, J.; Yu, C.; Mo, R.; Zhu, Z.; Dong, Z.; Hu, X.; Zhuang, C.; Deng, W. Physiological and Transcriptome Analyses of Photosynthesis in Three Mulberry Cultivars within Two Propagation Methods (Cutting and Grafting) under Waterlogging Stress. *Plants* **2023**, *12*, 2066. [CrossRef] [PubMed]
8. Wang, Y.; Zhu, W.; Ren, F.; Zhao, N.; Xu, S.; Sun, P. Transcriptional Memory in *Taraxacum mongolicum* in Response to Long-Term Different Grazing Intensities. *Plants* **2022**, *11*, 2251. [CrossRef] [PubMed]
9. Jayawardhane, J.; Wijesinghe, M.K.P.S.; Bykova, N.V.; Igamberdiev, A.U. Metabolic Changes in Seed Embryos of Hypoxia-Tolerant Rice and Hypoxia-Sensitive Barley at the Onset of Germination. *Plants* **2021**, *10*, 2456. [CrossRef] [PubMed]
10. Jeyasri, R.; Muthuramalingam, P.; Satish, L.; Pandian, S.K.; Chen, J.-T.; Ahmar, S.; Wang, X.; Mora-Poblete, F.; Ramesh, M. An Overview of Abiotic Stress in Cereal Crops: Negative Impacts, Regulation, Biotechnology and Integrated Omics. *Plants* **2021**, *10*, 1472. [CrossRef] [PubMed]
11. Nykiel, M.; Gietler, M.; Fidler, J.; Prabucka, B.; Rybarczyk-Płońska, A.; Graska, J.; Boguszewska-Mańkowska, D.; Muszyńska, E.; Morkunas, I.; Labudda, M. Signal Transduction in Cereal Plants Struggling with Environmental Stresses: From Perception to Response. *Plants* **2022**, *11*, 1009. [CrossRef] [PubMed]
12. Radani, Y.; Li, R.; Korboe, H.M.; Ma, H.; Yang, L. Transcriptional and Post-Translational Regulation of Plant bHLH Transcription Factors during the Response to Environmental Stresses. *Plants* **2023**, *12*, 2113. [CrossRef] [PubMed]

plants MDPI

Review

An Overview of Abiotic Stress in Cereal Crops: Negative Impacts, Regulation, Biotechnology and Integrated Omics

Rajendran Jeyasri [1,†], Pandiyan Muthuramalingam [1,2,†], Lakkakula Satish [1,3], Shunmugiah Karutha Pandian [1], Jen-Tsung Chen [4], Sunny Ahmar [5], Xiukang Wang [6], Freddy Mora-Poblete [5,*] and Manikandan Ramesh [1,*]

1 Department of Biotechnology, Science Campus, Alagappa University, Karaikudi 630003, India; jeyasri8220@gmail.com (R.J.); pandianmuthuramalingam@gmail.com (P.M.); lsatish@post.bgu.ac.il (L.S.); pandiansk@gmail.com (S.K.P.)
2 Department of Biotechnology, Sri Shakthi Institute of Engineering and Technology, Coimbatore 641062, India
3 Department of Biotechnology Engineering, Ben-Gurion University of the Negev, Beer Sheva 84105, Israel
4 Department of Life Sciences, National University of Kaohsiung, Kaohsiung 81148, Taiwan; jentsung@nuk.edu.tw
5 Institute of Biological Sciences, University of Talca, 2 Norte 685, Talca 3460000, Chile; sunnyahmar13@gmail.com
6 College of Life Sciences, Yan'an University, Yan'an 716000, China; wangxiukang@yau.edu.cn
* Correspondence: morapoblete@gmail.com (F.M.-P.); mrbiotech.alu@gmail.com (M.R.)
† These authors contributed equally to this review.

Citation: Jeyasri, R.; Muthuramalingam, P.; Satish, L.; Pandian, S.K.; Chen, J.-T.; Ahmar, S.; Wang, X.; Mora-Poblete, F.; Ramesh, M. An Overview of Abiotic Stress in Cereal Crops: Negative Impacts, Regulation, Biotechnology and Integrated Omics. *Plants* **2021**, *10*, 1472. https://doi.org/10.3390/plants10071472

Academic Editors: Małgorzata Nykiel, Mateusz Labudda, Beata Prabucka, Marta Gietler and Justyna Fidler

Received: 1 June 2021
Accepted: 16 July 2021
Published: 19 July 2021

Abstract: Abiotic stresses (AbS), such as drought, salinity, and thermal stresses, could highly affect the growth and development of plants. For decades, researchers have attempted to unravel the mechanisms of AbS for enhancing the corresponding tolerance of plants, especially for crop production in agriculture. In the present communication, we summarized the significant factors (atmosphere, soil and water) of AbS, their regulations, and integrated omics in the most important cereal crops in the world, especially rice, wheat, sorghum, and maize. It has been suggested that using systems biology and advanced sequencing approaches in genomics could help solve the AbS response in cereals. An emphasis was given to holistic approaches such as, bioinformatics and functional omics, gene mining and agronomic traits, genome-wide association studies (GWAS), and transcription factors (TFs) family with respect to AbS. In addition, the development of omics studies has improved to address the identification of AbS responsive genes and it enables the interaction between signaling pathways, molecular insights, novel traits and their significance in cereal crops. This review compares AbS mechanisms to omics and bioinformatics resources to provide a comprehensive view of the mechanisms. Moreover, further studies are needed to obtain the information from the integrated omics databases to understand the AbS mechanisms for the development of large spectrum AbS-tolerant crop production.

Keywords: abiotic stress; GWAS; *Oryza sativa* L.; plant omics; *Triticum aestivum* L.; *Sorghum bicolor* L.; transcription factors; *Zea mays* L.

1. Introduction

Cereals are grasses (a monocot family Poaceae, also known as Gramineae) cultivated for the edible components of the grain. The most important staple cereal crops are wheat, rice, maize, sorghum, barley, oats, and millet. These cereals are cultivated for the edible components of their caryopsis, composed of the endosperm, germ, and bran. These plants have evolved to live in environments where they are often exposed to various stressors such as high temperature (HT), drought, salinity and mineral toxicity, and the water deficiency [1,2]. Cereals are widely utilized crops in world agriculture, with an overall production of 2500 million tonnes being harvested globally in 2011. On a worldwide basis, rice, wheat and maize are the three most important cereal crops, which together comprise at least 75% of the world's grain production. Also, in 2011 723, 704, and 883 million tonnes of rice, wheat, and maize were harvested, respectively. Cereals contain major nutritional and energy sources such as proteins, carbohydrates, minerals, amino acids, fiber, and micronutrients such as vitamins, magnesium, and zinc for the global population [3,4]. Asia, America, and Europe produce 80% of the world's cereal grains. Rice, sorghum, millet and wheat are widely produced in Asia; likewise, corn and sorghum in America and barley, rye, and oats in Europe. Cereals are a pivotal nutrient source in both developed and developing countries, however, the utilization pattern of these cereal grains differs. In developed countries, more than 70% of total cereal production is fed to the animals, whereas in underdeveloped countries, 68 to 98% of the cereal production is used for human consumption [5].

Abiotic stresses (AbS) (predominantly drought, cold, salinity, and heat), adversely affect diverse plant developmental stages. They are highly complex and affect the various plant dynamisms at the transcriptome, cellular, and physiological processes such as flowering, grain filling, and maturation [6,7]. The ability of cereal crops to tolerate dominant AbS comprises water deficit (drought), flood (anoxia), salinity, high/low temperature, and other osmotic stresses is an essential aspect of yield resilience, and its improvement has long been a target for plant breeders and researchers [8]. The rapidly changing global climate is affecting crop productivity and food availability due to the ever-increasing population, resulting in a demand for stress-tolerant crop varieties that have never been greater [9,10]. Understanding the molecular cross-talk of plant responses to various stresses is crucial in providing opportunities for the development of broad-spectrum stress-tolerant crops. As a result, it is crucial to understand the AbS tolerance dynamic while also devising a new and improved approach for dealing with their detrimental influence on the agricultural sector. The recent game-changing advances in bioinformatics and integrating omics technologies could serve as the most immediate and prospective strategies for improving AbS tolerance in cereal crops. Omics approaches lead to understanding the stress tolerance mechanisms at the molecular level including genomics, functional genomics, genetic engineering, gene expression, protein or metabolite profile(s), and their overall phenotypic effects. In crop breeding techniques, the identification and characterization of the genes and the specific genetic regions associated with both the quantitative and qualitative agronomic traits have been a major challenge. In recent breeding programs, a high-throughput marker-assisted system is extensively being used to enhance selection accuracy and efficiency.

2. Cereal Crops

2.1. Comparative Nutritive Values of Cereal Crops

Cereal grains have low protein content when compared to food legumes and oilseeds, with rice being the lowest. Among the essential amino acids for humans, lysine is the most limiting in all cereal grains. Most cereal proteins are rich in cysteine, methionine and sulfur-containing amino acids. The biological value (BV) ranges from 55 to 77.7%, protein digestibility (TD) 77 to 99.7%, and net protein utilization (NPU) 50 to 73.8% in different cereal diets fed to growing rats. Barley contains relatively more amounts of lysine compared with other cereal crops. The utilization of legumes is known to be affected by the presence of several antinutrients, such as metal chelates, antivitamins, goitrogens, cyanogens, inhibitors of proteases and amylases, toxic phenolic glycosides, and amino acid derivatives [11]. Hence, proper processing of cereal-legume mixture is required to minimize these antinutrients before consumption. Compared with animal foods cereal grains products are inferior in both nutritional and sensory qualities. Physical, chemical, biological, and/or physiological modifications can improve both the nutritional and evident qualities of the grains [12]. Moreover, natural processes like fermentation and controlled germination with natural microflora are highly beneficial in improving the quality of cereal-based food. Important cereal crops and their comparative nutritive values are given in Table 1.

Table 1. Comparative nutritive value of cereal grains.

Factor	Wheat	Maize	Rice	Barley	Sorghum	Oat	Millet	Rye
Available CHO (%)	69.7	63.6	64.3	55.8	62.9	62.9	63.4	71.8
Energy (kJ/100 g)	1570	1660	1610	1630	1610	1640	1650	1570
Digestible energy (%)	86.4	87.2	96.3	81.0	79.9	70.6	87.2	85.0
Vitamins (mg/100 g)								
Thiamin	0.45	0.32	0.29	0.10	0.33	0.60	0.63	0.66
Riboflavin	0.10	0.10	0.04	0.04	0.13	0.14	0.33	0.25
Niacin	3.7	1.9	4.0	2.7	3.4	1.3	2.0	1.3
Amino acids (g/16 g N)								
Lysine	2.3	2.5	3.8	3.2	2.7	4.0	2.7	3.7
Threonine	2.8	3.2	3.6	2.9	3.3	3.6	3.2	3.3
Met. & Cys.	3.6	3.9	3.9	3.9	2.8	4.8	3.6	3.7
Tryptophan	1.0	0.6	1.1	1.7	1.0	0.9	1.3	1.0
Protein quality (%)								
True digestibility	96.0	95.0	99.7	88.0	84.8	84.1	93.0	77.0
Biological value	55.0	61.0	74.0	70.0	59.2	70.4	60.0	77.7
Net protein utilization	53.0	58.0	73.8	62.0	50.0	59.1	56.0	59.0
Utilization protein	5.6	5.7	5.4	6.8	4.2	5.5	6.4	5.1

2.2. Rice (*Oryza sativa* L.)

Rice is the second most widely consumed cereal, serving as a staple food for more than half of the world population and 90% of Asians. Rice, known as the grain of life, contains 80% carbohydrates, 7–8% protein, 3% fat, and 3% fiber [13]. In most countries, rice being the most dominant cereal crop can improve the health condition of millions of people who consume it. It plays an important role in health benefits and Lifestyle-related disease prevention such as high blood pressure, cancer prevention, Alzheimer's diseases, skincare, diabetes, heart disease, and dysentery [14–16]. However, rice plants are severely affected by various AbS, major stressors such as drought, cold, salinity, and high temperature [17].

Excess salinity in the soil is one of the major abiotic stress factors that affect the growth and productivity of a wide variety of crops including rice. Generally, rice can tolerate a modest amount of saltwater without affecting its growth and yield. However, it highly depends on the types and species of rice used as well as their growth stage [18]. According to Lee et al. [19], indica has a higher tolerance level than japonica at the seedlings stage. Rice is grouped as a salinity-sensitive cereal at an early stage of growth, which limits its production efficiency at the mature stage [20–22].

2.3. Maize (*Zea mays* L.)

Maize or corn is an important cereal crop grown in diverse agro-ecological zones and farming systems and socio-economic backgrounds in sub-Saharan Africa (SSA) [23]. After rice and wheat, maize is the world's third most important crop, and it is known as the "Queen of Cereals" because it has the highest production potential of all the cereals [24]. It is a predominant source of nutrition as well as phytochemical compounds such as carotenoids, phenolic compounds, and phytosterols [25]. Phytochemicals are naturally occurring bioactive compounds in plants that provide human health benefits while also preventing the risk of major chronic diseases [25,26]. Maize is believed to have potential anti-HIV activity due to the presence of *Galanthus nivalis* agglutinin (GNA) lectin or GNA-maize [25]. The studies revealed that phytochemicals in grains due to their potent antioxidant activities demonstrate significantly reducing the risk of many diseases such as cardiovascular disease, type 2 diabetes, Diet-related disorders, and cancers [27].

The growth and productivity of maize are severely affected by several abiotic stresses such as salinity drought, waterlogging, cold, and nitrogen stress. Among them, salinity stress causes several biochemical and physiological changes in maize, such as disruption of cellular homeostasis, ionic imbalance, nitrogen fixation, respiration, inhibition of several metabolic enzymes such as photosynthetic enzymes etc (enzyme toxicity) [28]. Drought is one of the most detrimental AbS which are seriously affecting the productivity of cereal crops. However, while maize is known as the "Queen of Cereals" across the world, it is susceptible to drought. Drought can affect kernel weight following silking stage up to maturity. Severe drought can reduce the yields of maize during this period by 20 to 30%. During the grain filling stage because of reduced photosynthesis accelerated leaf senescence takes place [29]. More recently, new biotechnological tools have emerged to further accelerate the grain selection and improvement for enhancing the tolerance to drought conditions.

2.4. *Wheat (Triticum aestivum* L.*)*

Wheat is a major staple crop for several countries, grown on around 10 million ha in Africa and it comes from a type of grass (*T. aestivum* L.) that is grown in countless varieties worldwide. As a result of a growing population, changing food preferences, and socioeconomic change associated with urbanization and industrialization, wheat consumption steadily increased during the past two decades in all African countries [30,31]. Wheat consumption provides up to 50% of daily calories and proteins. Wheat is often considered to be a source of energy (carbohydrate) and also contains a significant amount of other important nutrients including proteins, fiber, and minor components such as lipids, vitamins, minerals, and phytochemicals that may significantly contribute to the individual diet [32,33]. It reduces the risk of cardiovascular disease, type 2 diabetes, and forms of cancer (notably colorectal cancer) [34,35]. Dietary fiber components also have high heritability and are amenable to manipulation by breeding. Therefore, plant breeders should be able to select plants with enhanced health benefits in addition to increased crop productivity [36,37].

However, wheat plants are severely affected by different AbS, such as salinity, drought, cold, and heat. Heat stress affects wheat growth and yield, particularly at grain developmental stages. The effect of a 3-day heat shock on biomass production was less than the pre- and post-treatment growing temperature [38]. Salinity stress also affects the growth and productivity of wheat. The increasing salinity of irrigation water had a significant adverse effect on the yield of wheat. Domestic wheat lines could be grown at salinity concentrations ranging from 4 and 8 g/L with minimal reduction in biological and grain yield. Moreover, many physiological and biochemical approaches have been developed in plants to survive at high salt concentration. The most effective approach to solve the salt problem is to improve wheat adaptation under salinity stress conditions and enhance its grain yield. Various biotechnological approaches are required to understand the genetic and physiological mechanisms of natural differences in salinity tolerance of wheat and to obtain methods to explore the inherent genetic differences, to get new candidate genes for improving salt tolerance in wheat [39,40].

2.5. *Sorghum (Sorghum bicolor* L.*)*

Sorghum is the fifth most important food and feed crop in the world. It is the main cereal food for semi-arid tropical regions of Africa, Asia, and Latin America. Sorghum species (*S. vulgare* and *S. bicolor*) are members of the grass family. Sorghum is resistant to drought and water-logging and is grown in different soil conditions [41]. It is mainly composed of starch, protein and unsaturated fatty acids and is an essential source of some vitamins and minerals [42]. Sorghum is the most abundant and omnivorous cereal secondary metabolites of plants, including up to 6% of 3-deoxyanthocyanidine, phenolic acid, flavonoids, and tannins [43,44]. Phenolic and soluble compounds play an important role in balancing or stabilizing the intestinal microbiota and the parameters associated with obesity, oxidative stress, inflammation, diabetes, dyslipidemia, hypertension, and cancer [45].

Among the various AbS, drought and temperature stress are of foremost important that limits sorghum production. Severe drought causes considerable yield loss in sorghum, and it has a greater impact on grain filling and flowering compared to vegetative stage.

Drought adversely affects various physiological functions, inflorescence development and leaf growth of the sorghum [46,47]. Interestingly, genome and transcriptome sequencing, annotation projects and recent literatures paved the way to identify the candidate genes through omics databases, which are predicted to be involved in individual and combined abiotic stress (CAbS) responses. This claim supports to understanding the stress tolerance mechanism and environmental adaptation, as well as promoting the green revolution of all other food crops [47,48].

3. Abiotic Stress (AbS) Dynamism on Cereal Crops

The atmosphere, soil, water and their associated factors are the major abiotic stressors affecting modern agricultural systems [49].

3.1. Atmospheric Factors

3.1.1. Rainfall

In semiarid regions, rainfall is one of the primary AbS factors affecting soil erosion and crop production in rain-fed agriculture. It controls soil salinity and acidic properties. Sulfur dioxide (SO_2) and nitrogen oxides (NOx) from fossil fuel burning merge with water and oxygen in the atmosphere resulting in acid rain. It also reduces the soil pH and removes nutrients and minerals from the soil that can be harmful to plants [50].

3.1.2. Temperature

The plant underwent a high temperature that caused certain mechanical damages including expansion-induced-lysis, phase transitions and fracture lesions in membranes, and physical damage [51]. Depending on the temperature plant species have been classified into three groups: chilling-sensitive, freezing-sensitive, and freezing-resistant plants [52]. Freezing may alter the growth and can cause frost-hardening/cold hardening and also induces the production of reactive oxygen species (ROS), which damages membrane components, and results in protein denaturation [53,54]. Chilling stress including reduced leaf expansion and wilting, affects the reproductive development of plants, chlorosis, and may lead to necrosis.

3.1.3. Gases

An increase in the level of greenhouse gases in the atmosphere such as carbon dioxide (CO_2), methane (CH_4), nitrous oxide (N_2O), ozone (O_3), water vapor (H_2O), and some artificial chemicals such as chlorofluorocarbons (CFCs) increases the atmospheric temperature. As CO_2 levels increase in the atmosphere, nitrogen concentration is decreased in plants. Thus, decreasing the protein levels and affects the growth ability of plants [55].

3.1.4. Radiation

Ultraviolet (UV) and ionizing radiations affect plant growth and development in many ways. Radiations may disrupt stomatal resistance, damage plant cells, increases cell mutations, prevents seed growth, and reduce plant fertility [56,57]. Photon irradiation (such as H_2O, CO_2, CH_4, N_2O, and O_3) generates cellular damages in root and leaf tissue.

3.1.5. Wind

Plant tissues can be damaged during hot, humid, hazy weather with little wind. Wind decreases the phytohormonal content of roots and shoots in cereal crops. The direction and velocity of the wind will affect plant growth and development [49].

3.2. Soil Factors

3.2.1. Soil Properties

Salt stress (salinity) causes multifarious effects in plants such as ionic effect, osmotic effect, nutrient and hormonal imbalance, and production of ROS [58]. Plant growth and productivity are severely affected by the accumulation of sodium (Na^+) and chloride (Cl^-)

ions that leads to creating ionic, osmotic, and oxidative stresses [59–63]. Sodium-ion (Na^+) has been detected to intervene in multiple metabolic processes such as protein translation, transcription, and enzyme activity that ultimately led to osmotic stress. Functional genomic approaches provide new opportunities to unravel the functional roles of AbS responsive genes, enabling the identification of genes and generation of stress-tolerant plants especially on important cereal crops [64]. Plants can generally tolerate salinity stress mainly by three mechanisms, which include ion exclusion, osmotic tolerance and tissue tolerance. Osmotic tolerance is regulated by long-distance signaling waves that reduce cell expansion in root tips, leaves, and stomatal conductance [65,66]. Ion exclusion mainly involves the transport of Na^+ and Cl^- into roots which prevent the reduction of Na^+ accumulation in shoots. Tissue tolerance involves the exposure of tissues to the accumulated Na^+ and Cl^- at cellular and intracellular levels, synthesis of compatible solutes, and production of the enzyme catalyzing detoxification of ROS [67].

3.2.2. Pollution

Soil contamination was widespread as a result of the rapid application of some harmful pesticide compounds which can infiltrate the natural environment in two ways depending on their solubility. The pesticide-contaminated soil may affect the availability of nutrients in plants. Rapid industrialization near agricultural lands may affect crop growth and production. Heavy metals such as Fe, Mn, Zn, Cu, Mg, Mo, B, and Ni in the soil of a particular area significantly affect the morphological, metabolic, and physiological anomalies in plants. It ranges from chlorosis of shoot to lipid peroxidation and protein degradation [68]. Among them, 'B' is an essential element for plant growth and simultaneously affects the growth and yield at higher levels.

3.2.3. Degradation

Nutrient deficiency has been considered as the main cause of poor crop productivity. Of the global land area of 13.5 billion ha, among that only 3.03 billion ha (22%) is cultivable and about 2 billion ha is degraded. Disposal of oil shales, heavy metal contamination of soil, and spillage of crude oils adversely affect the severe root damage that governs nutrient irregularity [69].

3.3. *Water Factor*

3.3.1. Suboptimal

The most important constraints for agriculture are the water limitation, declining rainfall, and increasing temperature which significantly affects the growth of crop plant areas. A wide range of strategies such as genetical, physiological, biochemical, and molecular levels are well defined in plants. However, recent advancements are also available to obtain important drought-tolerant crops using conventional, marker-assisted breeding, and genetic engineering [70,71].

3.3.2. Supraoptimal Salinity

High salinity can be toxic for many crops, but halophytes have adapted to the worse salt conditions, which are found in several coastal areas, such as salt marshes, and inland arid areas. These zones, having a high rate of evaporation tends to concentrate salts in the mineral content of the soil. Those halophytes are morpho-physiologically and reproductively adapted to saline, waterlogged and anaerobic conditions. Nutrient deficiency or the presence of toxic substances such as heavy metals in the soil can also result in the AbS [49].

3.3.3. Waterlogging

Flood water usually causes a water-logged situation in the field. In waterlogged soils, within 24 h the oxygen concentration drops down to zero because water replaces most of the air in the soil pore space. Roots need oxygen for their respiration and cell viability. Water-logging limits oxygen supply to the roots, if any remaining oxygen is used up by the roots from flooded or waterlogged soils, then the normal function of the roots will be arrested. Therefore, the leaves and stems are unable to obtain enough minerals and nutrients hence the roots start to die off in water-logging conditions [72].

4. Bioinformatics and Functional Omics Approach to Explore the AbS Tolerance Mechanism

The process of gene regulatory dynamism in various cellular, physiological and biochemical mechanisms in plants was altered by environmental factors. Therefore, to analyze the processes involved in this regulatory mechanism, several functional omics projects, have been initiated across the world in recent years [49,73,74]. Moreover, multiple omics and bioinformatics approaches have been used for developing crop plants that are tolerant to AbS through molecular breeding and genetic engineering and also through advancements and increasing knowledge in genetics, genomics, and molecular physiology [75,76]. Hence, the development of new functional omics and computational biology software and tools paved the way to identify the AbS responsible candidate genes from gene pools. In addition, the use of high-throughput techniques has been employed such as expression reads by RNA-Seq, random and targeted mutagenesis, gene shifting, complementation, and synthetic promoter trapping approaches make many avenues for functional analyses of AbS responsive genes and tolerance mechanisms [77]. Transcription factors (TFs) are crucially important in knowing the appropriate molecular processes and pathways that are involved in plant growth and survival under AbS conditions [78–80]. AbS are the quantitative and multiple genes associated in nature, these stress molecular cross-talks and their pathway interconnections are found under AbS conditions. Again, understanding the post-translational modifications (PTM) degradation of proteins and non-coding miRNA interactions allow the modulation of the target proteins. Similarly, some of the siRNAs play an important role as stress-inducers and affect protein synthesis including alternative splicing. The genome-wide association studies (GWAS) have become popular to provide novel strategies to identify and characterize the unique stress-responsive genes, which are introduced into crop plants and used in building up the tolerance against various AbS conditions [80]. In this record, the identification, and characterization of specific stress-responsive genes along with their promoters are analyzed for specificity. Integrated omics and bioinformatics approaches have been used to study the AbS responsible genes, their corresponding growth regulations and their encoding global metabolic network. With the advent of newly developed functional omics can be broadly categorized into two potential approaches used in manipulating gene-pool for enhancing the AbS tolerance, which contains: (i) identification of stress-responsive genes followed by genetic engineering to improve the stress tolerant cereals development and (ii) mining of marker associated with agronomically important genes and their use in marker-assisted breeding programs.

Importance of Omics in Enhancing Nutritive Values

In the era of post-genomics, revealing the interconnected functional attributes between genes transcripts, metabolites, proteins and nutritional biology remains a major challenge. New advances in omics such as genomics, transcriptomics, proteomics and metabolomics will have an essential advantage in the bio-fortification processes in plants.

5. Gene Mining

Functionally integrated omics and computational biology tools for AbS tolerance include reconnaissance of novel AbS responsive genes and the expression levels, that may induce the AbS response. These omics approaches are used to understand the functional role of AbS -responsive genes and the generation of stress-tolerant transgenic lines. In addition, they also pave the way for guiding genetic and metabolic engineering studies. In cereals, large numbers of expressed sequence tags (ESTs) have been generated from different cDNA libraries, collected from AbS treated tissues at different developmental stages, and are considered as a significant functional genomic approach to impute the AbS responsive genes. The expressed sequence tag database (dbEST) provides information about the number and type of different AbS tolerant species. Genomics-based approaches were started in all the crop plants such as barley, rice, maize, wheat, and sorghum (stress-genomics.org). A total of 13,022 AbS related ESTs were reported from *Hordeum vulgare*, 13,058 genes from *O. sativa*, 17,189 from *S. bicolor*, 2641 from *Secale cereale*, 20,846 from *T. aestivum* and 5695 regulators from *Z. mays* using gene index of TIGR database (http://www.tigr.org/tdb/tgi/ accessed on 16 April 2021) [81]. However, publicly available ESTs libraries with AbS data are few. For this reason, the sequence-specific approach that is based on cDNA libraries from cereal stress resistant genotypes may be improved in different developmental periods to encompass a larger range of tissue types and organs of many species. Functional characterizations of the different AbS treated genes are studied using BlastX [82] by comparing the Swissprot dataset (http://www.uniprot.org/ accessed on 24 April 2021) [83]. Moreover, the EST clustering results provides consensus sequences which are more informative than typical EST data. Drought-responsive genes have a lot of features that include metallothionein-like proteins, late embryonic abundant (LEA) proteins, heat-shock proteins, cytochrome P450 enzymes, catalases, peroxidases, kinases, phosphatases, and TFs (DREB, MYB, MYC, AP2-EREBP, ZF-HD, NAC, WRKY, and bHLH protein) that were abundantly expressed during drought stress [78,84]. Recent research was carried out to identify ESTs in the NaCl-treated cDNA library of *Thellungiella halophila* and also from monocots like barley, wheat, maize, and rice [85]. Furthermore, novel studies are needed to be conducted on all food crops for future food security and production.

6. Transcript Signature Usage in Stress Responsive Gene Mining

Several computational approaches are employed to quantify the expression intensities of EST on its programs in cereal crops such as rice [86], barley [87], and maize [88]. The microarray technology is used for transcript expression profiling, Mission Planning and Scheduling System (MPSS), Serial Analysis of Gene Expression (SAGE), and quantitative real-time PCR (qRT-PCR) these are further modern approaches for the quantification under the controlled and stressed tissues of large number of gene expression studies [89]. Microarray-based transcript profiling was carried out in *Arabidopsis* [90] as well as in important cereal species [91,92] have comprehensively analyzed the gene expression signature in response to unique and multiple AbS. An EST-cDNA array technology provides a pivotal tool to compare the relative expression levels between normal and treated plants to unveil the functions of AbS responsive candidates. In addition to that, it is used to identify and understand the genes involved in transcriptional reprogramming and signal transduction pathways.

TFs, are the protein-containing domains that imply the DNA binding and transcript regulatory activities. The TF families were classified into more than 50, based on the presence of highly conserved motifs (Table 2). Computational omics studies were used to identify and understand the molecular cross-talks of plant developmental stages, expression profiling, and physicochemical properties [84,93]. The TFs are mainly involved in diverse stress tolerance and plant growth metabolisms such as cold, salinity, drought, metal, submergence, heat, low/high temperature, light, UV, O_3, osmotic, oxidative stress, and signaling, tissues development, regulation of plant physiological metabolisms respectively. It is the essential key factor to unravel the candidate gene functions and their dynamism.

Table 2. C3 and C4 crop species transcription factor families and their number of family members. (Source: https://grassius.org/grasstfdb.php/ accessed on 19 May 2021) [94] (http://planttfdb.cbi.pku.edu.cn/ accessed on 19 May 2021) [95]).

S. No	Transcription Factors (TFs)	Rice	Sorghum	Maize	Wheat
1	ABI3-VP1	55	60	51	-
2	BBR/BPC	04	05	04	05
3	C2C2-GATA	23	27	36	-
4	CCAAT-HAP2	11	09	16	-
5	G2-like	44	41	59	100
6	HSF	25	24	29	53

Table 2. *Cont.*

S. No	Transcription Factors (TFs)	Rice	Sorghum	Maize	Wheat
7	Orphans	185	177	339	-
8	WHIRLY	02	03	02	07
9	Alfin-like	09	13	19	-
10	bHLH	135	143	175	-
11	C2C2-YABBY	08	08	13	-
12	CCAAT-HAP3	12	-	04	-
13	GeBP	13	15	21	12
14	MADS	70	76	77	-
15	SBP	19	18	32	37
16	WRKY	103	94	125	171
17	AP2-EREBP	164	156	212	-
18	bZIP	92	91	128	186
19	C2H2	09	07	10	224
20	CCAAT-HAP5	22	-	18	-
21	GRAS	60	76	86	121
22	MYB	114	113	167	263
23	TCP	22	28	44	28
24	ZF-HD	15	15	21	20
25	ARF	27	27	38	45
26	BZR	06	08	10	-
27	C3H	46	44	54	100
28	CPP	11	08	13	24
29	GRF	12	10	15	16
30	MYB-related	71	80	116	227
31	Trihelix	17	19	43	46
32	ZIM	21	19	36	-
33	ARID	06	07	10	-
34	C2C2-CO-like	08	09	14	-
35	CAMTA	06	06	05	20
36	E2F-DP	09	10	19	24
37	Homeobox	95	83	133	-
38	NAC	144	123	134	263
39	TUB	15	13	15	-
40	ARR-B	06	10	08	22
41	C2C2-Dof	30	29	46	-
42	CCAAT-DR1	01	-	17	-
43	EIL	09	07	09	16
44	HRT	01	01	-	-
45	NLP	13	13	17	-
46	VOZ	02	02	05	06
47	CCAAT	-	32	-	-
48	mTERF	-	-	30	-
49	OVATE	-	-	43	-
50	Sigma70-like	-	-	09	-
51	PLATZ	-	-	15	-
52	FAR1-like	-	-	15	-
53	Rcd1-like	-	-	10	-
54	FLO/ LFY	-	-	02	-
55	S1Fa-like	-	-	02	03
56	CSD	-	-	04	-
57	LBD	-	-	44	61
58	DBP	-	-	04	-
59	SHI/STY (SRS)	-	-	09	-
60	AP2	-	-	-	43
61	BES1	-	-	-	10
62	ERF	-	-	-	181
63	HRT-like	-	-	-	03
64	M-type-MADS	-	-	-	77
65	NF-X1	-	-	-	02
66	RAV	-	-	-	08

Table 2. *Cont.*

S. No	Transcription Factors (TFs)	Rice	Sorghum	Maize	Wheat
67	TALE	-	-	-	52
68	DBB	-	-	-	17
69	FAR1	-	-	-	201
70	MIKC_MADS	-	-	-	51
71	NF-YA	-	-	-	22
72	Dof	-	-	-	52
73	HB-PHD	-	-	-	06
74	NF-YB	-	-	-	34
75	YABBY	-	-	-	25
76	B3	-	-	-	140
77	GATA	-	-	-	48
78	HB-other	-	-	-	44
79	LFY	-	-	-	02
80	NF-YC	-	-	-	20
81	SRS	-	-	-	05
82	CO-like	-	-	-	07
83	HD-ZIP	-	-	-	62
84	LSD	-	-	-	13
85	Nin-like	-	-	-	29
86	STAT	-	-	-	02
87	WOX	-	-	-	26

'-' indicates that TF family was absent in particular crop.

Diverse literary information revealed the importance of plant stress mechanisms analyzed by gene expression signature. Two kinds of pathways involved in this type of gene expression studies are (i) desiccation tolerance in an ABA-dependent manner by ABA-responsive element binding factors (ABF), MYC and MYB TFs, and (ii) ABA-independent and associated with drought-responsive element binding factors (DREB) [96]. Recent studies are providing evidence, that other than the ABA-associated pathways, there exists an interlinked relationship between cold and salinity stress signaling pathways.

In cereal crop species, microarray and RNA-Seq based stress-regulated transcripts were used for large spectrum gene expression signature analysis in rice, up-regulation of few candidates that encodes cell division, 40S ribosomal proteins, glycine-rich proteins, elongation factor, and induce the phytohormone regulating genes. The genes which are generally downregulated in the sensitive rice cultivators are glycine-rich proteins, ABA and stress-induced proteins, metallothionein-like proteins, glutathione S-transferase, ascorbate peroxidase, water channel protein isoforms, subtilisin inhibitor, tyrosine inhibitor, and so on [97,98].

During long-term AbS treatments, protease inhibitors, stress proteins, aquaporins, antioxidant components, and some unknown genes were induced and are expected to impart tolerance. A few transcriptional expression signature analyses were conducted in important cereal crops. In *Arabidopsis*, the multiple stress interactions were studied using functional genomics approaches [10,85]. Multiple stress interactions of AbS treatments were investigated to identify the key players having a pivotal role in multiple individual stresses such as drought, cold, flood, salt, UV, and temperature responses [99]. Using 1300 full-length clones, only 44 genes were identified, which are directly induced either by drought or cold stress dynamisms were reported [100]. By using 7000 whole clone inserts, 213 salinity responsible genes, 299 drought responsive genes, 245 ABA-regulating key genes, and 54 cold-regulating genes were identified [101]. Functional omics and bioinformatics tools to identify gene expression patterns related to multiple stress interactions have been considered as a significant aspect of modern plant stress biology research.

To study gene interaction and downstream elements, the analysis and characterization of the transcriptional responses in knock-out mutants or transgenic plants under abiotic stress tolerance are considered a differential method. Three members of the CBF/DREB1 family, CBF1, CBF2, and CBF3 (DREB1b, DREB1c, and DREB1a respectively) quantification of 41 downstream genes as CBF targets were identified by Fowler and Thomashow, [102]. Comparative genome analysis of the AbS responses among diverse tolerant species is extensively considered as an important approach to reconnaissance of evolutionarily conserved and unique stress defense mechanisms [103]. Various promoters of a group of the abiotic stress-responsive genes from different species of maize, rice, and *Arabidopsis* harboring DRE, GCC-box, ABF, and w-box *cis*-elements were reported [104]. Computational omics gene mining and profiling led to the novel way to understand the huge number of genes involved in AbS responses given in Table 3.

Table 3. Genes encoding enzymes/proteins associated with abiotic stress response in cereals.

Gene Category	Gene	Cellular Response	Species	Reference
Osmolyte compounds				
Glycine betaine	*BADH*	Heave metal stress	Rice	[105]
	CodA	Salt, Cold and drought stress	Rice	[106]
	bet A	Cold, Drought stress	Maize	[107]
Proline	*P5CS*	Drought	Wheat	[108]
Regulatory genes				
bZIP	*bZIP4*	Salinity stress	Maize	[109]
	HBP1b	Drought, Salt, cold	Rice	[110]
	bZIP16	Dehydration, salt and ABA	Rice	[111]
Transporters				
Na+-H+-dependent K+ transporter	ZmHKT1	Salt stress	Maize	[112]
Na+-K+-symporter	*HKT1*	Salt stress	Wheat	[113]
	HKT1	Salt stress	Rice	[114]
Stress-responsive genes				
Transcription factors	*SAP7*	Abiotic stress	Rice	[115]
	DREB	Abiotic stress	Maize	[116]
	MYB6	Drought and Salt	Rice	[117]
WCS genes	*WCS19*	Cold stress	Wheat	[118]
	WCS120	Cold stress	Wheat	[118]
Thaumatin-like protein	*TLP14*	Cold stress	Barley	[119]
Heat shock protein	*HSFA7*	Salt and Droght	Rice	[120]
	HSP20	Heat stress	Wheat	[121]
RAB genes	*RAB7*	Drought and Heat stress	Rice	[122]
	RAB11	Salt stress	Rice	[123]
LEA proteins	*HVA1*	Drought stress	Barley	[124]
	HVA1	Salt, Cold and dehydration stress	Rice	[125]
	HVA1	Salt and Drought tolerance	Barley	[126]
Antioxidants				
Ascorbate peroxidase	*APX*	Drought, Salt and Cold	Rice	[127]
Catalase	*CAT*	Drought stress	Wheat	[128]
Superoxide dismutase	*MnSOD*	Abiotic stress	Rice	[129]
	ZnSOD	Salt stress	Rice	[130]
	FeSOD	Drought stress	Rice	[131]

7. Identification of Genes and Their Agronomic Traits

A case study has been carried out on the evaluation of seven well-known candidate genes for their effects on improving drought resistance of transgenic rice under field conditions [132]. Several AbS affects the growth and yield of cereal crops such as drought, extremely high temperature, low water availability, mineral deficiencies or toxicities, and salinity [133,134]. In recent years, the initiative of developing drought-resistant cereal crops has been well recognized as the most promising and effective strategy for food security against drought and water deficit. In addition, over-expression of certain stress-responsive genes or TFs regulating the multiple stress proteins were shown to confer the increased tolerance to drought as well as in salinity and freezing stresses [117]. Poor management of agricultural water resources, soil degradation, and community pressures are all the prominent stressors that play a pivotal role in the agriculture perspectives across the world. Extensive genetic studies have indicated a huge variation for AbS tolerance but it has failed to attain its maximum goal due to the relatively poor knowledge from the molecular basis for stress-tolerant cereal crop plants. In this case, functional omics and bioinformatics play a crucial role by providing several tools for dissecting AbS responses in cereal crops especially in rice, barley, and wheat through interrogation of genes that may be useful for improving resistance to AbS.

Xiao et al. [132] have selected seven genes (*CBF3, SOS2, NCED2, NPK1, LOS5, ZAT10*, and *NHX1*) in drought resistance breeding and transformed them into rice cultivar Zhonghua 11 (*O. sativa* L. ssp. *Japonica*) under the control of a constitutive promoter (from rice *Actin1* gene) and an inducible promoter (from a rice homologous gene of *HVA22*), respectively. The developed transgenic rice was responsible for drought resistance under natural field conditions. These traits (Table 4) are an example that can be a useful reference for drought resistance engineering in cereal crops and also pave the way to address the other stress-related issues.

Table 4. Transgenic crop plants developed for AbS tolerance details.

Gene	Cellular Response	Species	Reference
Osmolyte compounds			
Pyrroline carboxylate synthase (p5cs)	Drought tolerance	Wheat	[108]
Choline dehydrogenase	Drought, Salt tolerance	Rice	[135]
Arginine decarboxylase	Drought, Heat, Freezing, Salinity tolerance	Rice	[136]
Glutamine synthetase	Oxidative stress tolerance	Rice	[137]
Trehalose-6-P-synthase	Salt, Drought, Cold tolerance	Rice	[138]
Mannitol dehydrogenase	Drought, Salt tolerance	Wheat	[139]
Regulatory genes			
Calcium dependent protein kinase	Drought tolerance	Rice	[140]
DREB1A	Drought tolerance	Rice	[141]
Transporters/symporter			
Na+/H+ antiporter	Salt tolerance	Rice	[142]
Potassium transporter (HKT1)	Salt tolerance	Rice	[143]
Stress-responsive genes			
HVA1	Drought, Salt tolerance	Rice	[125]
Alcohol dehydrogenase	Submergence tolerance	Rice	[144]
Ferritin	Heat tolerance	Wheat	[145]
HVA1	Salt and Drought tolerance	Barley	[126]
Pyruvate decarboxylase1	Submergence tolerance	Rice	[146]
Antioxidants			
Fe-superoxide dismutase	Drought tolerance	Rice	[147]
Mn-superoxide dismutase	Salt tolerance	Rice	[148]

8. Genome-Wide Association Studies (GWAS)

By the use of an ultra- high throughput genotyping technology, GWAS became available as a powerful alternative for dissecting quantitative traits in crop plants [149]. GWAS provides an understanding of transcriptional regulation, metabolic response of rice and other cereal crops to diverse environmental conditions and also in significant relation to AbS [150]. Genome-wide association mapping identifies major QTL regions without the need of constructing a mapping population.

Currently, GWAS was categorized into two types, population-based association studies and a family-based approach [151]. In population-based association studies, it was revealed that unrelated individuals are used to examine genome-wide associations between single nucleotide polymorphisms (SNPs) and their associated phenotypic traits. Family-based mapping studies can be applied to complex pedigrees derived from the crosses among different genotypes. Both approaches have complementary pros and cons.

Population-based GWAS takes advantage of more recombination events that have accumulated over time of generations in historical populations, with the disadvantage of finding false positives or false negatives results applied in *Arabidopsis* [152]. Family-based GWAS can eliminate the effects of population structure and therefore escape the false- positives and false-negatives, but recombination accumulated over a few generations during pedigree development used in *Z. mays* NAM (nested association mapping) population was developed to characterize flowering regulation [151]. The maize NAM population of 200 recombinant inbred lines from 25 parents were crossed to the fully sequenced genotype (B73). Therefore, GWAS is an alternative and complementary approach for understanding the trait and molecular level mechanisms in plants. This approach paves the un-paralleled way for the history of plant stress and molecular biology.

9. Conclusions and Future Perspectives

Ever-increasing advances in multiple analytical stages are paving the way to address the diverse omics scale outcomes, fortifying the researcher's knowledge and long-standing questions in plant biology. Plants are acclimatizing to the environment by altering their genome, metabolome, transcriptome, lipidome, proteome, secretome and miRNAome. Even in this post-genome era, the complete understanding of plant molecular responses to the stress tolerance mechanism is not yet achieved, as the list of genes involved in stress response is increasing rapidly. The identity of these proteins and their functions are close to being determined. Post-transcriptional studies including splicing and RNA silencing and posttranslational mechanisms such as SUMOylation, phosphorylation, and ubiquitination lead to a prompt response in plants against stress. Thus, employing the developing omics approaches and GWAS will contribute to better understanding the AbS response.

At the same time, we are majorly facing multiple global issues such as climate changes, global warming, water, food, and energy security. Further, voluminous research is highly needed to unravel the specific molecular function of the plants. AbS tolerance cross-talks that are speculating in particular plant species, significantly in C_3 and C_4 grass species, are also essential in identifying the candidates to improve the diverse molecular and biochemical functions. In addition, these molecular studies are also subjected and compared to advanced omics datasets from C_3 and C_4 genome species, which are helping to improve the applied or translational research, to unveil the plant molecular systems in response to stress biologist, molecular biologist, and plant physiologist, and also the ever-increasing command of mankind. This eagle's eye review can open the penstocks to budding scientists.

Author Contributions: Conceptualization, R.J., P.M. and M.R.; Methodology, R.J., P.M., and L.S.; Investigation, R.J., P.M., L.S., S.A. and X.W., Formal analysis, R.J. and P.M.; Project administration, R.J. and P.M.; Resources, R.J., P.M., L.S., F.M.-P. and M.R.; Validation, L.S., S.K.P., J.-T.C., S.A. and X.W.; Writing—Original draft, R.J. and P.M.; Writing—review & editing, S.K.P., J.-T.C., F.M.-P. and M.R. All authors have read and agreed to the published version of the manuscript.

Funding: The APC was supported by the Chilean National Fund for Scientific and Technological Development (FONDECYT) grant number 1201973.

Institutional Review Board Statement: Not applicable.

Informed Consent Statement: Not applicable.

Data Availability Statement: Not applicable.

Acknowledgments: The authors gratefully acknowledge the use of Bioinformatics Infrastructure Facility, Alagappa University funded by Department of Biotechnology, Ministry of Science and technology, Government of India grant (No.BT/BI/25/015/2012). The authors also thankfully acknowledge DST-FIST (Grant No. SR/FST/LSI-639/2015(C)), UGC-SAP (Grant No.F.5-1/2018/DRS-II (SAP-II)) and DST-PURSE (Grant No. SR/PURSE Phase 2/38 (G)) for providing lab facilities. The authors also thank RUSA 2.0 [F. 24-51/2014-U, Policy (TN Multi-Gen), Dept of Edn, GoI].

Conflicts of Interest: Authors have no conflict of interest.

References

1. Giordano, M.; Petropoulos, S.; Rouphael, Y. Response and Defence Mechanisms of Vegetable Crops against Drought, Heat and Salinity Stress. *Agriculture* **2021**, *11*, 463. [CrossRef]
2. Kumari, V.V.; Roy, A.; Vijayan, R.; Banerjee, P.; Verma, V.C.; Nalia, A.; Pramanik, M.; Mukherjee, B.; Ghosh, A.; Reja, H.; et al. Drought and Heat Stress in Cool-Season Food Legumes in Sub-Tropical Regions: Consequences, Adaptation, and Mitigation Strategies. *Plants* **2021**, *10*, 1038. [CrossRef]
3. O'Neil, C.E.; Nicklas, T.A.; Zanovec, M.; Cho, S. Whole-Grain Consumption Is Associated with Diet Quality and Nutrient Intake in Adults: The National Health and Nutrition Examination Survey, 1999–2004. *J. Am. Diet. Assoc.* **2010**, *110*, 1461–1468. [CrossRef] [PubMed]
4. Papanikolaou, Y.; Fulgoni, V.L. Certain grain foods can be meaningful contributors to nutrient density in the diets of US chil-dren and adolescents: Data from the National Health and Nutrition Examination Survey, 2009–2012. *Nutrients* **2017**, *9*, 160. [CrossRef]
5. Olugbire, O.O.; Olorunfemi, S.; Oke, D.O. Global utilisation of cereals: Sustainability and environmental issues. *Agro-Science* **2021**, *20*, 9–14. [CrossRef]
6. Atkinson, N.J.; Urwin, P.E. The interaction of plant biotic and abiotic stresses: From genes to the field. *J. Exp. Bot.* **2012**, *63*, 3523–3543. [CrossRef] [PubMed]
7. Maiti, R.; Satya, P. Research advances in major cereal crops for adaptation to abiotic stresses. *GM Crop. Food* **2014**, *5*, 259–279. [CrossRef] [PubMed]
8. Halford, N.G.; Curtis, T.Y.; Chen, Z.; Huang, J. Effects of abiotic stress and crop management on cereal grain composition: Implications for food quality and safety. *J. Exp. Bot.* **2014**, *66*, 1145–1156. [CrossRef]
9. Takeda, S.; Matsuoka, M. Genetic approaches to crop improvement: Responding to environmental and population changes. *Nat. Rev. Genet.* **2008**, *9*, 444–457. [CrossRef]
10. Newton, A.C.; Johnson, S.N.; Gregory, P.J. Implications of climate change for diseases, crop yields and food security. *Euphytica* **2011**, *179*, 3–18. [CrossRef]
11. Mohan, V.; Tresina, P.; Daffodil, E. Antinutritional Factors in Legume Seeds: Characteristics and Determination. *Encycl. Food Health* **2016**, 211–220. [CrossRef]
12. Piltz, J.W.; Rodham, C.A.; Wilkins, J.F.; Hackney, B.F. A Comparison of Cereal and Cereal/Vetch Crops for Fodder Conser-vation. *Agriculture* **2021**, *11*, 459. [CrossRef]
13. Chaudhari, P.R.; Tamrakar, N.; Singh, L.; Tandon, A.; Sharma, D. Rice nutritional and medicinal properties. *J. Pharmacogn. Phytochem.* **2018**, *7*, 150–156.
14. Muraki, I.; Wu, H.; Imamura, F.; Laden, F.; Rimm, E.B.; Hu, F.B.; Willett, W.C.; Sun, Q. Rice consumption and risk of cardio-vascular disease: Results from a pooled analysis of 3 US cohorts. *Am. J. Clin. Nutr.* **2015**, *101*, 164–172. [CrossRef] [PubMed]
15. Tan, B.L.; Norhaizan, M.E. Scientific evidence of rice by-products for cancer prevention: Chemopreventive properties of waste products from rice milling on carcinogenesis in vitro and in vivo. *Biomed. Res. Int.* **2017**, *2017*, 9017902. [CrossRef]
16. Okuda, M.; Fujita, Y.; Katsube, T.; Tabata, H.; Yoshino, K.; Hashimoto, M.; Sugimoto, H. Highly water pressurized brown rice improves cognitive dysfunction in senescence-accelerated mouse prone 8 and reduces amyloid beta in the brain. *BMC Complement. Altern. Med.* **2018**, *18*, 110. [CrossRef] [PubMed]
17. Almeida, D.M.; Almadanim, M.C.; Lourenço, T.; Abreu, I.A.; Saibo, N.J.M.; Oliveira, M.M. Screening for Abiotic Stress Tolerance in Rice: Salt, Cold, and Drought. In *Environmental Responses in Plants*; Humana Press: New York, NY, USA, 2016; pp. 155–182.
18. Hasanuzzaman, M.; Nahar, K.; Fujita, M.; Ahmad, P.; Chandna, R.; Prasad, M.N.V.; Ozturk, M. Enhancing plant productivity under salt stress: Relevance of poly-omics. In *Salt Stress in Plants*; Springer: New York, NY, USA, 2013; pp. 113–156.
19. Lee, K.S.; Choi, W.Y.; Ko, J.C.; Kim, T.S.; Gregoria, G.B. Salinity tolerance of japonica and indica rice (*Oryza sativa* L.) at the seedling stage. *Planta* **2003**, *216*, 1043–1046. [CrossRef]
20. Todaka, D.; Nakashima, K.; Shinozaki, K.; Yamaguchi-Shinozaki, K. Towards understanding transcriptional regulatory net-works in abiotic stress responses and tolerance in rice. *Rice* **2012**, *5*, 1–9. [CrossRef] [PubMed]
21. Das, P.; Nutan, K.K.; Singla-Pareek, S.N.; Pareek, A. Understanding salinity responses and adopting 'omics-based' approaches to generate salinity tolerant cultivars of rice. *Front. Plant Sci.* **2015**, *6*, 712. [CrossRef]
22. Alam, M.; Bell, R.W.; Hasanuzzaman, M.; Salahin, N.; Rashid, M.H.; Akter, N.; Akhter, S.; Islam, M.S.; Islam, S.; Naznin, S.; et al. Rice (*Oryza sativa* L.) establishment techniques and their implications for soil properties, global warming potential mitigation and crop yields. *Agronomy* **2020**, *10*, 888. [CrossRef]
23. Macauley, H.; Ramadjita, T. Cereal crops: Rice, maize, millet, sorghum, wheat. *Feed. Afr.* **2015**, *36*.
24. Sandhu, K.S.; Singh, N.; Malhi, N.S. Some properties of corn grains and their flours I: Physicochemical, functional and chapati-making properties of flours. *Food Chem.* **2007**, *101*, 938–946. [CrossRef]
25. Shah, T.R.; Prasad, K.; Kumar, P. Maize—A potential source of human nutrition and health: A review. *Cogent Food Agric.* **2016**, *2*, 1166995.
26. Liu, R.H. Potential synergy of phytochemicals in cancer prevention: Mechanism of action. *J. Nutr.* **2004**, *134*, 3479–3485. [CrossRef]
27. Díaz-Gómez, J.L.; Castorena-Torres, F.; Preciado-Ortiz, R.E.; García-Lara, S. Anti-cancer activity of maize bioactive peptides. *Front. Chem.* **2017**, *5*, 44. [CrossRef]
28. Iqbal, S.; Hussain, S.; Qayyaum, M.A.; Ashraf, M. The Response of Maize Physiology under Salinity Stress and Its Coping Strategies. In *Plant Stress Physiology*; IntechOpen: London, UK, 2020.

29. Sade, N.; del Mar Rubio-Wilhelmi, M.; Umnajkitikorn, K.; Blumwald, E. Stress-induced senescence and plant tolerance to abiotic stress. *J. Exp. Bot.* **2018**, *69*, 845–853. [CrossRef] [PubMed]
30. Shewry, P.R.; Hey, S.J. The contribution of wheat to human diet and health. *Food Energy Secur.* **2015**, *4*, 178–202. [CrossRef]
31. Hura, T. Wheat and Barley: Acclimatization to Abiotic and Biotic Stress. *Int. J. Mol. Sci.* **2020**, *21*, 7423. [CrossRef]
32. Luthria, D.L.; Lu, Y.; John, K.M. Bioactive phytochemicals in wheat: Extraction, analysis, processing, and functional properties. *J. Funct. Foods* **2015**, *18*, 910–925. [CrossRef]
33. Barros, L.; Fernandes, Â.; C.F.R. Ferreira, I.; Callejo, M.; Matallana-González, M.; Fernández-Ruiz, V.; Morales, P.; Carrillo, J.M. Potential health claims of durum and bread wheat flours as functional ingredients. *Nutrients* **2020**, *12*, 504.
34. Aune, D.; Keum, N.; Giovannucci, E.; Fadnes, L.T.; Boffetta, P.; Greenwood, D.C.; Tonstad, S.; Vatten, L.J.; Riboli, E.; Norat, T. Whole grain consumption and risk of cardiovascular disease, cancer, and all cause and cause specific mortality: Systematic review and dose-response meta-analysis of prospective studies. *BMJ* **2016**, *353*, i2716. [CrossRef] [PubMed]
35. Della Pepa, G.; Vetrani, C.; Vitale, M.; Riccardi, G. Wholegrain intake and risk of type 2 diabetes: Evidence from epidemiological and intervention studies. *Nutrients* **2018**, *10*, 1288. [CrossRef] [PubMed]
36. Bedő, Z.; Láng, L.; Rakszegi, M. Breeding for grain-quality traits. In *Cereal Grains*; Woodhead Publishing: Cambridge, UK, 2017; pp. 425–452.
37. Loskutov, I.G.; Khlestkina, E.K. Wheat, Barley, and Oat Breeding for Health Benefit Components in Grain. *Plants* **2021**, *10*, 86. [CrossRef] [PubMed]
38. Schapendonk, A.H.C.M.; Xu, H.Y.; Van Der Putten, P.E.L.; Spiertz, J.H.J. Heat-shock effects on photosynthesis and sink-source dynamics in wheat (*Triticum aestivum* L.). *NJAS-Wagening. J. Life Sci.* **2007**, *55*, 37–54. [CrossRef]
39. Srivastava, N. Biochemical and molecular responses in higher plants under salt stress. In *Plant Adaptation Strategies in Changing Environment*; Springer: Singapore, 2017; pp. 117–151.
40. Shah, T.; Xu, J.; Zou, X.; Cheng, Y.; Nasir, M.; Zhang, X. Omics approaches for engineering wheat production under abiotic stresses. *Int. J. Mol. Sci.* **2018**, *19*, 2390. [CrossRef]
41. Calone, R.; Sanoubar, R.; Lambertini, C.; Speranza, M.; Antisari, L.V.; Vianello, G.; Barbanti, L. Salt tolerance and Na allocation in Sorghum bicolor under variable soil and water salinity. *Plants* **2020**, *9*, 561. [CrossRef] [PubMed]
42. Ramatoulaye, F.; Mady, C.; Fallou, S. Production and use sorghum: A literature review. *J. Nutr. Health Food Sci.* **2016**, *4*, 1–4.
43. Rao, S.; Santhakumar, A.B.; Chinkwo, K.A.; Wu, G.; Johnson, S.K.; Blanchard, C.L. Characterization of phenolic compounds and antioxidant activity in sorghum grains. *J. Cereal Sci.* **2008**, *84*, 103–111. [CrossRef]
44. Xiong, Y.; Zhang, P.; Warner, R.D.; Fang, Z. Sorghum grain: From genotype, nutrition, and phenolic profile to its health benefits and food applications. *Compr. Rev. Food Sci. Food Saf.* **2019**, *18*, 2025–2046. [CrossRef] [PubMed]
45. Ba, K.; Tine, E.; Destain, J.; Cisse, N.; Thonart, P. Comparative study of phenolic compounds, the antioxidant power of various Senegalese sorghum varieties anda amylolytic enzymes of their malt. *Biotechnol. Agron. Société Environ.* **2010**, *14*, 131–139.
46. Djanaguiraman, M.; Prasad, P.V.; Ciampitti, I.A.; Talwar, H.S. Impacts of Abiotic Stresses on Sorghum Physiology. In *Sorghum in the 21st Century: Food–Fodder–Feed–Fuel for a Rapidly Changing World*; Springer: Singapore, 2020; pp. 157–188.
47. Abdel-Ghany, S.E.; Ullah, F.; Ben-Hur, A.; Reddy, A.S. Transcriptome analysis of drought-resistant and drought-sensitive sorghum (Sorghum bicolor) genotypes in response to peg-induced drought stress. *Int. J. Mol. Sci.* **2020**, *21*, 772. [CrossRef]
48. Muthuramalingam, P.; Jeyasri, R.; Kalaiyarasi, D.; Pandian, S.; Krishnan, S.R.; Satish, L.; Pandian, S.K.; Ramesh, M. Emerging advances in computational omics tools for systems analysis of gramineae family grass species and their abiotic stress responsive functions. *OMICS-Based Approach Plant Biotechnol.* **2019**, *185*, 185.
49. Sahu, M.; Dehury, B.; Modi, M.K.; Barooah, M. Functional Genomics and Bioinformatics Approach to Understand Regulation of Abiotic Stress in Cereal Crops. In *Crop Improvement in the Era of Climate Change*; I.K. International Publishing House Pvt. Ltd.: Delhi, India, 2014; p. 205.
50. Gong, Y.; Hao, Y.; Li, J.; Li, H.; Shen, Z.; Wang, W.; Wang, S. The effects of rainfall runoff pollutants on plant physiology in a bioretention system based on pilot experiments. *Sustainability* **2019**, *11*, 6402. [CrossRef]
51. Tomás, D.; Rodrigues, J.C.; Viegas, W.; Silva, M. Assessment of high temperature effects on grain yield and composition in bread wheat commercial varieties. *Agronomy* **2020**, *10*, 499. [CrossRef]
52. Kai, H.; Iba, K. Temperature stress in plants. In *eLS*; John Wiley & Sons, Ltd: Chichester, UK, 2014.
53. Beck, E.H.; Heim, R.; Hansen, J. Plant resistance to cold stress: Mechanisms and environmental signals triggering frost hardening and dehardening. *J. Biosci.* **2004**, *29*, 449–459. [CrossRef]
54. Baek, K.H.; Skinner, D.Z. Production of reactive oxygen species by freezing stress and the protective roles of antioxidant enzymes in plants. *J. Agric. Chem. Environ.* **2012**, *1*, 34–40. [CrossRef]
55. Cassia, R.; Nocioni, M.; Correa-Aragunde, N.; Lamattina, L. Climate change and the impact of greenhouse gasses: CO2 and NO, friends and foes of plant oxidative stress. *Front. Plant Sci.* **2018**, *9*, 273. [CrossRef] [PubMed]
56. Foroughbakhch Pournavab, R.; Bacópulos Mejía, E.; Benavides Mendoza, A.; Salas Cruz, L.R.; Ngangyo Heya, M. Ultraviolet radiation effect on seed germination and seedling growth of common species from Northeastern Mexico. *Agronomy* **2019**, *9*, 269. [CrossRef]
57. Metwally, S.A.; Shoaib, R.M.; Hashish, K.I.; El-Tayeb, T.A. In vitro ultraviolet radiation effects on growth, chemical constituents and molecular aspects of Spathiphyllum plant. *Bull. Natl. Res. Cent.* **2019**, *43*, 94. [CrossRef]

58. Rao, M.P.N.; Dong, Z.Y.; Xiao, M.; Li, W.J. Effect of salt stress on plants and role of microbes in promoting plant growth under salt stress. In *Microorganisms in Saline Environments: Strategies and Functions*; Springer: Cham, Switzerland, 2019; pp. 423–435.
59. Borsani, O.; Dıaz, P.; Agius, M.F.; Valpuesta, V.; Monza, J. Water stress generates an oxidative stress through the induction of a specific Cu/Zn superoxide dismutase in Lotus corniculatus leaves. *Plant Sci.* **2001**, *161*, 757–763. [CrossRef]
60. Tarakcioglu, C.; Inal, A. Changes induced by salinity, demarcating specific ion ratio (Na/Cl) and osmolality in ion and proline accumulation, nitrate reductase activity, and growth performance of lettuce. *J. Plant Nutr.* **2002**, *25*, 27–41. [CrossRef]
61. Eraslan, F.; Inal, A.; Gunes, A.; Alpaslan, M. Impact of exogenous salicylic acid on the growth, antioxidant activity and physiology of carrot plants subjected to combined salinity and boron toxicity. *Sci. Hortic.* **2007**, *113*, 120–128. [CrossRef]
62. Ahmad, P.; Prasad, M.N.V. *Abiotic Stress Responses in Plants: Metabolism, Productivity and Sustainability*; Springer Science & Business Media: New York, USA, 2011.
63. Yildiz, M.; Poyraz, İ.; Çavdar, A.; Özgen, Y.; Beyaz, R. Plant Responses to Salt Stress. In *Plant Breeding-Current and Future Views*; IntechOpen: London, UK, 2020.
64. Shelden, M.C.; Roessner, U. Advances in functional genomics for investigating salinity stress tolerance mechanisms in cereals. *Front. Plant Sci.* **2013**, *4*, 123. [CrossRef]
65. Rajendran, K.; Tester, M.; Roy, S.J. Quantifying the three main components of salinity tolerance in cereals. *Plant Cell Environ.* **2009**, *32*, 237–249. [CrossRef] [PubMed]
66. Roy, S.J.; Negrão, S.; Tester, M. Salt resistant crop plants. *Curr. Opin. Biotechnol.* **2014**, *26*, 115–124. [CrossRef]
67. Reddy, I.N.B.L.; Kim, B.K.; Yoon, I.S.; Kim, K.H.; Kwon, T.R. Salt tolerance in rice: Focus on mechanisms and approaches. *Rice Sci.* **2017**, *24*, 123–144. [CrossRef]
68. Emamverdian, A.; Ding, Y.; Mokhberdoran, F.; Xie, Y. Heavy metal stress and some mechanisms of plant defense response. *Sci. World J.* **2015**, *2015*, 75612. [CrossRef] [PubMed]
69. Shah, F.; Wu, W. Soil and crop management strategies to ensure higher crop productivity within sustainable environments. *Sustainability* **2019**, *11*, 1485. [CrossRef]
70. Oladosu, Y.; Rafii, M.Y.; Samuel, C.; Fatai, A.; Magaji, U.; Kareem, I.; Kamarudin, Z.S.; Muhammad, I.I.; Kolapo, K. Drought resistance in rice from conventional to molecular breeding: A review. *Int. J. Mol. Sci.* **2019**, *20*, 3519. [CrossRef]
71. Rosero, A.; Berdugo-Cely, J.A.; Šamajová, O.; Šamaj, J.; Cerkal, R. A Dual Strategy of Breeding for Drought Tolerance and Introducing Drought-Tolerant, Underutilized Crops into Production Systems to Enhance Their Resilience to Water Deficiency. *Plants* **2020**, *9*, 1263. [CrossRef]
72. Liliane, T.N.; Charles, M.S. Factors Affecting Yield of Crops. In *Agronomy-Climate Change & Food Security*; IntechOpen: London, UK, 2020; p. 9.
73. Chaudhary, J.; Khatri, P.; Singla, P.; Kumawat, S.; Kumari, A.; Vikram, A.; Jindal, S.K.; Kardile, H.; Kumar, R.; Sonah, H.; et al. Advances in omics approaches for abiotic stress tolerance in tomato. *Biology* **2019**, *8*, 90. [CrossRef] [PubMed]
74. Razzaq, M.K.; Aleem, M.; Mansoor, S.; Khan, M.A.; Rauf, S.; Iqbal, S.; Siddique, K.H. Omics and CRISPR-Cas9 Approaches for Molecular Insight, Functional Gene Analysis, and Stress Tolerance Development in Crops. *Int. J. Mol. Sci.* **2021**, *22*, 1292. [CrossRef] [PubMed]
75. Ismail, A.M.; Horie, T. Genomics, physiology, and molecular breeding approaches for improving salt tolerance. *Annu. Rev. Plant Biol.* **2017**, *68*, 405–434. [CrossRef] [PubMed]
76. Tiwari, J.K.; Plett, D.; Garnett, T.; Chakrabarti, S.K.; Singh, R.K. Integrated genomics, physiology and breeding approaches for improving nitrogen use efficiency in potato: Translating knowledge from other crops. *Funct. Plant Biol.* **2018**, *45*, 587–605. [CrossRef]
77. Chantre Nongpiur, R.; Lata Singla-Pareek, S.; Pareek, A. Genomics approaches for improving salinity stress tolerance in crop plants. *Curr. Genom.* **2016**, *17*, 343–357. [CrossRef] [PubMed]
78. Muthuramalingam, P.; Krishnan, S.R.; Saravanan, K.; Mareeswaran, N.; Kumar, R.; Ramesh, M. Genome-wide identification of major transcription factor superfamilies in rice identifies key candidates involved in abiotic stress dynamism. *J. Plant Biochem. Biotechnol.* **2018**, *27*, 300–317. [CrossRef]
79. Muthuramalingam, P.; Jeyasri, R.; Bharathi, R.K.A.S.; Suba, V.; Pandian, S.T.K.; Ramesh, M. Global integrated omics expression analyses of abiotic stress signaling HSF transcription factor genes in *Oryza sativa* L.: An in silico approach. *Genomics* **2020**, *112*, 908–918. [CrossRef]
80. Le, T.D.; Gathignol, F.; Vu, H.T.; Nguyen, K.L.; Tran, L.H.; Vu, H.T.T.; Dinh, T.X.; Lazennec, F.; Pham, X.H.; Véry, A.-A.; et al. Genome-Wide Association Mapping of Salinity Tolerance at the Seedling Stage in a Panel of Vietnamese Landraces Reveals New Valuable QTLs for Salinity Stress Tolerance Breeding in Rice. *Plants* **2021**, *10*, 1088. [CrossRef]
81. Chan, A.P.; Pertea, G.; Cheung, F.; Lee, D.; Zheng, L.; Whitelaw, C.; Pontaroli, A.C.; San Miguel, P.; Yuan, Y.; Bennetzen, J.; et al. The TIGR maize database. *Nucleic Acids Res.* **2006**, *34*, D771–D776. [CrossRef]
82. Altschul, S.F.; Gish, W.; Miller, W.; Myers, E.W.; Lipman, D.J. Basic local alignment search tool. *J. Mol. Biol* **1990**, *215*, 403–410. [CrossRef]
83. Bairoch, A.; Apweiler, R. The SWISS-PROT protein sequence database and its supplement TrEMBL in 2000. *Nucleic Acids Res.* **2000**, *28*, 45–48. [CrossRef]

84. Jeyasri, R.; Muthuramalingam, P.; Satish, L.; Adarshan, S.; Aishwarya Lakshmi, M.; Pandian, S.K.; Chen, J.T.; Ahmar, S.; Wang, X.; Freddy, M.P.; et al. The role of *OsWRKY* genes in rice when faced single and multiple abiotic stresses. *Agronomy* **2021**, *11*, 1301. [CrossRef]
85. Wang, Z.L.; Li, P.H.; Fredricksen, M.; Gong, Z.Z.; Kim, C.S.; Zhang, C.; Bohnert, H.J.; Zhu, J.K.; Bressan, R.A.; Hasegawa, P.M.; et al. Expressed sequence tags from Thellungiella halophila, a new model to study plant salt-tolerance. *Plant Sci.* **2004**, *166*, 609–616. [CrossRef]
86. Kawasaki, S.; Borchert, C.; Deyholos, M.; Wang, H.; Brazille, S.; Kawai, K.; Galbraith, D.; Bohnert, H.J. Gene expression profiles during the initial phase of salt stress in rice. *Plant Cell* **2001**, *13*, 889–905. [CrossRef] [PubMed]
87. Sreenivasulu, N.; Altschmied, L.; Radchuk, V.; Gubatz, S.; Wobus, U.; Weschke, W. Transcript profiles and deduced changes of metabolic pathways in maternal and filial tissues of developing barley grains. *Plant Jour* **2004**, *37*, 539–553. [CrossRef] [PubMed]
88. Lee, J.; Williams, M.E.; Tingey, S.V.; Rafalski, J.A. DNA array profiling of gene expression changes during maize embryo development. *Funct. Integr. Genom.* **2002**, *2*, 13–27. [CrossRef] [PubMed]
89. Buitink, J.; Leger, J.J.; Guisle, I.; Vu, B.L.; Wuilleme, S.; Lamirault, G.; Le Bars, A.; Le Meur, N.; Becker, A.; Küster, H.; et al. Transcriptome profiling uncovers metabolic and regulatory processes occurring during the transition from desiccation sensitive to desiccation-tolerant stages in Medicago truncatula seeds. *Plant J.* **2006**, *47*, 735–750. [CrossRef] [PubMed]
90. Liu, Y.; Ji, X.; Zheng, L.; Nie, X.; Wang, Y. Microarray analysis of transcriptional responses to abscisic acid and salt stress in Arabidopsis thaliana. *Int. J. Mol. Sci.* **2013**, *14*, 9979–9998. [CrossRef] [PubMed]
91. Rensink, W.A.; Buell, C.R. Microarray expression profiling resources for plant genomics. *Trends Plant Sci.* **2005**, *10*, 603–609. [CrossRef] [PubMed]
92. Li, L.; Deng, X.W. Microarray-based Approaches to Rice Transcriptome Analysis. In *Rice Biology in the Genomics Era*; Springer: Berlin/Heidelberg, Germany, 2008; pp. 37–51.
93. Kumar, R.; Sharma, V.; Suresh, S.; Ramrao, D.P.; Veershetty, A.; Kumar, S.; Priscilla, K.; Hangargi, B.; Narasanna, R.; Pandey, M.K.; et al. Understanding Omics Driven Plant Improvement and de novo Crop Domestication: Some Examples. *Front. Genet.* **2021**, *12*, 415. [CrossRef] [PubMed]
94. Yilmaz, A.; Nishiyama, M.Y., Jr.; Fuentes, B.G.; Souza, G.M.; Janies, D.; Gray, J.; Grotewold, E. GRASSIUS: A platform for comparative regulatory genomics across the grasses. *Plant Physiol.* **2009**, *149*, 171–180. [CrossRef] [PubMed]
95. Jin, J.; Tian, F.; Yang, D.C.; Meng, Y.Q.; Kong, L.; Luo, J.; Gao, G. PlantTFDB 4.0: Toward a central hub for transcription factors and regulatory interactions in plants. *Nucleic Acids Res.* **2016**, *45*, gkw982. [CrossRef]
96. Roychoudhury, A.; Paul, S.; Basu, S. Cross-talk between abscisic acid-dependent and abscisic acid-independent pathways during abiotic stress. *Plant Cell Rep.* **2013**, *32*, 985–1006. [CrossRef] [PubMed]
97. Czolpinska, M.; Rurek, M. Plant glycine-rich proteins in stress response: An emerging, still prospective story. *Front. Plant Sci.* **2018**, *9*, 302. [CrossRef]
98. Zenda, T.; Liu, S.; Dong, A.; Duan, H. Advances in Cereal Crop Genomics for Resilience under Climate Change. *Life* **2021**, *11*, 502. [CrossRef]
99. Koyro, H.W.; Ahmad, P.; Geissler, N. Abiotic stress responses in plants: An overview. In *Environmental Adaptations and Stress Tolerance of Plants in the Era of Climate Change*; Springer Science & Business Media: New York, NY, USA, 2012; pp. 1–28.
100. Seki, M.; Narusaka, M.; Abe, H.; Kasuga, M.; Yamaguchi-Shinozaki, K.; Carninci, P.; Hayashizaki, Y.; Shinozaki, K. Monitoring the expression pattern of 1300 *Arabidopsis* genes under drought and cold stresses by using a full-length cDNA microarray. *Plant Cell* **2001**, *13*, 61–72. [CrossRef] [PubMed]
101. Seki, M.; Narusaka, M.; Ishida, J.; Nanjo, T.; Fujita, M.; Oono, Y.; Kamiya, A.; Nakajima, M.; Enju, A.; Sakurai, T. Monitoring the expression profiles of 7000 *Arabidopsis* genes under drought, cold and high-salinity stresses using a full-length cDNA microarray. *Plant Jour* **2002**, *31*, 279–292. [CrossRef]
102. Fowler, S.; Thomashow, M.F. *Arabidopsis* transcriptome profiling indicates that multiple regulatory pathways are activated during cold acclimation in addition to the CBF cold response pathway. *Plant Cell* **2002**, *14*, 1675–1690. [CrossRef]
103. Lenka, S.K.; Katiyar, A.; Chinnusamy, V.; Bansal, K.C. Comparative analysis of drought-responsive transcriptome in Indica rice genotypes with contrasting drought tolerance. *Plant Biotechnol. J.* **2011**, *9*, 315–327. [CrossRef]
104. Dubouzet, J.G.; Sakuma, Y.; Ito, Y.; Kasuga, M.; Dubouzet, E.G.; Miura, S.; Seki, M.; Shinozaki, K.; Yamaguchi- Shinozaki, K. OsDREB genes in rice, *Oryza sativa* L., encode transcription activators that function in drought-, high-salt- and cold responsive gene expression. *Plant. J.* **2003**, *33*, 751–763. [CrossRef]
105. Paul, S.; Roychoudhury, A. Transcriptome profiling of abiotic stress-responsive genes during cadmium chloride-mediated stress in two indica rice varieties. *J. Plant Growth Regul.* **2018**, *37*, 657–667. [CrossRef]
106. Kathuria, H.; Giri, J.; Nataraja, K.N.; Murata, N.; Udayakumar, M.; Tyagi, A.K. Glycinebetaine-induced water-stress tolerance in codA-expressing transgenic indica rice is associated with up-regulation of several stress responsive genes. *Plant Biotechnol. J.* **2009**, *7*, 512–526. [CrossRef]
107. Quan, R.; Shang, M.; Zhang, H.; Zhao, Y.; Zhang, J. Engineering of enhanced glycine betaine synthesis improves drought tolerance in maize. *Plant Biotechnol. J.* **2004**, *2*, 477–486. [CrossRef] [PubMed]
108. Maghsoudi, K.; Emam, Y.; Niazi, A.; Pessarakli, M.; Arvin, M.J. P5CS expression level and proline accumulation in the sensitive and tolerant wheat cultivars under control and drought stress conditions in the presence/absence of silicon and salicylic acid. *J. Plant Interact.* **2018**, *13*, 461–471. [CrossRef]

109. Ma, H.; Liu, C.; Li, Z.; Ran, Q.; Xie, G.; Wang, B.; Fang, S.; Chu, J.; Zhang, J. ZmbZIP4 contributes to stress resistance in maize by regulating ABA synthesis and root development. *Plant Physiol.* **2018**, *178*, 753–770. [CrossRef]
110. Das, P.; Lakra, N.; Nutan, K.K.; Singla-Pareek, S.L.; Pareek, A. A unique bZIP transcription factor imparting multiple stress tolerance in Rice. *Rice* **2019**, *12*, 58. [CrossRef]
111. Pandey, A.S.; Sharma, E.; Jain, N.; Singh, B.; Burman, N.; Khurana, J.P. A rice bZIP transcription factor, OsbZIP16, regulates abiotic stress tolerance when over-expressed in Arabidopsis. *J. Plant Biochem. Biotechnol.* **2018**, *27*, 393–400. [CrossRef]
112. Jiang, Z.; Song, G.; Shan, X.; Wei, Z.; Liu, Y.; Jiang, C.; Jiang, Y.; Jin, F.; Li, Y. Association analysis and identification of ZmHKT1; 5 variation with salt-stress tolerance. *Front. Plant Sci.* **2018**, *9*, 1485. [CrossRef] [PubMed]
113. Xu, B.; Hrmova, M.; Gilliham, M. High affinity Na + transport by wheat HKT1;5 is blocked by K +. *Plant Direct* **2020**, *4*, e00275. [CrossRef]
114. Basu, S.; Roychoudhury, A. Expression profiling of abiotic stress-inducible genes in response to multiple stresses in rice (*Oryza sativa* L.) varieties with contrasting level of stress tolerance. *Biomed. Res. Int.* **2014**, *2014*, 706890. [CrossRef]
115. Muthuramalingam, P.; Jeyasri, R.; Selvaraj, A.; Kalaiyarasi, D.; Aruni, W.; Pandian, S.T.K.; Ramesh, M. Global transcriptome analysis of novel stress associated protein (SAP) genes expression dynamism of combined abiotic stresses in *Oryza sativa* (L.). *J. Biomol. Struct. Dyn.* **2020**, *39*, 2106–2117. [CrossRef]
116. Li, S.; Zhao, Q.; Zhu, D.; Yu, J. A DREB-like transcription factor from maize (*Zea mays*), ZmDREB4. 1, plays a negative role in plant growth and development. *Front. Plant Sci.* **2018**, *9*, 395. [CrossRef]
117. Tang, Y.; Bao, X.; Zhi, Y.; Wu, Q.; Guo, Y.; Yin, X.; Zeng, L.; Li, J.; Zhang, J.; He, W.; et al. Overexpression of a MYB family gene, OsMYB6, increases drought and salinity stress tolerance in transgenic rice. *Front. Plant Sci.* **2019**, *10*, 168. [CrossRef]
118. Tchagang, A.B.; Fauteux, F.; Tulpan, D.; Pan, Y. Bioinformatics identification of new targets for improving low temperature stress tolerance in spring and winter wheat. *BMC Bioinform.* **2017**, *18*, 1746. [CrossRef] [PubMed]
119. Iqbal, I.; Tripathi, R.K.; Wilkins, O.; Singh, J. Thaumatin-like Protein (TLP) gene family in barley: Genome-wide exploration and expression analysis during germination. *Genes* **2020**, *11*, 1080. [CrossRef]
120. Liu, A.L.; Zou, J.; Liu, C.F.; Zhou, X.Y.; Zhang, X.W.; Luo, G.Y.; Chen, X.B. Over-expression of OsHsfA7 enhanced salt and drought tolerance in transgenic rice. *BMB Rep.* **2013**, *46*, 31. [CrossRef]
121. Muthusamy, S.K.; Dalal, M.; Chinnusamy, V.; Bansal, K.C. Genome-wide identification and analysis of biotic and abiotic stress regulation of small heat shock protein (HSP20) family genes in bread wheat. *J. Plant Physiol.* **2017**, *211*, 100–113. [CrossRef] [PubMed]
122. El-Esawi, M.A.; Alayafi, A.A. Overexpression of rice Rab7 gene improves drought and heat tolerance and increases grain yield in rice (*Oryza sativa* L.). *Genes* **2019**, *10*, 56. [CrossRef] [PubMed]
123. Chen, C.; Heo, J.B. Overexpression of constitutively active OsRab11 in plants enhances tolerance to high salinity levels. *J. Plant. Biol.* **2018**, *61*, 169–176. [CrossRef]
124. Yao, X.; Wu, K.; Yao, Y.; Li, J.; Ren, Y.; Chi, D. The response mechanism of the HVA1 gene in hulless barley under drought stress. *Ital. J. Agron.* **2017**, *12*, 804. [CrossRef]
125. Chen, Y.S.; Lo, S.F.; Sun, P.K.; Lu, C.A.; Ho, T.H.D.; Yu, S.M. A late embryogenesis abundant protein HVA 1 regulated by an inducible promoter enhances root growth and abiotic stress tolerance in rice without yield penalty. *Plant Biotechnol. J.* **2015**, *13*, 105–116. [CrossRef]
126. Gürel, F.; Öztürk, Z.N.; Uçarlı, C.; Rosellini, D. Barley genes as tools to confer abiotic stress tolerance in crops. *Front. Plant Sci.* **2016**, *7*, 1137. [CrossRef] [PubMed]
127. Zhang, Z.; Zhang, Q.; Wu, J.; Zheng, X.; Zheng, S.; Sun, X.; Qiu, Q.; Lu, T. Gene knockout study reveals that cytosolic ascorbate peroxidase 2 (OsAPX2) plays a critical role in growth and reproduction in rice under drought, salt and cold stresses. *PLoS ONE* **2013**, *8*, e57472. [CrossRef]
128. Luna, C.M.; Pastori, G.M.; Driscoll, S.; Groten, K.; Bernard, S.; Foyer, C.H. Drought controls on H2O2 accumulation, catalase (CAT) activity and CAT gene expression in wheat. *J. Exp. Bot.* **2005**, *56*, 417–423. [CrossRef] [PubMed]
129. Shiraya, T.; Mori, T.; Maruyama, T.; Sasaki, M.; Takamatsu, T.; Oikawa, K.; Itoh, K.; Kaneko, K.; Ichikawa, H.; Mitsui, T. Golgi/plastid-type manganese superoxide dismutase involved in heat-stress tolerance during grain filling of rice. *Plant Biotechnol. J.* **2015**, *13*, 1251–1263. [CrossRef] [PubMed]
130. Guan, Q.; Liao, X.; He, M.; Li, X.; Wang, Z.; Ma, H.; Yu, S.; Liu, S. Tolerance analysis of chloroplast OsCu/Zn-SOD overexpressing rice under NaCl and NaHCO3 stress. *PLoS ONE* **2017**, *12*, e0186052. [CrossRef] [PubMed]
131. de Deus, K.E.; Lanna, A.C.; Abreu, F.R.M.; Silveira, R.D.D.; Pereira, W.J.; Brondani, C.; Vianello, R.P. Molecular and biochemical characterization of superoxide dismutase (SOD) in upland rice under drought. *Embrapa Arroz Feijão-Artigo Periódico Indexado (ALICE)* **2015**, *9*, 744–753.
132. Xiao, B.Z.; Chen, X.; Xiang, C.B.; Tang, N.; Zhang, Q.F.; Xiong, L.Z. Evaluation of seven function-known candidate genes for their effects on improving drought resistance of transgenic rice under field conditions. *Mol. Plant* **2009**, *2*, 73–83. [CrossRef]
133. Xie, G.; Kato, H.; Sasaki, K.; Imai, R. A cold-induced thioredoxinh of rice, OsTrx23, negatively regulates kinase activities of OsMPK3 and OsMPK6 in vitro. *FEBS Lett.* **2009**, *583*, 2734–2738. [CrossRef]
134. Menguer, P.K.; Sperotto, R.A.; Ricachenevsky, F.K. A walk on the wild side: Oryza species as source for rice abiotic stress tolerance. Genet. *Mol. Biol.* **2017**, *40*, 238–252. [CrossRef]

135. Mohanty, A.; Kathuria, H.; Ferjani, A.; Sakamoto, A.; Mohanty, P.; Murata, N.; Tyagi, A. Transgenics of an elite indica rice variety Pusa Basmati 1 harbouring the codA gene are highly tolerant to salt stress. *Theor. Appl. Genet.* **2002**, *106*, 51–57. [CrossRef]
136. Thomas, S.; Krishna, G.K.; Yadav, P.; Pal, M. Cloning and abiotic stress responsive expression analysis of Arginine decarboxylase genes in contrasting rice genotypes. *Indian J. Genet.* **2019**, *79*, 411–419. [CrossRef]
137. Lee, H.J.; Abdula, S.E.; Jang, D.W.; Park, S.H.; Yoon, U.H.; Jung, Y.J.; Kang, K.K.; Nou, I.S.; Cho, Y.G. Overexpression of the glutamine synthetase gene modulates oxidative stress response in rice after exposure to cadmium stress. *Plant Cell Rep.* **2013**, *32*, 1521–1529. [CrossRef]
138. Garg, A.K.; Kim, J.K.; Owens, T.G.; Ranwala, A.P.; Do Choi, Y.; Kochian, L.V.; Wu, R.J. Trehalose Accumulation in Rice Plants Confers High Tolerance Levels to Different Abiotic Stresses. *Proc. Natl. Acad. Sci. USA* **2002**, *99*, 15898–15903. [CrossRef]
139. Abebe, T.; Guenzi, A.C.; Martin, B.; Cushman, J.C. Tolerance of mannitol-accumulating transgenic wheat to water stress and salinity. *Plant Physiol.* **2003**, *131*, 1748–1755. [CrossRef] [PubMed]
140. Wei, S.; Hu, W.; Deng, X.; Zhang, Y.; Liu, X.; Zhao, X.; Luo, Q.; Jin, Z.; Li, Y.; Zhou, S.; et al. A rice calcium-dependent protein kinase OsCPK9 positively regulates drought stress tolerance and spikelet fertility. *BMC Plant Biol.* **2014**, *14*, 133. [CrossRef] [PubMed]
141. Geda, C.K.; Repalli, S.K.; Dash, G.K.; Swain, P.; Rao, G.J.N. Enhancement of Drought Tolerance in Rice through Introgression of Arabidopsis DREB1A through Transgenic Approach. *J. Rice Res.* **2019**, *7*, 2.
142. Amin, U.S.M.; Biswas, S.; Elias, S.M.; Razzaque, S.; Haque, T.; Malo, R.; Seraj, Z.I. Enhanced salt tolerance conferred by the complete 2.3 kb cDNA of the rice vacuolar Na+/H+ antiporter gene compared to 1.9 kb coding region with 5′ UTR in transgenic lines of rice. *Front. Plant Sci.* **2016**, *7*, 14. [CrossRef] [PubMed]
143. Khan, I.; Mohamed, S.; Regnault, T.; Mieulet, D.; Guiderdoni, E.; Sentenac, H.; Véry, A.A. Constitutive contribution by the rice OsHKT1; 4 Na+ transporter to xylem sap desalinization and low Na+ accumulation in young leaves under low as high external Na+ conditions. *Front. Plant Sci.* **2020**, *11*, 1130. [CrossRef] [PubMed]
144. Mohanty, B.; Takahashi, H.; Benildo, G.; Wijaya, E.; Nakazono, M.; Lee, D.Y. Transcriptional regulatory mechanism of alcohol dehydrogenase 1-deficient mutant of rice for cell survival under complete submergence. *Rice* **2016**, *9*, 51. [CrossRef]
145. Zang, X.; Geng, X.; Wang, F.; Liu, Z.; Zhang, L.; Zhao, Y.; Tian, X.; Ni, Z.; Yao, Y.; Xin, M.; et al. Overexpression of wheat ferritin gene TaFER-5B enhances tolerance to heat stress and other abiotic stresses associated with the ROS scavenging. *BMC Plant Biol.* **2017**, *17*, 14. [CrossRef]
146. Quimio, C.A.; Torrizo, L.B.; Setter, T.L.; Ellis, M.; Grover, A.; Abrigo, E.M.; Oliva, N.P.; Ella, E.S.; Carpena, A.L.; Ito, O.; et al. Enhancement of submergence tolerance in transgenic rice overproducing pyruvate decarboxylase. *J. Plant Physiol.* **2000**, *156*, 516–521. [CrossRef]
147. Wang, Y.; Deng, C.; Ai, P.; Zhang, Z. ALM1, encoding a Fe-superoxide dismutase, is critical for rice chloroplast biogenesis and drought stress response. *Crop J.* **2020**. [CrossRef]
148. Saibi, W.; Brini, F. Superoxide dismutase (SOD) and abiotic stress tolerance in plants: An overview. In *Superoxide Dismutase: Structure, Synthesis and Applications*; Magliozzi, S., Ed.; Nova Science Publishers, Inc.: Hauppauge, NY, USA, 2018; pp. 101–142.
149. Alqudah, A.M.; Sallam, A.; Baenziger, P.S.; Börner, A. GWAS: Fast-forwarding gene identification and characterization in temperate Cereals: Lessons from Barley–A review. *J. Adv. Res.* **2020**, *22*, 119–135. [CrossRef] [PubMed]
150. Muthuramalingam, P.; Krishnan, S.R.; Pandian, S.; Mareeswaran, N.; Aruni, W.; Pandian, S.K.; Ramesh, M. Global analysis of threonine metabolism genes unravel key players in rice to improve the abiotic stress tolerance. *Sci. Rep.* **2018**, *8*, 9270. [CrossRef] [PubMed]
151. Mitchell-Olds, T. Complex-trait analysis in plants. *Genome Biol.* **2010**, *11*, 113. [CrossRef]
152. Atwell, S.; Huang, Y.S.; Vilhjálmsson, B.J.; Willems, G.; Horton, M.; Li, Y.; Meng, D.; Platt, A.; Tarone, A.M.; Hu, T.T.; et al. Genome-wide association study of 107 phenotypes in Arabidopsis thaliana inbred lines. *Nature* **2010**, *465*, 627–631. [CrossRef]

plants

MDPI

Review

Signal Transduction in Cereal Plants Struggling with Environmental Stresses: From Perception to Response

Małgorzata Nykiel [1,*], Marta Gietler [1], Justyna Fidler [1], Beata Prabucka [1], Anna Rybarczyk-Płońska [1], Jakub Graska [1], Dominika Boguszewska-Mańkowska [2], Ewa Muszyńska [3], Iwona Morkunas [4] and Mateusz Labudda [1]

1 Department of Biochemistry and Microbiology, Institute of Biology, Warsaw University of Life Sciences-SGGW, 02-776 Warsaw, Poland; marta_gietler@sggw.edu.pl (M.G.); justyna_fidler@sggw.edu.pl (J.F.); beata_prabucka@sggw.edu.pl (B.P.); anna_rybarczyk_plonska@sggw.edu.pl (A.R.-P.); jakubgraska1@gmail.com (J.G.); mateusz_labudda@sggw.edu.pl (M.L.)
2 Plant Breeding and Acclimatization Institute-National Research Institute, 05-870 Radzików, Poland; d.boguszewska-mankowska@ihar.edu.pl
3 Department of Botany, Institute of Biology, Warsaw University of Life Sciences-SGGW, 02-776 Warsaw, Poland; ewa_muszynska@sggw.edu.pl
4 Department of Plant Physiology, Poznań University of Life Sciences, Wołyńska 35, 60-637 Poznań, Poland; iwona.morkunas@gmail.com
* Correspondence: malgorzata_nykiel@sggw.edu.pl; Tel.: +48-22-593-2575

Abstract: Cereal plants under abiotic or biotic stressors to survive unfavourable conditions and continue growth and development, rapidly and precisely identify external stimuli and activate complex molecular, biochemical, and physiological responses. To elicit a response to the stress factors, interactions between reactive oxygen and nitrogen species, calcium ions, mitogen-activated protein kinases, calcium-dependent protein kinases, calcineurin B-like interacting protein kinase, phytohormones and transcription factors occur. The integration of all these elements enables the change of gene expression, and the release of the antioxidant defence and protein repair systems. There are still numerous gaps in knowledge on these subjects in the literature caused by the multitude of signalling cascade components, simultaneous activation of multiple pathways and the intersection of their individual elements in response to both single and multiple stresses. Here, signal transduction pathways in cereal plants under drought, salinity, heavy metal stress, pathogen, and pest attack, as well as the crosstalk between the reactions during double stress responses are discussed. This article is a summary of the latest discoveries on signal transduction pathways and it integrates the available information to better outline the whole research problem for future research challenges as well as for the creative breeding of stress-tolerant cultivars of cereals.

Keywords: abiotic stress; biotic stress; cereal; crosstalk; drought; heavy metal; phytohormone; salinity; pathogen; pest

Citation: Nykiel, M.; Gietler, M.; Fidler, J.; Prabucka, B.; Rybarczyk-Płońska, A.; Graska, J.; Boguszewska-Mańkowska, D.; Muszyńska, E.; Morkunas, I.; Labudda, M. Signal Transduction in Cereal Plants Struggling with Environmental Stresses: From Perception to Response. *Plants* **2022**, *11*, 1009. https://doi.org/10.3390/plants11081009

Academic Editor: Francisco M. del Amor Saavedra

Received: 16 March 2022
Accepted: 6 April 2022
Published: 7 April 2022

1. Introduction

The climatic conditions have changed many times during the history of Earth, but now, these alterations are strongly intensified by heavy industrial activity. As agriculture is a branch of the economy most dependent on climatic conditions, the progressing climate changes create a completely new situation for agricultural activity, especially for plant production [1]. The rate of temperature changes causes many extreme atmospheric phenomena that have not occurred or have appeared very rarely so far. Noticeable changes in the quantity and quality of rainfall (an increase in the number of storms followed by periods without rainfall) increase the risk of both flooding and drought. In addition, the increase in temperature may favour the overwintering of plant pathogens and pests, which have not so far posed a threat to native crops [2]. Moreover, in the era of global warming, the mobility of pollutants, including heavy metals in the environment increases [3].

Plant responses to environmental factors are extraordinarily complex. They can be observed at various levels of plant organisation, ranging from changes at the cell level, i.e., the changes in the activity of basic biochemical processes such as DNA replication, respiration, and photosynthesis to morphological and anatomical changes in plant organs [4–7]. However, mentioned biochemical changes are preceded by the activation of an efficient signalling system that endures environmental fluctuations [8]. The presented review highlights issues related to stress factor recognition/stress factor perception, induction, and transmission of the signal, and subsequent signalling responses at molecular and metabolic levels in cereals under the influence of various stress factors caused by global warming. This work shows the latest research results in the context of the defence mechanism induction of cereals to different abiotic stress and tolerance/existence under stressful environments.

It is well known that there are four main common signal transduction pathways in plants, which can interact with each other, i.e., reactive species signalling, calcium-dependent signalling, plant hormone signalling and signalling based on phosphorylation/dephosphorylation of proteins by kinase cascades [9]. Plant hormones are important regulators of the tailored responses to different stresses. The coordination of regulatory mechanisms among different hormones, or the interaction of hormone signalling with other molecules, such as reactive oxygen species (ROS), reactive nitrogen species (RNS), and hydrogen sulphide (H_2S), is flexible and changes over time [10–12].

Abiotic and biotic stressors can cause a state of excess excitation energy in plant cells which leads to universal consequences: disturbances in electron transport, increased reduction in plastoquinone and uncontrolled generation of ROS, mainly superoxide anion (O_2^-), hydrogen peroxide (H_2O_2), hydroxyl radical (HO^-) and singlet oxygen ($_1O^2$). The reaction of ROS with nitric oxide leads also to the formation of RNS, and potentially the appearance of not only oxidative but also nitrosative stress [13]. As ROS and RNS can act as a double-edged sword, they also play an important role in redox signalling as secondary messengers or signalling molecules and take part in signal transmission to the nucleus through redox reactions using the mitogen-activated protein kinases (MAPK) pathway and control of the antioxidant system in the plant cell [10]. Emerging signals are also facilitated by a series of phosphorylation/dephosphorylation events. This protein modification can both activate or deactivate effective proteins, which adjust cell metabolism to current environmental conditions with low energy costs. Phosphorylation cascade can be calcium-dependent (calcium-dependent protein kinases; CDPKs, calcineurin B-like interacting protein kinase; CIPK) or calcium-independent, for example, MAPK.

All signalling pathways can lead to the activation of appropriate transcription factors (TFs), enabling the transcription of genes crucial for maintaining plant homeostasis under stress. Additionally, the gene expression may be regulated by the presence of microRNA (miRNAs)-key regulators of plant responses to abiotic and biotic stresses and plant development [14]. The end result is the translation of proteins whose role is to reduce stress (e.g., by sequestration of salt ions or inactivation of heavy metals) or to eliminate its negative consequences. The resulting metabolic adjustments are crucial for maintaining the balance between simultaneously occurring processes of growth and development and stress defence. In this review, we present, confront and discuss different recent views on signal transduction in cereal plants under various stresses.

2. Drought

During drought, disturbances in the water management of plants occur, leading, inter alia, to closing the stomata and limiting transpiration [15]. Overlapping structural and functional changes, including oxidative stress, inhibition of photosynthesis and alternations in the distribution of assimilation products lead to disruption of plant growth and development [5,16]. Under drought, assimilates move from leaves (donor organs) to the roots (acceptor organs), which are responsible for water and nutrient uptake, at the expense of the biomass of the aerial parts. Plants suffering from water deficit are usually smaller, light-coloured, lack turgor, and are more susceptible to disease and pest attacks. All of the above result in lesser nutrients supplementation, which reduces the yield of a crop [1].

Plants initially identify water scarcity conditions by the roots, which results in the initiation of several molecular signals that transfer from the roots to the shoots [17]. For this reason, the root is a key organ that determines the effectiveness of the plant's response to the stress of water shortage. The root shows high plasticity of its features and its reaction to drought can vary. There can be observed root elongation [18] or shortening [19], but also its length may not change [20]. The root system of cereals varies, e.g., wheat (*Triticum aestivum* L.) produces coarser, moderately branched roots, which allows for more efficient water management, while rice (*Oryza sativa* L.) forms thin, more branched underground organs which can better penetrate soil [21].

A special role in signalling is attributed to chloroplasts and mitochondria that are considered sensors of changes occurring in the environment. Chloroplasts and mitochondria generate ROS and transmit retrograde signals to the nucleus [22]. The direct signals of drought are transduced in plants through ROS, such as singlet oxygen ($^1O_2$), superoxide anion radical ($O_2^{\cdot -}$), hydroxyl radical ($HO^{\cdot}$) and hydrogen peroxide (H_2O_2) [23]. Signalling by ROS can take place through pathways based on susceptible proteins containing thiol groups which are subject to reversible oxidation [24]. Thiol reducing molecules, such as glutathione and specific isoforms of thiols reductases, thioredoxins (TRX) and glutaredoxins (GRX), were found in diverse nuclear subcompartments, further supporting the assumption that thiol-dependent systems are active in the nucleus [25]. Thioredoxins and glutaredoxins are not only responsible for the reduction of thiol groups of numerous metabolic enzymes and molecules belonging to ROS scavenging systems, but also regulate thiol-based post-transcriptional redox modifications of proteins [26]. In *T. aestivum*, TRX isoforms are accumulated in the nucleus upon oxidative stress. It is likely that the influence of ROS on the expression of nuclear genes may be based on the regulation of redox-sensitive TFs [27].

The MAPK cascade-mediated ROS removal is an important mechanism regulating drought stress tolerance [28]. The MAPK cascade leads to the activation of antioxidant enzymes such as superoxide dismutases (SOD), peroxidases (POX) and catalases (CAT) in many cereal plants [29,30]. A proteomic study of *T. aestivum* plants revealed the abundance of CAT and three isoforms of SOD (chloroplastic cytosolic Cu/Zn-SOD and mitochondrial Mn-SOD) in response to drought. These antioxidant enzymes were involved in the survival strategy of wheat by avoiding the excess generation of ROS [31]. Moreover, salicylic acid (SA), ethylene (ET), jasmonic acid (JA), cytokinins (CKs), gibberellins (GAs) and brassinosteroids (BRs) play a vital function in regulating various phenomena in cereal acclimatisation to drought stress [32]. For example, BR signalling regulates drought tolerance in wheat, which is partially achieved through brassinazole-resistant 2 (TaBZR2) TF. TaBZR2 increased the *glutathione S-transferase-1* (*TaGST1*) expression and a decrease in ROS level was observed [33]. Furthermore, maize calcium/calmodulin-dependent protein kinase (ZmCCaMK) was involved in BR signalling and it was required for BR-induced antioxidant defence [34].

The key step in cereals' response to drought is increased concentrations of abscisic acid (ABA) in the root, which may contribute to increased root hydraulic conductivity. By this mechanism, cereals adjust their cellular processes by triggering a network of long-distance signalling events that start with the perception of stress signals and lead through transduction of those signals to switch on acclimation cellular responses, such as changes in gene expression [35]. ABA regulates the expression of different stress-responsive genes involved in the accumulation of compatible osmolytes, synthesis of late embryogenesis abundant (LEA) proteins, dehydrins, chitinases, glucanases, as well as other protective proteins, such as the heat shock protein (HSP) [36]. The resulting osmotic adjustment helped to maintain higher leaf relative water content at low leaf water potential under drought. It enabled sustained growth while under reduced leaf water potential [37]. Recent evidence indicates that H_2S is actively involved in the regulation of ethylene-induced stomatal closure and also interacts with H_2O_2 by impacting the activities of the inward K^+ ion and anion channels [38]. Wheat can adapt to osmotic stress by H_2S production and activation of the antioxidant system [12]. It was proven that H_2S induced the ABA-triggered ascorbate–glutathione (AsA-GSH) cycle under osmotic stress. Obviously, H_2S was involved in the ABA-related closing of stomata in response to various environmental stresses, however, the interaction between them is still unclear and requires further research [11].

In response to osmotic stress occurring often with drought, levels of growth-stimulating hormones: GAs, CKs, sometimes indole acetic acid (IAA) decrease, while there is an observed increase in the level of hormones that usually inhibit cell elongation growth or accelerate maturation and/or aging of tissues: ABA, ET, JA, methyl jasmonate (MeJA) and BRs [39]. Here, ABA acts as the hub of the hormonal crosstalk between several stress signalling cascades [40]. Osmotic stress-responsive gene expression is regulated by ABA-dependent and ABA-independent pathways [41]. In the ABA-dependent pathway, numerous types of TFs, such as MYeloBlastosis (MYB), a basic helix–loop–helix (bHLH), the basic region leucine zipper (bZIP), ethylene response factor (ERF) and homeodomain TF are involved [42]. It was proven that overexpression of *JERF1* (ERF gene) significantly enhanced drought tolerance of transgenic rice [43]. According to Zhang et al. [43], the JERF1 activated the expression of stress-responsive genes and increased the synthesis of the osmolyte proline by regulating the expression of *OsP5CS*, encoding deltal-pyrroline-5-carboxylate synthetase (proline biosynthesis enzyme) [43]. JERF1 also triggered the expression of two rice genes encoding ABA biosynthesis enzymes, zeaxanthin epoxidase 2 (*OsABA2*) and xanthoxin dehydrogenase (*Os03g0810800*) [43].

Signal transduction of osmotic stress also depends on Ca^{2+}, nitric oxide (NO), reactive sulphur species which induce MAPK [28], calcium-dependent protein kinase [44], and calcineurin B-like interacting protein kinase families of protein kinases [45], or phospholipid signalling [13]. The MAPK cascade plays an important role in the drought stress response mainly by responding to ABA and regulating ROS production. Numerous components of MAPK cascades were described as responding to water deficiency in cereals. For example in rice, the transcripts of *OsMKK4*, *OsMKK1*, *OsMPK8*, *OsMPK7*, *OsMPK5* and *OsMPK4* were accumulated under drought stress [46–48]. In wheat, the expression levels of *TaMKKK16*, *TaMKK1* and *TaMPK8* changed in response to drought stress [49]. Ma et al. [50] found that *OsMKK10.2–OsMPK3* were responsible for conferring drought stress tolerance in rice via ABA signalling [50]. However, the exact relationship of the MAPK cascade with ABA has not yet been described [28]. In rice, MAPK5, MAPK7, MAPK8 and MAPK12 were induced by drought and MAPK4 was repressed under water shortage [51]. In rice, MAPK kinases regulated the activity of transcription factors such as OsWRKY30 and increased drought tolerance [52].

The CDPKs and CIPK are families of protein kinases. In rice, a Ca^{2+} dependent kinase such as OsCIPK12 increased the concentration of proline and soluble sugars, which may improve drought tolerance. Additionally, OsCDPK7 enhanced the expression of the gene whose product is the rab16A protein, potentially involved in drought tolerance [53]. In addition to the mentioned protein kinases, rice has a total of 74 heat shock proteins classified into four categories: sHSP, HSP70, HSP90 and HSP100 [54]. These HSPs are activated by ABA-dependent heat shock transcription factors (HSFs), but only some of them are activated by drought [55]. Furthermore, in wheat and barley, the expression of several dehydrins (Dhn) belonging to group two of LEA proteins was observed under drought [56]. Karami et al. [57] reported induction of several genes of Dhn, such as *Dhn1*, *Dhn3*, *Dhn5*, *Dhn7*, and *Dhn9*, in barley flag leaf under drought. Relative expression levels of *Dhn3* and *Dhn9* revealed positive correlations with chlorophyll a and b contents, osmotic adjustment, plant biomass and grain yield, and negative correlations with malondialdehyde (MDA), a marker of membrane oxidative lipids damage, and electrolyte leakage levels.

The majority of LEA proteins display a preponderance of hydrophilic and charged amino acid residues. On the basis of the literature, their function as antioxidants, membranes and protein stabilisers, and indirect participants as molecular shields in cell protection are considered [58]. *HVA1* gene encoding group three of LEA protein from barley (*Hordeum vulgare* L.) was transformed into rice and the tolerance to water deficit of the transgenic rice was improved under the greenhouse conditions [59]. The overexpression of the *HVA1* gene in the roots and leaves of wheat also tended to retain tolerance to drought stress. In wheat, in response to drought, the size of LEA proteins reached up to 200 kDa, therefore, these proteins were resistant to denaturation [60]. It was observed that overexpression of *OsLEA6* and *OsLEA3-1* led to enhanced drought tolerance of rice plants in the field [61].

As an important metabolic pathway, phosphatidylinositol metabolism generates signalling molecules that are essential for survival under drought [46]. Phospholipid molecules are involved in signalling processes leading to adjustments in root growth, pollen and vascular development, hormone effects and cell responses to environmental stimuli in plants [46]. Wang et al. [62] showed that the expression of maize *ZmPLC1*, encoding phospholipase C, was up-regulated under dehydration and it improved the drought tolerance of maize through the interaction with other signalling pathways in guard cells [62].

It is well known that the NO performs the signalling function in plant cells. Signal transduction by NO is mediated by cyclic guanosine monophosphate (cGMP) and activation of guanylate cyclase [63]. NO regulates the levels of cellular ROS content and toxicity through the activation of antioxidant enzymes [64]. Gan et al. [64] showed that NO (applied exogenously) increased drought resistance in barley. The application of NO not only increased the activity of antioxidant enzymes but also increased the content of proline. NO was also found to crosstalk with ABA, JA, SA and CKs to mitigate the adverse effect of drought stress [65]. There are also many studies showing the cooperation of H_2S and NO in response to drought [11].

3. Salinity

Salinity is one of the most important abiotic stress that negatively influences plant growth and productivity, especially rice, wheat and barley, which are the main food crops worldwide [66,67]. Soil salinity caused 25–30% of the irrigated area worldwide to be commercially unproductive and it is estimated that progressive salinity, expanding at a rate of 10% per year, will lead to nearly 50% of agricultural land by 2050 being unproductive [68]. The high concentration of salts in the soil may cause a reduction in water and nutrient uptake due to salt accumulation in the root zone (physiological drought), therefore inducing ion and nutrient imbalance, and water stress in plants. Salinity, due to the presence of NaCl in the soil, is the most common, hence the most harmful effect of salinity is the accumulation of Na^+ and Cl^- [69]. Excess Na^+ in the plant inhibits the uptake of essential micronutrients such as K^+ and Ca^{2+} from the soil, a shortage of the second one is especially crucial because it participates in the maintenance of cell membrane integrity, as well as in the synthesis of new cell walls [70]. Thus, overaccumulation of Na^+ leads to damage and enhanced permeability of membranes. Loss of membrane integrity can also lead to K^+ leakage from cells, which can affect enzymatic reactions since many enzymes require K^+ as a cofactor. Additionally, these enzymes are sensitive to high cytosolic Na^+ content. The accumulation of Na^+ also alters the activity of photosynthetic enzymes and it is harmful to other photosynthesis compounds, such as chlorophylls, and carotenoids [71]. What is more, it is assumed that disturbed ion homeostasis (excess of Na^+ and shortage of Ca^{2+} and K^+) might contribute to oxidative stress which is resulting in the overproduction of ROS and an inefficient ROS detoxification system. The following consequences are oxidative damage of various plant cellular components such as nucleic acids, proteins, sugars and lipids, and hence the inhibition of proper plant development and growth [72].

The response of plants to salinity occurs through the perception and transduction of a signal associated with the disruption of ion and osmotic homeostasis. It is considered that plant cells sense the increase in cytosol Na^+ levels through a sensor or a receptor. Nonetheless, no specific sensor or receptor was identified in plants so far. Therefore, it is not known how an excess of Na^+ is detected by plants, so it can be assumed that the perception of salt stress signal remains unrevealed [73]. However, the most common salt stress signalling pathways—the salt overly sensitive (SOS)—are well characterised in plants, including cereals. Additionally, the MAPK cascade, which transduces stress signals to a variety of transcription factors that further activate salt-responsive genes, plays an important role in salt stress signalling in plants [74].

SOS pathway genes encode proteins that are engaged in the active efflux of excess Na^+ from the cytosol (Figure 1). SOS1 is a plasma membrane Na^+/H^+ antiporter, activated through phosphorylation catalysed by the SOS2–SOS3 kinase complex. SOS3 is a Ca^{2+} sensor, which belongs to the calcineurin B-like signal protein family [75]. It perceives the cytosolic Ca^{2+} signal, which is triggered by a salt-induced excess of Na^+. ABA plays a key role in increasing Ca^{2+} content, which is released from intracellular storage compartments [76]. Then, SOS3 interacts with SOS2, which is a CIPK serine-threonine protein kinase. The SOS3/SOS2 kinase complex regulates the expression of *SOS1* genes therefore it can stimulate SOS1 Na^+/H^+ antiporter activity [77]. In seedlings of three bread wheat genotypes, which were characterised as highly tolerant, moderately tolerant and sensitive to salinity stress, the expression of *SOS1*, *SOS2* and *SOS3* genes was observed at a significantly higher level in the salt-tolerant genotype. What is more, both constitutive and salt-induced expression of *SOS1* was 2-fold higher in the leaf of this genotype. This was correlated with low Na^+ levels in tissue and better leaf K^+/Na^+ ratio in leaves, which was probably a result of the facilitated exclusion of toxic Na^+ into root apoplast [75]. Similarly in experiments by Jiang et al. [77], the expression of several genes belonging to the TaSOS1 gene family was up-regulated in response to salinity in the wheat-tolerant genotype after 1 day of salt treatment. What is more, overexpression of the wheat genes encoding TaSOS1 and TaSOS1-974 (with a deletion on the C-terminus) in tobacco resulted in improved Na^+ efflux and K^+ influx rates in the roots of the transgenic plant compared to wild-type (WT) tobacco upon salt stress. Among these three types of plants, the lowest content of MDA and electrolyte leakage was observed in *TaSOS1-974* transgenic plants while the highest was observed in WT tobacco. This indicates that the overexpression of *TaSOS1-974* might alleviate oxidative damage of the plasma membrane generated upon salinity [78]. Comparable results were obtained in Arabidopsis SOS1 mutant plants with the overexpression of durum wheat (*Triticum durum* Desf.) gene *TdSOS1Δ972* (with a deletion on the C-terminus). These plants showed greater water retention capacity and maintained a better K^+/Na^+ ratio in their shoots and roots, as well as their seeds, had a better germination rate upon salinity than in Arabidopsis *SOS1* mutant transformed with empty binary vector or *TdSOS1* (full-length) [79]. These results confirmed that in proteins belonging to the SOS1 family, the C-terminus function as an auto-inhibitory domain. Autoinhibition of SOS1 is released when the C-terminus domain is phosphorylated by activated SOS2 [80]. Additionally, in rice and barley, the involvement of SOS genes in response to salinity was observed. Fu et al. [81] showed that rice *OsSOS3* was significantly up-regulated in roots under salt stress. Additionally, the expression of *OsSOS2* and *OsSOS1* was markedly up-regulated and a high transcript level of these genes was maintained. In turn, barley *HvSOS3* was only slightly up-regulated in roots under stress. Other barley *SOS* genes, *HvSOS1* and *HvSOS3*, showed slight changes in roots during salt treatment. All tested rice genes showed higher absolute expression than barley genes. However, rice was more sensitive to salt stress than barley. In rice, a higher excess of Na^+ was observed in the shoots, which was harmful for physiological processes, e.g., protein degradation. On the other hand, in rice, the level of Na^+ in the roots was lower than in barley, which might be the result of Na^+ efflux through the SOS pathway. Despite this phenomenon, barley maintains normal metabolism. These results show the differences in salt tolerance between these two species [81].

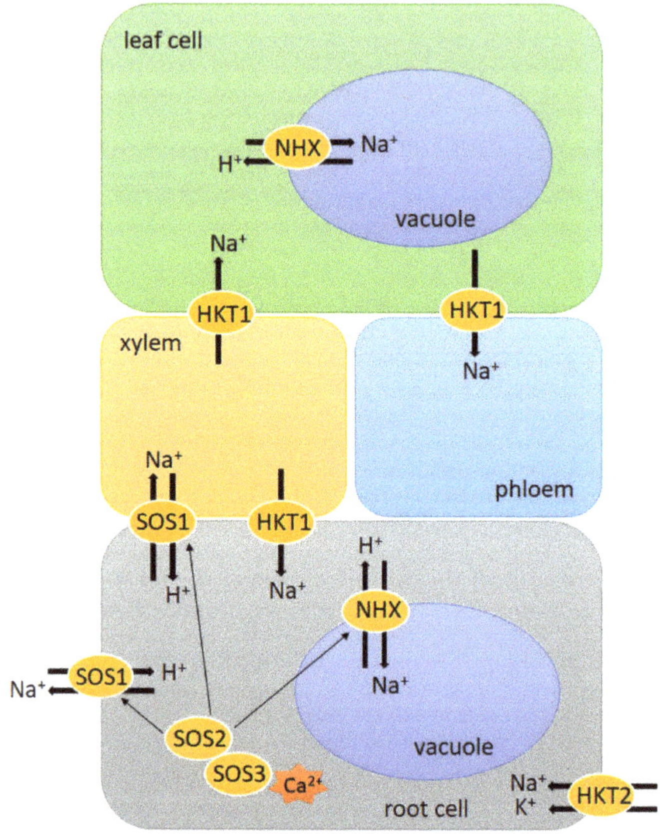

Figure 1. Na^+ transportation under salinity. Plants remove Na^+ from the cytoplasm using plasma membrane Na^+/H^+ antiporter (SOS1), which is activated through phosphorylation, catalysed by the SOS2–SOS3 kinase complex, SOS3 is a Ca^{2+} sensor. Compartmentation of Na^+ into vacuoles occurs by Na^+/H^+ antiporter (NHX), which is also activated by SOS2–SOS3 kinase complex. High-affinity K^+ transport (HKT) proteins, are Na^+ transporters (class 1) or Na^+/K^+ symporters (class 2). HKT1 proteins remove Na^+ from xylem. HKT2 play role in Na^+ uptake in the root. Details are described in Salinity paragraph.

Besides the exclusion of Na^+ from the cytosol, compartmentation of Na^+ into vacuoles by tonoplast Na^+/H^+ antiporter (NHX) is also another essential mechanism in salt stress response (Figure 1). The necessary proton gradient required for NHX activity is derived from vacuolar H^+-pyrophosphatase and H^+-ATPase. The SOS3/SOS2 kinase complex regulates both NHX and H^+-ATPase activity under salt stress. In wheat, the expression of the *NHX1* gene was markedly increased under saline conditions compared to the control [82]. Additionally, overexpression of the wheat *TaNHX2* gene in eggplant (*Solanum melongena* L.) and sunflower (*Helianthus annuus* L.) increased salinity tolerance in comparison to WT plants. Both transgenic species showed improved growth as well as reduced ROS and MDA contents, which correlated with the high activity of antioxidant enzymes such as SOD and ascorbate peroxidase (APX) [83,84]. A comparison of salt stress response in barley and rice showed that the expression of one of the *NHX* genes was significantly higher in barley (*HvNHX5*) than in rice (*OsNHX5*) in the roots treated with salt. However, the expression of rice *OsNHX1*, *OsNHX2* and *OsNHX4* in shoots was higher than in barley *HvNHXs*. This may indicate that a higher concentration of Na^+ in rice shoots is a result of the up-regulated expression of NHX genes [81]. Moreover, the overexpression of barley *HvNHX2* in Arabidopsis showed that, under salt conditions, transgenic plants grew normally, while WT plants were not able to. Additionally, transgenic plants had a higher concentration of Na^+ in the shoots and had longer roots than WT plants [85]. Similarly, the overexpression of *OsNHX1* in transgenic rice showed increased salt tolerance in transgenic plants and delayed appearance of negative effects connected with damage or death [86]. These results suggest that the vacuolar Na^+ compartmentalisation plays a beneficial role in improving cereals' salt tolerance.

Another element involved in the response to salinity is the family of high-affinity K^+ transport (HKT) proteins, which, contrary to their name, are Na^+ transporters (class 1) or Na^+/K^+ symporters (class 2) (Figure 1). HKT1 proteins remove Na^+ from xylem sap and sequestrate Na^+ into xylem parenchyma cells. The function of this mechanism is to confine toxic Na^+ to the roots, therefore, it prevents the accumulation of Na^+ in shoots and leaves, protecting the photosynthetic tissues from damage [58,87]. By contrast, SOS1 plays a role in the protection of the root since it exports Na^+ out of the root and facilitates its loading into the xylem. These two mechanisms function antagonistically, and it is not fully understood how they are activated and regulated to avoid Na^+ loading and unloading. The role of HKT proteins differs between species in response to salinity. For most species, Na^+ exclusion from the leaf blade is correlated with enhanced salinity tolerance and is due to HKT1. The comparison of two rice varieties with different sensitivity to salinity showed that, under salinity stress, the Na^+ concentration in the leaf blades was much lower in Ouukan383 (salinity tolerant) than in Kanniho (salinity sensitive). It is the result of a high expression level of *OsHKT1;4* in the leaf sheaths of in Ouukan383 cultivar, corresponding to higher Na^+ accumulation in the leaf sheaths and lower Na^+ accumulation in the leaf blades. What is more, under salinity conditions, the expression of the *OsHKT1;5* gene was induced in the roots of Ouukan383 but was repressed in the roots of Kanniho. These findings indicate that the expression of *OsHTS1s* might be correlated with better tolerance to salt stress [88]. In addition, a mutation in *OsHKT1;5* in rice showed that lack of OsHKT1;5 protein in roots leads to excess Na^+ accumulation in leaves in response to salt stress [89]. On the other hand, the expression of *ZmHKT1;5* in two maize genotypes (*Zea mays* L.), SC131 (more tolerant) and SC132 (less tolerant), was not significantly affected under salt stress. However, the expression of *ZmHKT2* was highly induced in SC132 while its transcripts were absent in SC131. It can be concluded that differences in the salinity tolerance in these maize genotypes might be the result of weaker Na^+ and K^+ translocation to the shoots due to high expression of *ZmHKT2* in the roots of SC132 since it is responsible for reduced leaf K^+ concentration, enhanced Na^+ uptake in the roots and later more translocation to the shoots [90].

Signalling through the MAPK cascade leads to cellular responses against various stresses. This pathway relies on successive phosphorylation reactions, thus maintaining proper cell phosphorus (P) content is crucial. During salt stress, Cl^- may reduce plant P content due to ionic competition. Therefore, salinity may negatively affect the MAPK pathway. However, activation of the components of this signalling cascade does not always function as a positive regulator in the stress response. Hao et al. [91] showed that wheat TaMPK4, one of the members of MAPK, was a positive regulator in salt stress response. Sense- and antisense-expressing of *TaMPK4* in tobacco strongly modified plant growth under salinity. *TaMPK4*-overexpressing plants were much larger and showed a larger dry mass, leaf number and leaf areas, while *TaMPK4*-knockout plants were much smaller and showed a lower dry mass, leaf number and leaf areas, compared to WT plants. What is more, under salinity, plants with overexpression of *TaMPK4* had higher K^+ and osmolyte contents and lower Na^+ content than the WT plants, unlike *TaMPK4*-knockout plants [91]. Similarly, Arabidopsis plants with overexpression of *ZmSIMK1*, maize MAPK member, had increased tolerance to salt stress. Seeds of transgenic lines germinated better on medium containing NaCl, as well as at seedling stage, their growth was not inhibited, as was observed in WT plants [92]. On the other hand, the overexpression of wheat *TMKP1*, mitogen-activated protein kinase phosphatase (MKP), which is a negative regulator in the MAPK signalling pathways in Arabidopsis, resulted in improved tolerance to NaCl. Seeds of transgenic plants had a better germination rate and seedlings had lower content of MDA and ROS compared to WT. Improved resistance to salt stress in *TMKP1*-overexpressing plants was correlated with increased antioxidant enzyme activities, which resulted in less damage to cell components [93]. Additionally, Seong-Kon et al. [94] showed that rice *OsMAPK33* could play a negative role in salt tolerance. The expression of *OsMAPK33* was down-regulated until 8 h after the induction of salt stress, indicating that this is a negative regulator in response to salinity. Moreover, the overexpression of *OsMAPK33* in rice enhanced sensitivity to salt stress. It was assumed that it was a consequence of disrupted ion homeostasis since transgenic plants had reduced expression of ion transporter genes, such as the K^+/H^+ antiporter [94].

It was also reported that H_2S might be an important player in plants' response to salinity. It was shown that exogenous application of H_2S improved salt tolerance in some cereals such as rice [95], wheat [96] and barley [97]. The protective role of H_2S was the result of maintaining ion homeostasis, as well as reducing oxidative stress, which was reflected in decreased ROS and MDA contents under salt stress. In addition, antioxidant enzyme activity was increased with H_2S application. Exogenous H_2S might also enhance photosynthetic capacity as well as improve primary and energy metabolism. As it was shown in rice under influence of exogenous H_2S, proteins related to glycolysis, tricarboxylic acid cycle and ATP synthesis were up-regulated in salt-treated plants [95]. Moreover, exogenous H_2S up-regulated transcript level of genes encoding proteins involved in the SOS pathway and the MAPK pathway, as was recently shown in wheat [96].

4. Heavy Metals

The impact of heavy metals (HMs) on plants depends not only on the concentration and type of xenobiotic elements but also on their availability to plants, which is related to such soil factors as pH, cation exchange capacity, organic matter content and adsorption by clays. HMs in high concentration affect membrane permeability, inhibit enzymes activity, inactivate photosystems and disturb mineral metabolism [98]. Furthermore, HMs cause secondary oxidative stress, which results in the oxidation of plant membranes, damage of nucleic acid, leading to mutations, oxidative modifications of proteins resulting in loss of their activity, disruption of pigment function, and finally, cell death [99]. The toxicity of a specific substance, including HMs, depends on a variety of factors, e.g., how much of the substance organisms are exposed to, how they are exposed and for how long. Understanding the mechanisms underlying plant resistance or tolerance of plants to abiotic and biotic stress factors is extremely important in the era of global warming, where the mobility of pollutants in the environment increases [3]. However, some HMs are necessary (in non-toxic concentrations) for the proper development and growth of cereals. This category includes, among others, copper (Cu), iron (Fe), cobalt (Co), zinc (Zn), molybdenum (Mo), manganese (Mn), boron (B) and nickel (Ni), the presence of which is required for the proper functioning of the plant. However, excessive concentrations of even these essential micronutrients can also stress the plants. There is also a group of particularly highly toxic HMs including Pb, Hg, As and Cd that are ranked as the first, second, third and sixth, respectively, in the list of the US Agency for Toxic Substances and Disease Registry (ATSDR) [100].

HMs negatively affect the plant cell on many levels. They can directly inhibit enzymes and cause an oxidative burst, leading to the overproduction of ROS and RNS, which changes the oxidative potential in the cell [101]. ROS and RNS not only damage proteins, which can lead to their degradation but also alter membrane permeability, which puts the integrity of the cell at risk. In addition, HMs can induce chloroplasts and mitochondria damage, which inhibits basic metabolic processes in the cell, such as photosynthesis and the respiration chain. What is more, HMs also influence the stomatal movements and subsequently affect the transpiration rate [102]. HM also caused damage to DNA and inhibition of transcription and translation, which hinders the synthesis of proteins that may be of fundamental importance in the survival of the cell. All these changes lead to the failure of cell division, which prevents the correct growth and development of crops [98]. However, it should be observed that each of the HMs can affect the plant in a slightly separate way. Cd causes a strong inhibition of cereal growth, browning of the roots and chlorotic changes in leaves. Cd particularly affects photosynthetic enzymes such as Fe (III) reductase. In turn, Hg blocks the flow of water in the plant by interacting with the water channels, thereby blocking them. The action of Pb focuses on changing the permeability of the cell membrane, disturbing the hormonal balance of the plant and inhibiting the activity of selected enzymes due to the interaction of Pb with their sulfhydryl groups. As has a similar effect on enzymes, as it also reacts with sulfhydryl-containing proteins, disrupting their function. As also binds to vicinal thiols present in dehydrogenases, which not only inhibits cellular respiration but also leads to overproduction of ROS [103].

Due to the different effects of individual HMs on cereals, the response of the plant to HM stress is multifaceted and is associated with the activation of several signalling pathways causing a change in the expression of the relevant TFs and/or genes: (a) calcium-dependent signalling; (b) signalling mediated by MAPK; (c) signalling via ROS; (d) hormone signalling [9]. Calcium signalling occurs through several sensors which include calmodulins (CaM), calmodulin-like proteins, calcineurin B-like proteins and CDPK. The activation of individual sensors depends on the concentration of Ca^{2+}. It was observed that both the recurrent and long-term Cr (VI) stress in rice increased the activity of CDPK [104]. The signalling cascade based on MPAK caused the phosphorylation of selected transcription factors (ABA-responsive element; ABRE, dehydration-responsive element binding; DREB, bZIP, MYB, MYC, NAC and WRKY-containing a conserved WRKYGQK domain and a zinc finger-like motif) resulting in the altered expression of genes related to the HM stress response [9]. Induction of OsMAPK2 and myelin basic protein kinase was recorded in Cd-treated rice. In response to the increased production of ROS, cereals improve the activity of their antioxidant system by increasing both enzymatic (SOD, CAT, APX, dehydroascorbate reductase) and non-enzymatic (betaines, proline and ascorbate) activities, which allows them to avoid or reduce oxidative damage to the plant cell, however, some redox imbalance is necessary for the induction of a proper stress response [105]. ROS and kinase-related pathways may cross with each other. In rice, the activation of MPAK by excessive accumulation of ROS was reported as a result of secondary oxidative stress induced by HM stress. What is more, ROS also influence changes in the plant's hormonal system, in particular, auxin (AUX), ET and JA and ABA signalling. Treatment of rice with JA was shown to increase the antioxidant response of rice to Cd [106]. Treatment of rice plants with As resulted in a change in ABA metabolism, which influenced the modulation of signal transduction and the plant defence stress response [107]. Besides those signal transduction pathways, miRNAs also play a crucial role in the response to HM stress. miRNAs are 20–24 nucleotide non-coding RNAs that regulate gene expression at the post-transcriptional level by targeting mRNA degradation or by translation repression [108]. Due to the different properties of individual HMs, their uptake pathways, as well as signal induction and transmission, differ from each other.

As (V), being the main form of As in the soil, is similar in structure to P ions and thus its uptake into the plant is possible via phosphate transporters. Under anaerobic conditions, As also reaches the cell via aquaporins (AQPs). AQPs include various family subclasses of proteins that can uptake As, including tonoplast intrinsic proteins, cell membrane intrinsic proteins, and nodulin-like proteins. In rice, As (III) ions can be taken up by silicon pathways and methylene forms of low silicon transporter proteins (Lsi1 and Lsi2), which have the ability to transport As (III) ions both from and into the cell [109]. Due to the similarity of As (V) to P ions, ATP synthesis in plant cells is disturbed. As (III) in turn reacts with thiol groups of proteins, including enzymes, leading to the disturbance of cell homeostasis. In rice, As caused the production of ROS and the activation of the MAPK-inducing phosphorylation cascade including MKK4, MPK3, MPK4 [110] and calcium-dependent signalling by CaM, CaM kinase and CaM-like protein [107,111]. Moreover, rice induces down-regulation of miR172 (miRNA) and up-regulation of miR393, miR397 and miR408. The last one (miR408) has a direct role in targeting Cu-containing proteins or SOD [112]. On the other hand, ROS down-regulated miR397 targeted laccase, which led to increased activity of the lignin biosynthesis pathway by the accumulation of laccase enzymes [113]. Additionally, one of the miRNAs, miR528, was crucial for As tolerance in rice [114].

Cd in the environment occurs in an ionic form (Cd^{2+}) and is bound into chelates. Cd^{2+} is taken up into the plant by non-specific HM transporters, whose levels depend on transpiration. The most important uptake routes for both Cd forms include Zn-regulated transporters, Fe-regulated transporters, hyperpolarisation-activated Ca^{2+} channels, depolarisation-activated Ca^{2+} channels, voltage-insensitive cation channels, yellow-stripe 1-like proteins (YSL) and the natural resistance-associated macrophage protein (NRAMP). Transport to the xylem occurs via apoplastic ATP-binding cassette (ABC) transports and P1B ATPase and H^+/Cd^{2+} antiports. During defence responses, cereals activate TFs such as DREB, APETALA2 (AP2) and bZIP [103]. Cd accumulation activated the MAPK pathway: MAPK2, MPK3, MPK6, MSRMK3, WJUMK in rice [110,115,116] and MPK3 [49] in maize. It also activated components of the hormonal pathway, mainly by auxins: MAPK3/6/7, YUCCA, PIN proteins, ARF (auxin response factors) and IAA [117]. Exposing rice to Cd stress led to the up-regulation of miR441, and down-regulation of 12 other miRNAs, including miR192, which targeted ABC transporters. Increased activity of ABC transporters enables Cd sequestration and stress alleviation [118]. Cd up-regulated the transcription factors belonging to MYB, AP2, DREB, WRKY and NAC at different time intervals in rice [119]. As for MYB, OsMYB45 was especially related to Cd toxicity, as its mutation increased H_2O_2 content in the leaves of mutant and decreased CAT activity compared to the WT plants [120], and OsARM1 (arsenite-responsive MYB1) regulated As-associated transporters genes *OsLsi1*, *OsLsi2* and *OsLsi6* [121].

In most plants, the occurrence of aluminium (Al) is limited to the roots, although the presence of Al-citrate in the xylem and Al-oxalate in the leaves of buckwheat was reported [122]. Al is excluded into the soil by organic acids aided secretion through transporters such as the Al-activated malate transporter (ALMT) family, ABC transporters family (STAR1 and STAR2), multidrug and toxic compound extrusion (MATE) family and aluminium transporter 1 (NRAMP/NRAT1) family [123]. In wheat, Al accumulation enabled pathways dependent on MAPK: 48 kDa MAPK, 42 kDa protein kinase [124], Ca: myosin, calpain, phospholipase C, phospholipase A2 [125], and ethylene: ALMT1, 1-aminocyclopropane-1-carboxylic acid (ACC) synthase (ACS), ACC oxidase (ACO) [126]. Similar to previously described HMs, Al also down-regulated most of the miRNAs in rice such as miR156, miR395, miR398, miR159 and only miR399, miR166, miR168 were up-regulated in response to Al [127]. This however is not true for all crops, as maize showed mostly miRNAs up-regulation with the exception of miR171 and miR396 [128]. MiR395 targets genes of ATP sulfurylase (APS) and SULTR2:1, which are crucial for GSH and phytochelatin (PCs) synthesis [129].

Another important HM is Hg. The bioavailable Hg compounds in the soil are Hg^{2+} and methylmercury. Hg with a hydrophilic character is easily trapped by the roots, transported to the shoots, and then released back into the atmosphere in gaseous form. Hg tends to accumulate in the roots and cannot be transferred to plant shoots. Transport of Hg in the plant is possible due to ABC transporters. They can pump Hg^{2+} conjugates to or from the vacuole of higher eukaryotes [130]. It was shown that an accumulation of Hg led to the activation of MAPK proteins in rice, especially MSRMK2, MSRMK3, WJUMK [115], and the ET pathway via OsACS2, OsACO1, OsACO2, OsACO5 and OsACO6, 5 MAPKKK, 1 MAPKK and 2 MAPK [131].

Pb in the form of a dipositive cation is passively aborted by root hairs. Its further transport is severely limited by its low solubility. Pb transport in the plant is accomplished by the apoplast of xylem tissues but is blocked in the Kasparian bands of the endoderm. It can then be sequestered via ABC transporters, P-type pumps, pleiotropic drug resistance (PDR1), inner membrane proteins of mitochondria, ATM1, leucine-rice repeat proteins (LRR), Ca^{2+} gated channels, cyclic nucleotide ion gated channels and K^{+} gated channels [132]. In rice, Pb activates 34 kDa, 40 kDa and 42 kDa MAPK, and a calcium-dependent pathway via CDPK-like kinase [133].

In order to limit the negative effects of HMs, the signal cascade causes adaptive changes in the plant cell, relying on detoxification to prevent the involvement of HMs in undesirable toxic reactions. Defence strategies include preventing or reducing the uptake by limiting the transport of metal ions to the apoplast by binding them to the cell wall or cell exudate, or by inhibiting long-distance transport [134]. To achieve that, activation of appropriate TFs and induction of the transcription of particular genes related to the HMs response is necessary. Some of the up-regulated genes are associated with the activation or amplification of selected signal transduction pathways. For example, As treatment of rice increased the expression of the ABA biosynthesis genes: *OsNCED2* and *OsNCED3* [135], while chromium treatment of rice increased the expression of four ET biosynthesis-related genes (*ACS1, ACS2, ACO4* and *ACO5*) [136,137], two genes associated with MAP cascades (*OsMPK3, OsCML31*), three protein kinase-related genes (*OsWAKL-Os, OsLRK10L-2, OsDUF26-If*) and two TF-related genes (*OsWRKY26, OsAP2/ERF-130*) [137]. Another group of genes expressed by the action of HMs are genes encoding phosphatases. Phosphorylation/dephosphorylation is the most common post-translational modification, whose role is to activate and deactivate selected proteins, which results in the adaptation of the metabolism to the plants' needs. In rice treated with chromium, increased expression of five families of genes encoding phosphatases (*OsLMWP, OsDSP, OsPP2A, OsPTP* and *OsPP2C*) was observed [137]. Due to the fact that one of the strategies for reducing the negative impact of HMs is their translocation, another group of genes up-regulated as a result of stress are those related to the transport of HMs. In rice, Cr strongly induced a number of genes involved in the vesicle trafficking pathway, including five OsExo70 genes (*Os01g0763700, Os06g0255900, Os01g0905300, Os01g0905200* and *Os11g0649900*) and one *Tom1* gene (*Os05g0475300*) [137]. In durum wheat, the exposure to Cd induced several vacuolar HM transporter genes, especially *ZIF1, ZIF-like* genes [138].

When HMs are present at elevated concentrations, cells activate a complex network of storage and detoxification strategies, such as chelating metal ions with phytochelatins (PC) and metallothioneins (MTs) in the cytosol, as well as transport and sequestration into the vacuole via vacuole transporters [139]. HMs activate the synthesis of phytochelatin synthase (PCS) and metallothionein, and then HM–PC and HM–MTs complexes of low molecular weight (LMW) are formed in the cytosol. LMW HM–PCs complexes are consistently transported across the tonoplast into the vacuole via the ATP binding cassette and the V-ATPase transporter (ABCC1/2). After compartmentalisation, the LMW complexes further integrate HMs and are generated by chloroplasts sulfide (S^{2-}) to eventually form HM–PC complexes of high molecular weight (HMW). MTs regulate cellular redox homeostasis independently and by stimulating the antioxidant system and stabilising high cellular GSH concentrations. It was well documented that the biosynthesis of PCs can be regulated at the post-translational level by metals in many plant species. However, the overexpression of the phytochelatin synthase (PCS) gene in plants does not always result in enhanced tolerance to HM stress [140]. Moreover, MTs not only bind HM but also partake in the elevation of oxidative stress by acting as ROS scavengers, thus, integrating those two pathways [141]. MTs are tissue-specific. For example, the *OsMT2c* gene encoding for type 2 MT was expressed in the roots, leaf sheathes and leaves of rice, but was almost absent in seeds [142]. Moreover, to protect proteins against HM stress, HSP proteins are also synthesised, belonging to HSPs70, HSPs60, HSPs90, HSPs100 and HSPs classes. HSPs70 were induced in rice by As, Ag, Cu, Cd and Cr (HSP70, BiP), HSPs60s by Hg ($cpn60^2$), HSPs90 by Cu, As and Cd (HSP81-2, HSP82, HSP81-1), HSPs100 by As, Cu and Co (HSP101, ClpB-C), and HSPs by Cu, Cd, Fe, Al and Zn (HSP17.4, HSP23.9, HSP78.3) [140].

5. Biotic Stress

Plants are exposed to a wide variety of pathogens and pests, the life cycle of which and the impact on plants differ significantly. Therefore, it is difficult to identify one common signalling pathway associated with the biotic stress response. The plant–parasite relationship is quite specific and depends on both the defence mechanisms and the structure of the plant itself, as well as those of pathogen, therefore the signal transduction pathway is multifaceted and quite strongly individualised. Research on the subject is fairly limited, but in this review, we attempted to describe its known elements.

Plants have an innate immune system able to recognise evolutionarily conserved microbe/pathogen-associated molecular patterns or herbivore-associated molecular patterns [143,144]. The presence of transmembrane pattern recognition receptors and intracellular proteins of the nucleotide-binding domain and leucine-rich repeat superfamily enables the identification of pathogens/herbivores by plant cells which leads to induction of defence reactions including the synthesis of signalling molecules such as SA, ABA, JA, ET, H_2O_2 and NO [145]. The activation of those signalling patterns can cause alterations in gene expression, leading to specific defence responses. Both pathogens and insects can act locally and systematically [145].

Plants launch defence responses to shield themselves against pathogens and pests. Those responses are regulated by the infestation-induced production of hormones. SA, JA, ET and ABA are vital players in induced mechanisms against biotic stresses [146]. SA-dependent responses are usually efficacious against biotrophs, while JA-dependent responses are successful against necrotrophs and phytophagous insects [147]. Defence signalling of SA depends on the transcriptional co-factor called non-expresser of pathogenesis-related gene 1 (NPR1), ultimately leading to the activation of anti-microbial pathogenesis-related (PR) genes [148]. Following pathogen infection/insect infestation, molecules such as ABA, JA, SA, ET, H_2O_2 and NO are accumulated at different time points and convergence of signalling pathways can occur in a plant [149,150].

The biotrophic barley powdery mildew *Blumeria graminis* and the hemibiotrophic *Bipolaris sorokiniana* are economically significant pathogens of *H. vulgare*. To assess the barley defence responses to these pathogens, alternations in SA and genes of SA-dependent responses (*PR1*, *PR2*, *PR3* and *PR5*) were studied, which revealed that the level of SA was significantly enhanced in infected barley plants (both resistant and susceptible) at 24 h post-inoculation compared to control plants. Furthermore, time-course experiments showed a clear contradiction in patterns of expression of SA-dependent genes upon barley inoculation with *B. graminis* and *B. sorokiniana*. These studies also showed that the expression of *PR1* and *PR2* genes was induced in resistant barley inoculated with *B. sorokiniana* contrary to *B. graminis* infestation, indicating different SA-dependent responses in barley plants infested with fungal pathogens with different lifestyles [2].

MYB transcription factors play a vital role in cereal plant defence including responses to fungal pathogens. Wei et al. [151] presented results on characterisation of the *TaPIMP2* gene encoding a pathogen-activated MYB protein in *T. aestivum*. The expression of *TaPIMP2* was altered to a different extent and speed upon inoculation with *B. sorokiniana* or *Rhizoctonia cerealis*. In addition, different expression patterns of *TaPIMP2* were observed after *T. aestivum* plants were sprayed with ABA, 1-aminocyclopropane-1-carboxylic acid (ACC, precursor of ethylene) or SA. Silencing of *TaPIMP2* decreased the resistance of *B. sorokiniana*-resistant wheat to *B. sorokiniana* infection but did not change the resistance of *R. cerealis*-resistant wheat to *R. cerealis* infection. On the other hand, the overexpression of *TaPIMP2* remarkably increased resistance to *B. sorokiniana* rather than *R. cerealis* in transgenic wheat. Moreover, it was observed that TaPIMP2 is engaged in wheat resistance to *B. sorokiniana* due to stimulation of the expression of *PR1a*, *PR2*, *PR5* and *PR10*.

After the plant is mechanically injured or infested with necrotrophic pathogens or insects, the accumulation of JA and its derivatives—oxylipins (called jasmonates)—occurs [152]. For example, infestation of maize with a lepidopteran pest, the beet armyworm caterpillars (*Spodoptera exigua*) induced synthesis of JA, MeJA and jasmonoyl-L-isoleucine in infested-maize leaves [153]. There are two separate branches of the JA signalling that have a negative influence on each other: the ERF branch and the MYC branch [154]. The ERF branch is induced upon infestation with necrotrophs and is controlled by the AP2/ERF-domain transcription factors such as ERF1 and octadecanoid-responsive AP2/ERF 59 (ORA59). Furthermore, the ERF branch is co-regulated by ET and triggers the expression of many ERF-branch genes including the marker gene encoding plant defensin 1.2 (PDF1.2) [155]. Dong et al. [156] identified and characterised *B. sorokiniana*-induced defence gene (*TaPIEP1*) from the ERF branch (B-3c subgroup) of wheat. The mRNA level of *TaPIEP1* was induced upon both inoculations with *B. sorokiniana* and treatments with ET, JA, and ABA. Transgenic *T. aestivum* plants overexpressing *TaPIEP1* showed enhanced resistance to *B. sorokiniana*. The increased resistance of transgenic wheat lines showed also increased transcript levels of defence-associated genes from the ET/JA pathways. Wheat is one of the main cereals crucial for food production worldwide, therefore its pathogens should be one of the main focuses in biotic stress studies. Besides *B. sorokiniana*, *Puccinia striiformis*, which also causes stripe rust, is an important wheat pathogen. In response to *P. striiformis* reactive oxygen species burst is observed. Early accumulation of ROS leads to an increase in chlorophyll a and b levels, as well as to activation of antioxidative enzymes. It contributes to plant resistance to this pathogen [157].

Jisha et al. [158] proposed a model for the role of the AP2/ERF transcription factor, OsEREBP1, during the response of rice plants to infection with the bacterium *Xanthomonas oryzae* pv. *oryzae*. The authors suggested that enhanced expression of *OsEREBP1* can lead to accumulation of JA, which mediates activation of the helix–loop–helix transcription regulator RERJ1 and induces linalool synthase activity so that volatile monoterpene linalool molecules are accumulated resulting in improved tolerance to *X. oryzae* pv. *oryzae* infection.

The brown planthopper (*Nilaparvata lugens*) is a hemipteran pest infesting rice plants. This insect injures plants through feeding, and it also transmits rice grassy stunt virus and rice ragged stunt virus [159]. Xylanase inhibitors were described as players participating in plant defence. Zhan et al. [160] presented that infestation with *imagines* of *N. lugens*, wounding or MeJA treatment increased transcript and protein levels of OsXIP (an XIP-type rice xylanase inhibitor). By studying 5′ deletion in *OsXIP* promoter in rice mutant plants invaded by *N. lugens*, a 562 bp region was shown as crucial for the response to stress induced by pest feeding. Furthermore, a basic helix–loop–helix protein (OsbHLH59) and an AP2/ERF-transcription factor OsERF71 directly reacted with 562 bp sequence to induce the expression of *OsXIP*. The expression of genes *OsbHLH59* and *OsERF71* was also stimulated in rice roots and shoots by wounding and submerging in MeJA.

Fusarium head blight induced by *Fusarium* species such as *F. graminearum* is a globally important fungal disease of wheat. Transcriptional profiling of moderately resistant and susceptible to *F. graminearum* winter wheat cultivars have shown 2169 differentially expressed genes, induced by jasmonate and ethylene, e.g., encoding thionin, lipid-transfer protein, defensin and GDSL-like lipase. Moreover, defence-activated genes encoding jasmonate-dependent proteins were up-regulated in response to infection with *F. graminearum*, such as, for example, the subfamily of mannose-specific jacalin-like lectin-containing proteins [161].

During an infestation, pathogens and pests secrete effectors into host plant tissues. These effectors interact with plant defence systems, which may lead to effective colonisation and the spread of the infection [162]. Darino et al. [163] performed functional characterisation of the biotrophic fungus *Ustilago maydis* (causing smut disease on maize plants) effector jasmonate/ethylene signalling inducer 1 (Jsi1). Jsi1 reacts with members of the plant corepressor protein family Topless/Topless-related (TPL/TPR). It was shown that the increased expression of *Jsi1* in maize led to activation of the ERF-branch pathway by an ET-responsive element-binding factor-associated amphiphilic repression (EAR) motif, which takes after EAR motifs from plant ERF transcription factors interacting with TPL/TPR proteins. Interestingly, phytopathogen effector candidates with EAR motifs were also found to be secreted by an ascomycete fungus *Magnaporthe oryzae* (affecting rice) and a Basidiomycota fungus *Sporisorium reilianum* (affecting maize and sorghum) [163].

In winter wheat field studies, it was shown that JA application induced resistance to cereal aphids (*Metopolophium dirhodum, Sitobion avenae, Rhopalosiphum padi*) and thrips (*Limothrips denticornis* and *Thrips angusticeps*). JA at first caused a significant decrease in the number of pests, which, even though it increased in time, remained lower on wheat treated with JA [164].

The MYC branch is induced upon mechanical injury or feeding by insects. This branch is controlled by basic helix–loop–helix leucine zipper transcription factors MYC2, MYC3 and MYC4, and it is also coordinated by ABA [165]. The MYC-branch activation results in the induction of JA-responsive gene expression including marker genes of the MYC-branch such as *vegetative storage protein 1* and *2* (*VSP1* and *VSP2*) [166].

The rice water weevil (*Lissorhoptrus oryzophilus*) is the most harmful coleopteran pest of *O. sativa* plants. It was proven that the treatment of rice seeds with jasmonates led to resistance against *L. oryzophilus* but rice growth and fitness were reduced. Jasmonates caused delayed emergence and heading, and after full development of plants, lower yield in comparison to plants grown from untreated seeds [167,168]. Therefore, it can be stated that plant fitness is decreased upon activation of JA-dependent defence responses, however, other hormones including ABA, SA, GAs, AUX and BRs are also substantial regulators of the immune–fitness balance caused by phytopathogens [169,170]. In addition, the decrease in plant growth elicited by JA is most probably regulated via signalling crosstalk with AUX, SA, BRs, GAs and CKs [171].

The crosstalk between hormonal pathways promotes the induction of efficient responses against pathogens and pests [172]. Many observations of the mutual interaction between the SA and JA pathways were made [173]. Pharmacological studies showed that the expression of *PDF1.2* and *VSP2* is sensitive to SA treatment. The opposed influence of SA on JA-depended responses was observed. It was shown that exogenous treatment with SA decreased the expression of the JA-responsive genes (*PDF1.2* and *VSP2*) activated by MeJA, the necrotrophic fungi *Alternaria brassicicola* and *Botrytis cinerea*, and the western flower thrips (*Frankliniella occidentalis*) and *P. rapae*. However, infestation with the biotrophic oomycete *Hyaloperonospora parasitica* leading to SA-activation defence antagonised MeJA-dependent expression of *PDF1.2* and *VSP2* and infection with *H. parasitica* diminished *P. rapae*-activated expression of *VSP2* [174]. Moreover, it was documented that this effect (induced by SA exogenous exposition) persists in the next plant generation [175]. The antagonism between SA and JA pathways can change resistance to biotic stressors. It was observed that activation of the SA signalling by exogenous exposition to SA or infestation with the hemibiotrophic bacterium *Pseudomonas syringae*, made the plants more susceptible to *A. brassicicola* [176,177]. Moreover, decreased SA responses in transgenic plants expressing a bacterial salicylate hydroxylase gene (*nahG*) and *npr1* mutant plants were interdependent with attenuated feeding by the cabbage looper (*Trichoplusia ni*) caterpillars [178].

Similar antagonism is present in the ERF and the MYC branch. For example, it was proved that inducing the MYC2 branch in plants inhibits the ERF branch activated by *P. rapae* feeding, hence they are less alluring to the herbivore. Moreover, caterpillars of *P. rapae* preferred to feed on *jin1* (MYC2 transcription factor) mutants and *ORA59*-overexpressing ones more than on WT plants, showing that the ERF and the MYC branch crosstalk changes host–insect herbivore interactions [154]. This antagonism between the ERF and the MYC branch can also change resistance to necrotrophic pathogens. The ERF branch was elevated in plants with *MYC2*-mutated *jin1* and ABA biosynthesis mutant (*aba2-1*), leading to increased resistance to necrotrophs (*B. cinerea, Plectosphaerella cucumerina, Fusarium oxysporum*) [166,179–182].

Vast crosstalk between hormonal signalling pathways permits the plant under biotic stress for precise regulation of defence responses at various levels of plant organisation [183]. As elicitation of parasite-inducible responses is not without metabolic cost, trade-offs between immune defence and growth and development are clearly noticeable in plant organisms [146,184–186]. Hormonal crosstalk is sometimes discussed as an evolutionary cost-limiting strategy. Some researchers argue that this crosstalk may have evolved as a countermeasure to lessen energy costs by retardation of ineffective defence responses against specific invaders [187,188]. This hypothesis also seems to be confirmed by Vos et al. [189]. The authors analysed the effect of hormonal crosstalk on biotic stress resistance and host fitness upon multi-species infestation. Activation of SA- or JA/ABA-mediated responses by the biotrophs *Hyaloperonospora arabidopsidis* or *P. rapae*, respectively, decreased the level of induced JA/ET-response against the following infestation with *B. cinerea*. Notwithstanding, although there was increased susceptibility to this second invader, no long-term negative consequences were observed on host fitness when plants had been infected by multiple parasites. The authors concluded that host hormonal crosstalk during multi-parasite interactions gives the plants an opportunity to put their defence in order of importance while decreasing the energy fitness costs linked to activation of immune responses. This issue is extremely interesting and requires further research, especially in the context of crop plants including cereals.

6. Crosstalk Signalling between Abiotic and Biotic Stress

Current research on biotic and abiotic stress response pathways in plants suggests that there are significant similarities between them. The responses of cereals plants to biotic and abiotic stress are a complex web of interactions between secondary messengers, ROS, phytohormones, antioxidants, photosynthetic pigments, secondary metabolites, protein kinases, TFs, photosynthesis efficiency and chlorophyll a fluorescence parameters and ultra-structural adjustments [190–192]. Plants subjected to abiotic stress, e.g., high temperature, drought, salinity, are often more sensitive to subsequent attacks by pathogens [193]. There are reports about the decrease in the disease resistance of crops due to high humidity and high temperature [194]. Both types of stress factors cause the increase in such parameters as Ca^{2+}, ROS, and pH levels in the apoplast. MAPK kinases are activated, which is a common response to both stresses [195]. For example, OsMPK5 kinase in rice is an ortholog of AtMPK3 in Arabidopsis and NtWIPK in tobacco, which are well known to be activated by both different pathogens and abiotic environmental stimuli [195]. ABA triggered a signal and it negatively imposed on the signalling of defence hormones, e.g., SA. ABA/SA interaction is two-sided, as activation of SA signalling by pathogens lowers ABA concentration [194]. On the other hand, positive interactions were observed for JA/ET signalling in response to double stress. ABA can act as a molecular switch between both responses and plays a dominant role in the response to stress [196]. It can take place through the ABA-inducible genes *ERD15* and *ATAF1*, which may activate ABA-dependent biotic stress responses at the expense of abiotic responses [197]. A scheme for the interaction interface and overlapping signalling pathways of abiotic and biotic stress at the cellular level is presented in Figure 2.

CDPK families are also involved in crosstalk between biotic and abiotic stresses [198]. They are involved in various processes such as osmotic homeostasis, cell protection and root growth [44]. Some studies have reported that the CDPK genes not only behaved as positive regulators of abiotic or biotic stress signalling but also as negative regulators [44]. Overexpression of *OsCDPK12* in rice led to positive regulation of salt tolerance and negative regulation of blast resistance [199].

Phytohormones regulate the activity of transcription factors such as WRKY, MYB, ERF, NAC and the HSF family, which respond to both biotic and abiotic stress and play a vital role in the plant's response to simultaneously occurring stresses. WRKY30 and WRKY13 have a dualistic function in response to drought, salinity, cold and pathogen attack in rice [200]. Some WRKY such as *OsWRKY76* antagonistically regulated the response of rice to blast disease and cold stress [201] but *OsWRKY82* improved resistance against pathogens and tolerance against abiotic stress via the JA and ET pathways [202]. The rice *OsWRKY45* is induced in response to ABA in various abiotic stress and also by infection with *Pyricularia oryzae* Cav. and *Xanthomonas oryzae* pv. *oryzae*. In a study by Qiu and Yu [203], it was shown that constitutive overexpression of the *OsWRKY45* led to a significant increase in the expression of PR genes, resistance to bacterial pathogens, as well as tolerance to salt and drought stresses.

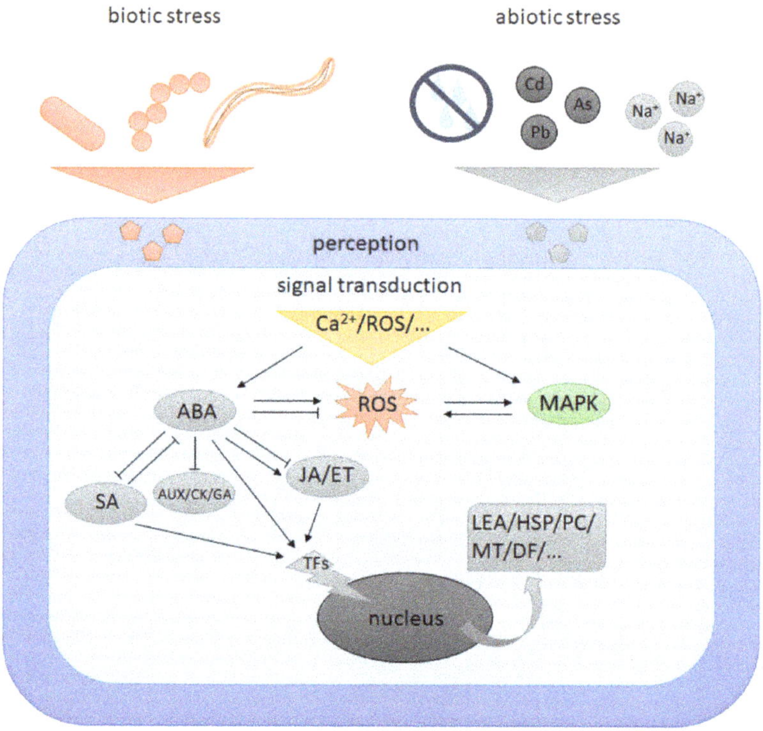

Figure 2. Scheme for the crosstalk signalling between abiotic and biotic stress. Both stress factors are first recognised by plant cells and then information is transduced through chemical signals such as Ca^{2+}, reactive oxygen species (ROS), as well as mitogen-activated protein kinases (MAPK) cascades. Abscisic acid (ABA) is mostly involved in abiotic stress acclimation, while salicylic acid (SA) and jasmonate/ethylene (JA/ET) are responsible for the reaction to abiotic as well as biotic stresses. Finally, phytohormones up-regulate transcription factors (TFs), which then contribute to expression of genes related to stress response, e.g., late embryogenesis abundant proteins (LEA), heat shock proteins (HSP), phytochelatins (PC), metallothioneins (MT), defensis (DF).

MYB transcription factors also may be common element of the response to various stresses. The MYB factor TaPIMP1 from wheat confers tolerance to drought and salt and pathogens stress when overexpressed in tobacco [204]. Another one of the candidates for common TF for multiple stresses response is JAmyb. *JAmyb* expression in response to salinity and osmotic stress was observed in rice seedlings. Microarray analysis showed that *JAmyb* overexpression stimulated the induction of several defence-related genes, some of which are predicted to be involved in osmosis regulation, ROS removal and ion homeostasis [205]. Additionally, transgenic rice plants overexpressing *JAmyb* exhibited improved resistance to blast [206]. A study by Yokotani et al. [205], showed that *JAmyb* expression was induced by H_2O_2 and paraquat. However, it is known that ROS overproduction is a common response to biotic and abiotic stress and could overlap with other stress responses. It is suggested that *JAmyb* might play a role in the crosstalk between JA and ROS-signal transduction pathways in dual stresses [205].

Another important TF is NAC. NAC are plant-specific TFs induced in various developmental stages and under abiotic and biotic stress [207]. The enhanced expression of the *TaNAC4* gene in wheat was observed under the fungus, salinity, wounding and cold stress [208]. Expression of *OsNAC6* in rice was induced by abiotic stresses, including cold, drought and high salinity, as well as by biotic stresses, such as wounding and blast disease [207]. OsNAC6, among others, increases the activity of peroxidase, which elevates oxidative stress.

It was shown that genes encoding cold-responsive/late embryogenesis abundant (COR/LEA) proteins, participate in improving cold resistance and protection of cells from dehydration and low-temperature [209]. It is known that the ABA participates in the regulation of *COR* gene (*WRAB15* and *WRAB18*) expression in wheat. Studies by Talanova et al. [210] showed enhanced expression of *WRAB15* and *WRAB18* genes in wheat leaves caused by the Cd, hardening or their combination. This may indicate the participation of these genes in the protective and adaptive responses of plants to different stress factors [210].

As mentioned above, the effect of one stress can make plants more sensitive to the next stress. On the other hand, exposure of plants to one stress affects their response during the next stress leading to enhanced defence mechanisms to later stress. This phenomenon called "priming" results in a faster and stronger induction of basal defence mechanisms upon subsequent biotic stress factors [3]. "Metabolic memory" in higher plants requires less energy expenditure than defence directly induced by insect feeding or infection caused by pathogens.

A list of genes that may be crucial in signalling the response to biotic and abiotic stresses is given in Tables 1 and 2.

Table 1. The list of genes with a potential role under abiotic and biotic stress signalling pathways.

Gene	Plant	Changes in Expression Level/Physiological Effect	References
		Drought stress	
JERF1	*Oryza sativa* L.	activates the expression of stress-responsive genes and increases the synthesis of the proline	[43]
OsABA2	*Oryza sativa* L.	increases ABA synthesis	[43]
Os03g0810800	*Oryza sativa* L.	increases ABA synthesis	[43]
TaGST1	*Triticum aestivum* L.	decreases ROSs	[33]
ZmCCaMK	*Zea mays* L.	participates in BR-induced antioxidant defence	[34]
OsMKK1 *OsMKK4*	*Oryza sativa* L.	increases under drought	[48]
OsMPK4 *OsMPK5* *OsMPK7* *OsMPK8*	*Oryza sativa* L.	increases under drought	[48]
TaMKK1 *TaMKKK16*	*Triticum aestivum* L.	increases under drought	[49]
TaMPK8	*Triticum aestivum* L.	increases under drought	[49]
OsMKK10.2 *OsMPK3*	*Oryza sativa* L.	increases drought stress tolerance in rice via ABA signalling	[50]
Dhn1 *Dhn3* *Dhn5* *Dhn7* *Dhn9*	*Hordeum vulgare* L.	increases under drought; show positive correlations with chlorophyll a and b contents; participates in osmotic adjustment; increases plant biomass and grain yield	[57]

Table 1. *Cont.*

Gene	Plant	Changes in Expression Level/Physiological Effect	References
OsLEA3-1 *OsLEA6*	*Oryza sativa* L.	enhances drought tolerance	[61]
ZmPLC1	*Zea mays* L.	partakes in interaction with other signalling pathways in guard cell; improves the drought tolerance	[62]
		Salinity stress	
SOS1 *SOS2* *SOS3*	*Triticum aestivum* L.	facilitates exclusion of toxic Na^+ into root apoplast; significantly higher level in salt-tolerant genotype	[77]
OsSOS1 *OsSOS2* *OsSOS3*	*Oryza sativa* L.	increases under salinity	[81]
HvSOS1	*Hordeum vulgare* L.	increases under salinity	[81]
NHX1	*Triticum aestivum* L.	increases under salinity	[82]
HvNHX5	*Hordeum vulgare* L.	increases under salinity in roots	[81]
OsNHX1 *OsNHX2* *OsNHX4* *OsNHX5*	*Oryza sativa* L.	increases under salinity in roots	[81,85,86]
OsHKT1;4 *OsHKT1;5*	*Oryza sativa* L.	increases salinity tolerance; decreases Na^+ accumulation	[88,89]
ZmHKT1;5 *ZmHKT2*	*Zea mays* L.	increases under salinity; reduces leaf K^+ concentration; enhances Na^+ uptake in the root; increases its translocation to the shoot	[90]
OsMAPK33	*Oryza sativa* L.	decreases under salt stress-negative regulator in salinity response	[94]
		Heavy metals	
OsMYB45	*Oryza sativa* L.	decreases H_2O_2 content in the leaves; increases CAT activity	[120]
OsLsi1 *OsLsi2* *OsLsi6*	*Oryza sativa* L.	participates in As transport	[121]
OsNCED2 *OsNCED3*	*Oryza sativa* L.	increases ABA biosynthesis	[135]
ACS1 *ACS2* *ACO4* *ACO5*	*Oryza sativa* L.	increases ET biosynthesis	[136,137]
OsMPK3 *OsCML31* *OsWAKL-Os* *OsLRK10L-2* *OsDUF26-If* *OsWRKY26* *OsAP2/ERF-130* *OsLMWP* *OsDSP* *OsPP2A* *OsPTP* *OsPP2C*	*Oryza sativa* L.	increases under Cr toxicity	[137]

Table 1. *Cont.*

Gene	Plant	Changes in Expression Level/Physiological Effect	References
OsExo70 *(Os01g0763700 Os06g0255900 Os01g0905300 Os01g0905200 Os11g0649900)*	*Oryza sativa* L.	increases under Cr toxicity; participates in vesicle trafficking pathway	[137]
Tom1 (Os05g0475300)	*Oryza sativa* L.	increases under Cr toxicity; participates in vesicle trafficking pathway	[137]
ZIF1 ZIF-like	*Triticum durum* Desf.	increases to Cd toxicity; participates in metal transport	[138]
YSL2	*Triticum durum* Desf.	increases to Cd toxicity; participates in metal transport	[138]
		Biotic stress	
PR1 PR2	*Hordeum vulgare* L.	increases expression under *B. sorokiniana* and decreases under *B. graminis* infestation	[2]
Jsi1	*Zea mays* L.	led to activation of the ERF-branch pathway by an ET-responsive element binding-factor-associated amphiphilic repression (EAR) motif	[163]
OsEREBP1	*Oryza sativa* L.	cause accumulation of JA	[158]
OsERF71	*Oryza sativa* L.	increases in roots and shoots as a result of wounding and submerging in MeJA	[160]
		Multi-stress	
OsWRKY76 OSWRKY82	*Oryza sativa* L.	antagonistically regulates the response of rice to blast disease and cold stress; increases resistance against pathogens and tolerance against abiotic stress via the jasmonic acid and ethylene pathways	[201,202]
OsNAC6	*Oryza sativa* L.	activates the expression peroxidase	[207]
WRAB15 WRAB18	*Triticum aestivum* L.	increases under cadmium, hardening temperature, or their combination; protective and adaptive functions	[210]

Table 2. The list of mutants and transgenic plants with changed stress tolerance under abiotic and biotic stress.

Gene	Species	Type of Manipulation	Effect	Reference
HVA1	Barley	Overexpresion of *HVA1* in rice and wheat	Improves tolerance to water deficit	[59,60]
TaSOS1-974	Wheat	Overexpresion of *TaSOS1-974* in tobacco	Improves Na^+ efflux and K^+ influx rates in the roots, decreases oxidative damage of plasma membrane generated upon salinity	[78]
TdSOS1Δ972	Durum wheat	Overexpresion of *TdSOS1Δ972* in Arabidopsis	Increases water retention capacity and germination rate upon salinity	[79]
TaNHX2	Wheat	Overexpression of *TaNHX2* in *Solanum melongena* L. and *Helianthus annuus* L.	Increases salinity tolerance, improves growth, reduces ROS and MDA content	[83,84]
HvNHX2	Barley	Overexpression of *HvNHX2* in Arabidopsis	Improves growth under salinity	[85]

Table 2. *Cont.*

Gene	Species	Type of Manipulation	Effect	Reference
OsNHX1	Rice	Overexpression of *OsNHX1* in rice	Increases salt tolerance, delays appearance of negative effects connected with damages or death	[86]
OsHKT1;5	Rice	Mutation in *OsHKT1;5* in rice	Excesses Na^+ accumulation in leaves under salinity	[89]
TaMPK4	Wheat	Overexpression of *TaMPK4* in wheat	Improves salinity tolerance, increases K^+ and osmolyte contents and decreases Na^+ content	[91]
ZmSIMK1	Maize	Overexpression of *ZmSIMK1* in Arabidopsis	Increases tolerance to salt stress	[92]
TMKP1	Wheat	Overexpression of *TMKP1* in Arabidopsis	Improves salinity tolerance, increases seeds germination rate, decreases ROS and MDA content under stress	[93]
OsMAPK33	Rice	Overexpression of *OsMAPK33* in rice	Enhances sensitivity to salt stress, disturbs ion homeostasis	[94]
OsMYB45	Rice	Mutation in *OsMYB45* in rice	Reduces resistance to Cd stress, increases H_2O_2 content, decreases CAT activity	[120]
TaPIMP2	Wheat	Overexpression of *TaPIMP2* in wheat	Increased resistance to *Bipolaris sorokiniana*	[151]
TaPIEP1	Wheat	Overexpression of *TaPIEP1* in wheat	Increased resistance to *Bipolaris sorokiniana*	[156]
TaPIMP1	Wheat	Overexpression of *TaPIMP1* in tobacco	Confers tolerance to drought, salt and pathogens stresses	[204]
OsXIP	Rice	Mutation in *OsXIP* in rice	Decreases response to stress induced by *Nilaparvata lugens*	[160]
OsCDPK12	Rice	Overexpression of *OsCDPK12* in rice	Increases salt tolerance, decreases blast resistance	[199]
JAmyb	Rice	Overexpression of *JAmyb* in rice	Improves resistance to blast	[205]
OsWRKY45	Rice	Overexpression of *OsWRKY45* in Arabidopsis	Increases resistance against pathogens and tolerance against abiotic stress	[203]

7. Conclusions

Different stresses affect plants in various ways, therefore proper plant acclimation enabling plant survival is dependent on the crop's ability to recognise the stress factor and its intensity, as well as on the ability to transmit the signal to the appropriate parts of both the cell and the plant in order to trigger an adequate response. While some plant defence mechanisms (such as ROS signalling) are not specific and occur under most stresses, others are strictly dependent on the specific stress factor (e.g., SOS). When cereals struggle to survive only with drought or with the presence of HMs, the situation is quite simple and well recognised in the literature. The problem appears when the same plant is affected by various stress factors at the same time or in short time intervals. In this case, the triggered defence mechanisms can be opposed to each other, which makes resistance to stress difficult. Therefore, learning about the signalling pathways and, more importantly, the interactions between them is crucial in plant cultivation, where multi-stress is common. It should be emphasised that these signal transduction pathways not only intersect with each other but are often opposed (ABA and SA), especially when both abiotic and biotic stress are present in the environment at the same time, which is of paramount importance for plant survival. By activating only selected response elements, and silencing others, it is possible to limit cereals' energy expenditure on ineffective acclimatisation mechanisms. Reducing unnecessary energy consumption allows the plant to continue to develop and grow despite the presence of the stress factor, however, the same mechanism may lead to increase susceptibility to one stress when others occur. Therefore, in the near future, research should focus on signalling pathways crosstalk and multi-stress response.

Author Contributions: Conceptualisation, M.N.; writing—original draft preparation, M.N., J.F., M.G., M.L., B.P., I.M. and E.M.; writing—review and editing, M.N., J.F., M.G., M.L., A.R.-P., D.B.-M., J.G., I.M. and E.M.; supervision, M.N. All authors have read and agreed to the published version of the manuscript.

Funding: This research received no external funding. The APC was funded by the Multidisciplinary Digital Publishing Institute (MDPI) publication discount voucher granted to Małgorzata Nykiel.

Institutional Review Board Statement: Not applicable.

Informed Consent Statement: Not applicable.

Data Availability Statement: Not applicable.

Conflicts of Interest: The authors declare no conflict of interest.

References

1. Raza, A.; Razzaq, A.; Mehmood, S.; Zou, X.; Zhang, X.; Lv, Y.; Xu, J. Impact of Climate Change on Crops Adaptation and Strategies to Tackle Its Outcome: A Review. *Plants* **2019**, *8*, 34. [CrossRef] [PubMed]
2. Al-Daoude, A.; Al-Shehadah, E.; Shoaib, A.; Jawhar, M.; Arabi, M.I.E. Salicylic acid pathway changes in barley plants challenged with either a biotrophic or a necrotrophic pathogen. *Cereal Res. Commun.* **2019**, *47*, 324–333. [CrossRef]
3. Morkunas, I.; Woźniak, A.; Mai, V.; Rucińska-Sobkowiak, R.; Jeandet, P. The Role of Heavy Metals in Plant Response to Biotic Stress. *Molecules* **2018**, *23*, 2320. [CrossRef] [PubMed]
4. Laxa, M.; Liebthal, M.; Telman, W.; Chibani, K.; Dietz, K.-J. The Role of the Plant Antioxidant System in Drought Tolerance. *Antioxidants* **2019**, *8*, 94. [CrossRef]
5. Abdelaal, K.; Attia, K.A.; Niedbała, G.; Wojciechowski, T.; Hafez, Y.; Alamery, S.; Alateeq, T.K.; Arafa, S.A. Mitigation of Drought Damages by Exogenous Chitosan and Yeast Extract with Modulating the Photosynthetic Pigments, Antioxidant Defense System and Improving the Productivity of Garlic Plants. *Horticulturae* **2021**, *7*, 510. [CrossRef]
6. Abdelaal, K.A.A.; Attia, K.A.; Alamery, S.F.; El-Afry, M.M.; Ghazy, A.I.; Tantawy, D.S.; Al-Doss, A.A.; El-Shawy, E.-S.E.; Abu-Elsaoud, A.M.; Hafez, Y.M. Exogenous Application of Proline and Salicylic Acid can Mitigate the Injurious Impacts of Drought Stress on Barley Plants Associated with Physiological and Histological Characters. *Sustainability* **2020**, *12*, 1736. [CrossRef]
7. Moghanm, F.S.; El-Banna, A.; El-Esawi, M.A.; Abdel-Daim, M.M.; Mosa, A.; Abdelaal, K.A.A. Genotoxic and Anatomical Deteriorations Associated with Potentially Toxic Elements Accumulation in Water Hyacinth Grown in Drainage Water Resources. *Sustainability* **2020**, *12*, 2147. [CrossRef]
8. Mahmood, T.; Khalid, S.; Abdullah, M.; Ahmed, Z.; Shah, M.K.N.; Ghafoor, A.; Du, X. Insights into Drought Stress Signaling in Plants and the Molecular Genetic Basis of Cotton Drought Tolerance. *Cells* **2019**, *9*, 105. [CrossRef]
9. Tiwari, S.; Lata, C. Heavy Metal Stress, Signaling, and Tolerance Due to Plant-Associated Microbes: An Overview. *Front. Plant Sci.* **2018**, *9*, 452. [CrossRef]
10. Hasanuzzaman, M.; Bhuyan, M.H.M.; Zulfiqar, F.; Raza, A.; Mohsin, S.; Mahmud, J.; Fujita, M.; Fotopoulos, V. Reactive Oxygen Species and Antioxidant Defense in Plants under Abiotic Stress: Revisiting the Crucial Role of a Universal Defense Regulator. *Antioxidants* **2020**, *9*, 681. [CrossRef]
11. Wang, C.; Deng, Y.; Liu, Z.; Liao, W. Hydrogen Sulfide in Plants: Crosstalk with Other Signal Molecules in Response to Abiotic Stresses. *Int. J. Mol. Sci.* **2021**, *22*, 12068. [CrossRef] [PubMed]
12. Shan, C.; Zhang, S.; Zhou, Y. Hydrogen sulfide is involved in the regulation of ascorbate-glutathione cycle by exogenous ABA in wheat seedling leaves under osmotic stress. *Cereal Res. Commun.* **2017**, *45*, 411–420. [CrossRef]
13. Sachdev, S.; Ansari, S.A.; Ansari, M.I.; Fujita, M.; Hasanuzzaman, M. Abiotic Stress and Reactive Oxygen Species: Generation, Signaling, and Defense Mechanisms. *Antioxidants* **2021**, *10*, 277. [CrossRef]
14. Dolata, J.; Zielezinski, A.; Stepien, A.; Kruszka, K.; Bielewicz, D.; Pacak, A.; Jarmolowski, A.; Karlowski, W.; Szweykowska-Kulinska, Z. Quantitative Analysis of Plant Primary Transcripts. In *RNA Abundance Analysis*; Jin, H., Kaloshian, I., Eds.; Methods in Molecular Biology; Springer: New York, NY, USA, 2021; Volume 2170, pp. 53–77. ISBN 978-1-07-160742-8.
15. Agurla, S.; Gahir, S.; Munemasa, S.; Murata, Y.; Raghavendra, A.S. Mechanism of Stomatal Closure in Plants Exposed to Drought and Cold Stress. In *Survival Strategies in Extreme Cold and Desiccation*; Iwaya-Inoue, M., Sakurai, M., Uemura, M., Eds.; Advances in Experimental Medicine and Biology; Springer: Singapore, 2018; Volume 1081, pp. 215–232. ISBN 9789811312434.
16. Arafa, S.A.; Attia, K.A.; Niedbała, G.; Piekutowska, M.; Alamery, S.; Abdelaal, K.; Alateeq, T.K.; Ali, M.A.M.; Elkelish, A.; Attallah, S.Y. Seed Priming Boost Adaptation in Pea Plants under Drought Stress. *Plants* **2021**, *10*, 2201. [CrossRef] [PubMed]

17. Hsiao, T.C.; Xu, L. Sensitivity of growth of roots versus leaves to water stress: Biophysical analysis and relation to water transport. *J. Exp. Bot.* **2000**, *51*, 1595–1616. [CrossRef] [PubMed]
18. Xu, B.; Gao, Z.; Wang, J.; Xu, W.; Huang, J. Morphological changes in roots of *Bothriochloa ischaemum* intercropped with *Lespedeza davurica* following phosphorus application and water stress. *Plant Biosyst.-Int. J. Deal. All Asp. Plant Biol.* **2015**, *149*, 298–306. [CrossRef]
19. Zhou, G.; Zhou, X.; Nie, Y.; Bai, S.H.; Zhou, L.; Shao, J.; Cheng, W.; Wang, J.; Hu, F.; Fu, Y. Drought-induced changes in root biomass largely result from altered root morphological traits: Evidence from a synthesis of global field trials: Effects of drought on root traits. *Plant Cell Environ.* **2018**, *41*, 2589–2599. [CrossRef] [PubMed]
20. Mori, M.; Inagaki, M. Root development and water-uptake under water deficit stress in drought-adaptive wheat genotypes. *Cereal Res. Commun.* **2012**, *40*, 44–52. [CrossRef]
21. Kadam, N.N.; Yin, X.; Bindraban, P.S.; Struik, P.C.; Jagadish, K.S.V. Does Morphological and Anatomical Plasticity during the Vegetative Stage Make Wheat More Tolerant of Water Deficit Stress Than Rice? *Plant Physiol.* **2015**, *167*, 1389–1401. [CrossRef]
22. Kleine, T.; Leister, D. Retrograde signaling: Organelles go networking. *Biochim. Et Biophys. Acta (BBA)-Bioenerg.* **2016**, *1857*, 1313–1325. [CrossRef]
23. Reczek, C.R.; Chandel, N.S. ROS-dependent signal transduction. *Curr. Opin. Cell Biol.* **2015**, *33*, 8–13. [CrossRef] [PubMed]
24. Waszczak, C.; Akter, S.; Jacques, S.; Huang, J.; Messens, J.; Van Breusegem, F. Oxidative post-translational modifications of cysteine residues in plant signal transduction. *J. Exp. Bot.* **2015**, *66*, 2923–2934. [CrossRef] [PubMed]
25. Martins, L.; Trujillo-Hernandez, J.A.; Reichheld, J.-P. Thiol Based Redox Signaling in Plant Nucleus. *Front. Plant Sci.* **2018**, *9*, 705. [CrossRef] [PubMed]
26. Meyer, Y.; Buchanan, B.B.; Vignols, F.; Reichheld, J.-P. Thioredoxins and Glutaredoxins: Unifying Elements in Redox Biology. *Annu. Rev. Genet.* **2009**, *43*, 335–367. [CrossRef]
27. Waszczak, C.; Akter, S.; Eeckhout, D.; Persiau, G.; Wahni, K.; Bodra, N.; Van Molle, I.; De Smet, B.; Vertommen, D.; Gevaert, K.; et al. Sulfenome mining in Arabidopsis thaliana. *Proc. Natl. Acad. Sci. USA* **2014**, *111*, 11545–11550. [CrossRef]
28. Lin, L.; Wu, J.; Jiang, M.; Wang, Y. Plant Mitogen-Activated Protein Kinase Cascades in Environmental Stresses. *Int. J. Mol. Sci.* **2021**, *22*, 1543. [CrossRef]
29. Caverzan, A.; Casassola, A.; Brammer, S.P. Antioxidant responses of wheat plants under stress. *Genet. Mol. Biol.* **2016**, *39*, 1–6. [CrossRef]
30. Kapoor, D.; Bhardwaj, S.; Landi, M.; Sharma, A.; Ramakrishnan, M.; Sharma, A. The Impact of Drought in Plant Metabolism: How to Exploit Tolerance Mechanisms to Increase Crop Production. *Appl. Sci.* **2020**, *10*, 5692. [CrossRef]
31. Ford, K.L.; Cassin, A.; Bacic, A. Quantitative Proteomic Analysis of Wheat Cultivars with Differing Drought Stress Tolerance. *Front. Plant Sci.* **2011**, *2*, 44. [CrossRef]
32. Ullah, A.; Manghwar, H.; Shaban, M.; Khan, A.H.; Akbar, A.; Ali, U.; Ali, E.; Fahad, S. Phytohormones enhanced drought tolerance in plants: A coping strategy. *Environ. Sci. Pollut. Res.* **2018**, *25*, 33103–33118. [CrossRef]
33. Cui, X.-Y.; Gao, Y.; Guo, J.; Yu, T.-F.; Zheng, W.-J.; Liu, Y.-W.; Chen, J.; Xu, Z.-S.; Ma, Y.-Z. BES/BZR Transcription Factor TaBZR2 Positively Regulates Drought Responses by Activation of *TaGST1*. *Plant Physiol.* **2019**, *180*, 605–620. [CrossRef] [PubMed]
34. Yan, J.; Guan, L.; Sun, Y.; Zhu, Y.; Liu, L.; Lu, R.; Jiang, M.; Tan, M.; Zhang, A. Calcium and ZmCCaMK are involved in brassinosteroid-induced antioxidant defense in maize leaves. *Plant Cell Physiol.* **2015**, *56*, 883–896. [CrossRef] [PubMed]
35. Siddiqui, M.N.; Léon, J.; Naz, A.A.; Ballvora, A. Genetics and genomics of root system variation in adaptation to drought stress in cereal crops. *J. Exp. Bot.* **2021**, *72*, 1007–1019. [CrossRef] [PubMed]
36. Verslues, P.E.; Agarwal, M.; Katiyar-Agarwal, S.; Zhu, J.; Zhu, J.-K. Methods and concepts in quantifying resistance to drought, salt and freezing, abiotic stresses that affect plant water status. *Plant J.* **2006**, *45*, 523–539. [CrossRef]
37. Blum, A. Drought resistance, water-use efficiency, and yield potential—Are they compatible, dissonant, or mutually exclusive? *Aust. J. Agric. Res.* **2005**, *56*, 1159. [CrossRef]
38. Jin, Z.; Wang, Z.; Ma, Q.; Sun, L.; Zhang, L.; Liu, Z.; Liu, D.; Hao, X.; Pei, Y. Hydrogen sulfide mediates ion fluxes inducing stomatal closure in response to drought stress in Arabidopsis thaliana. *Plant Soil* **2017**, *419*, 141–152. [CrossRef]
39. Gietler, M.; Fidler, J.; Labudda, M.; Nykiel, M. Abscisic Acid—Enemy or Savior in the Response of Cereals to Abiotic and Biotic Stresses? *Int. J. Mol. Sci.* **2020**, *21*, 4607. [CrossRef]
40. Sah, S.K.; Reddy, K.R.; Li, J. Abscisic Acid and Abiotic Stress Tolerance in Crop Plants. *Front. Plant Sci.* **2016**, *7*, 571. [CrossRef]
41. Yoshida, T.; Mogami, J.; Yamaguchi-Shinozaki, K. ABA-dependent and ABA-independent signaling in response to osmotic stress in plants. *Curr. Opin. Plant Biol.* **2014**, *21*, 133–139. [CrossRef]
42. Zhu, Q.; Zhang, J.; Gao, X.; Tong, J.; Xiao, L.; Li, W.; Zhang, H. The Arabidopsis AP2/ERF transcription factor RAP2.6 participates in ABA, salt and osmotic stress responses. *Gene* **2010**, *457*, 1–12. [CrossRef]
43. Zhang, Z.; Li, F.; Li, D.; Zhang, H.; Huang, R. Expression of ethylene response factor JERF1 in rice improves tolerance to drought. *Planta* **2010**, *232*, 765–774. [CrossRef] [PubMed]
44. Mittal, S.; Mallikarjuna, M.G.; Rao, A.R.; Jain, P.A.; Dash, P.K.; Thirunavukkarasu, N. Comparative Analysis of CDPK Family in Maize, Arabidopsis, Rice, and Sorghum Revealed Potential Targets for Drought Tolerance Improvement. *Front. Chem.* **2017**, *5*, 115. [CrossRef] [PubMed]

45. Lu, L.; Chen, X.; Wang, P.; Lu, Y.; Zhang, J.; Yang, X.; Cheng, T.; Shi, J.; Chen, J. CIPK11: A calcineurin B-like protein-interacting protein kinase from Nitraria tangutorum, confers tolerance to salt and drought in Arabidopsis. *BMC Plant Biol.* **2021**, *21*, 123. [CrossRef] [PubMed]
46. Xue, H.-W.; Chen, X.; Mei, Y. Function and regulation of phospholipid signalling in plants. *Biochem. J.* **2009**, *421*, 145–156. [CrossRef]
47. Yoo, S.J.; Kim, S.H.; Kim, M.J.; Ryu, C.M.; Kim, Y.C.; Cho, B.H.; Yang, K.Y. Involvement of the OsMKK4-OsMPK1 Cascade and its Downstream Transcription Factor OsWRKY53 in the Wounding Response in Rice. *Plant Pathol. J.* **2014**, *30*, 168–177. [CrossRef]
48. Kumar, K.; Rao, K.P.; Sharma, P.; Sinha, A.K. Differential regulation of rice mitogen activated protein kinase kinase (MKK) by abiotic stress. *Plant Physiol. Biochem.* **2008**, *46*, 891–897. [CrossRef]
49. Wang, M.; Yue, H.; Feng, K.; Deng, P.; Song, W.; Nie, X. Genome-wide identification, phylogeny and expressional profiles of mitogen activated protein kinase kinase kinase (MAPKKK) gene family in bread wheat (*Triticum aestivum* L.). *BMC Genom.* **2016**, *17*, 668. [CrossRef]
50. Ma, H.; Chen, J.; Zhang, Z.; Ma, L.; Yang, Z.; Zhang, Q.; Li, X.; Xiao, J.; Wang, S. MAPK kinase 10.2 promotes disease resistance and drought tolerance by activating different MAPKs in rice. *Plant J.* **2017**, *92*, 557–570. [CrossRef]
51. Hadiarto, T.; Tran, L.-S.P. Progress studies of drought-responsive genes in rice. *Plant Cell Rep.* **2011**, *30*, 297–310. [CrossRef]
52. Shen, H.; Liu, C.; Zhang, Y.; Meng, X.; Zhou, X.; Chu, C.; Wang, X. OsWRKY30 is activated by MAP kinases to confer drought tolerance in rice. *Plant Mol. Biol.* **2012**, *80*, 241–253. [CrossRef]
53. Saijo, Y.; Hata, S.; Kyozuka, J.; Shimamoto, K.; Izui, K. Over-expression of a single Ca $^{2+}$-dependent protein kinase confers both cold and salt/drought tolerance on rice plants. *Plant J.* **2000**, *23*, 319–327. [CrossRef] [PubMed]
54. Xiang, J.; Ran, J.; Zou, J.; Zhou, X.; Liu, A.; Zhang, X.; Peng, Y.; Tang, N.; Luo, G.; Chen, X. Heat shock factor OsHsfB2b negatively regulates drought and salt tolerance in rice. *Plant Cell Rep.* **2013**, *32*, 1795–1806. [CrossRef] [PubMed]
55. Chauhan, H.; Khurana, N.; Agarwal, P.; Khurana, P. Heat shock factors in rice (*Oryza sativa* L.): Genome-wide expression analysis during reproductive development and abiotic stress. *Mol. Genet. Genom.* **2011**, *286*, 171. [CrossRef] [PubMed]
56. Kosová, K.; Vítámvás, P.; Prášil, I.T. Wheat and barley dehydrins under cold, drought, and salinity–what can LEA-II proteins tell us about plant stress response? *Front. Plant Sci.* **2014**, *5*, 343. [CrossRef]
57. Karami, A.; Shahbazi, M.; Niknam, V.; Shobbar, Z.S.; Tafreshi, R.S.; Abedini, R.; Mabood, H.E. Expression analysis of dehydrin multigene family across tolerant and susceptible barley (*Hordeum vulgare* L.) genotypes in response to terminal drought stress. *Acta Physiol. Plant* **2013**, *35*, 2289–2297. [CrossRef]
58. Ali, A.; Raddatz, N.; Pardo, J.M.; Yun, D. HKT sodium and potassium transporters in *Arabidopsis thaliana* and related halophyte species. *Physiol. Plant.* **2021**, *171*, 546–558. [CrossRef]
59. Xu, D.; Duan, X.; Wang, B.; Hong, B.; Ho, T.; Wu, R. Expression of a Late Embryogenesis Abundant Protein Gene, HVA1, from Barley Confers Tolerance to Water Deficit and Salt Stress in Transgenic Rice. *Plant Physiol.* **1996**, *110*, 249–257. [CrossRef]
60. Hand, S.C.; Menze, M.A.; Toner, M.; Boswell, L.; Moore, D. LEA Proteins During Water Stress: Not Just for Plants Anymore. *Annu. Rev. Physiol.* **2011**, *73*, 115–134. [CrossRef]
61. Rodríguez-Valentín, R.; Campos, F.; Battaglia, M.; Solórzano, R.M.; Rosales, M.A.; Covarrubias, A.A. Group 6 Late Embryogenesis Abundant (LEA) Proteins in Monocotyledonous Plants: Genomic Organization and Transcript Accumulation Patterns in Response to Stress in Oryza sativa. *Plant Mol. Biol. Rep.* **2014**, *32*, 198–208. [CrossRef]
62. Wang, C.-R.; Yang, A.-F.; Yue, G.-D.; Gao, Q.; Yin, H.-Y.; Zhang, J.-R. Enhanced expression of phospholipase C 1 (ZmPLC1) improves drought tolerance in transgenic maize. *Planta* **2008**, *227*, 1127–1140. [CrossRef]
63. Gehring, C.; Turek, I.S. Cyclic Nucleotide Monophosphates and Their Cyclases in Plant Signaling. *Front. Plant Sci.* **2017**, *8*, 1704. [CrossRef] [PubMed]
64. Gan, L.; Wu, X.; Zhong, Y. Exogenously Applied Nitric Oxide Enhances the Drought Tolerance in Hulless Barley. *Plant Prod. Sci.* **2015**, *18*, 52–56. [CrossRef]
65. Lau, S.-E.; Hamdan, M.F.; Pua, T.-L.; Saidi, N.B.; Tan, B.C. Plant Nitric Oxide Signaling under Drought Stress. *Plants* **2021**, *10*, 360. [CrossRef] [PubMed]
66. Chang, J.; Cheong, B.E.; Natera, S.; Roessner, U. Morphological and metabolic responses to salt stress of rice (*Oryza sativa* L.) cultivars which differ in salinity tolerance. *Plant Physiol. Biochem.* **2019**, *144*, 427–435. [CrossRef] [PubMed]
67. Alnusairi, G.S.H.; Mazrou, Y.S.A.; Qari, S.H.; Elkelish, A.A.; Soliman, M.H.; Eweis, M.; Abdelaal, K.; El-Samad, G.A.; Ibrahim, M.F.M.; ElNahhas, N. Exogenous Nitric Oxide Reinforces Photosynthetic Efficiency, Osmolyte, Mineral Uptake, Antioxidant, Expression of Stress-Responsive Genes and Ameliorates the Effects of Salinity Stress in Wheat. *Plants* **2021**, *10*, 1693. [CrossRef] [PubMed]
68. Abiala, M.A.; Abdelrahman, M.; Burritt, D.J.; Tran, L.P. Salt stress tolerance mechanisms and potential applications of legumes for sustainable reclamation of salt-degraded soils. *Land Degrad. Dev.* **2018**, *29*, 3812–3822. [CrossRef]
69. Hussain, S.; Zhang, J.; Zhong, C.; Zhu, L.; Cao, X.; Yu, S.; Allen Bohr, J.; Hu, J.; Jin, Q. Effects of salt stress on rice growth, development characteristics, and the regulating ways: A review. *J. Integr. Agric.* **2017**, *16*, 2357–2374. [CrossRef]
70. Roy, P.R.; Tahjib-Ul-Arif, M.; Polash, M.A.S.; Hossen, M.Z.; Hossain, M.A. Physiological mechanisms of exogenous calcium on alleviating salinity-induced stress in rice (*Oryza sativa* L.). *Physiol. Mol. Biol. Plants* **2019**, *25*, 611–624. [CrossRef]

71. Riaz, M.; Arif, M.S.; Ashraf, M.A.; Mahmood, R.; Yasmeen, T.; Shakoor, M.B.; Shahzad, S.M.; Ali, M.; Saleem, I.; Arif, M.; et al. A Comprehensive Review on Rice Responses and Tolerance to Salt Stress. In *Advances in Rice Research for Abiotic Stress Tolerance*; Elsevier: Amsterdam, The Netherlands, 2019; pp. 133–158. ISBN 978-0-12-814332-2.
72. Hasanuzzaman, M.; Oku, H.; Nahar, K.; Bhuyan, M.H.M.B.; Mahmud, J.A.; Baluska, F.; Fujita, M. Nitric oxide-induced salt stress tolerance in plants: ROS metabolism, signaling, and molecular interactions. *Plant Biotechnol. Rep.* **2018**, *12*, 77–92. [CrossRef]
73. Yang, Y.; Guo, Y. Unraveling salt stress signaling in plants: Salt stress signaling. *J. Integr. Plant Biol.* **2018**, *60*, 796–804. [CrossRef]
74. Choudhary, P.; Pramitha, L.; Rana, S.; Verma, S.; Aggarwal, P.R.; Muthamilarasan, M. Hormonal crosstalk in regulating salinity stress tolerance in graminaceous crops. *Physiol. Plant.* **2021**, *173*, 1587–1596. [CrossRef] [PubMed]
75. Sathee, L.; Sairam, R.K.; Chinnusamy, V.; Jha, S.K. Differential transcript abundance of salt overly sensitive (SOS) pathway genes is a determinant of salinity stress tolerance of wheat. *Acta Physiol. Plant* **2015**, *37*, 169. [CrossRef]
76. Seifikalhor, M.; Aliniaeifard, S.; Shomali, A.; Azad, N.; Hassani, B.; Lastochkina, O.; Li, T. Calcium signaling and salt tolerance are diversely entwined in plants. *Plant Signal. Behav.* **2019**, *14*, 1665455. [CrossRef] [PubMed]
77. Jiang, W.; Pan, R.; Buitrago, S.; Wu, C.; Abou-Elwafa, S.F.; Xu, Y.; Zhang, W. Conservation and divergence of the TaSOS1 gene family in salt stress response in wheat (*Triticum aestivum* L.). *Physiol. Mol. Biol. Plants* **2021**, *27*, 1245–1260. [CrossRef] [PubMed]
78. Zhou, Y.; Lai, Z.; Yin, X.; Yu, S.; Xu, Y.; Wang, X.; Cong, X.; Luo, Y.; Xu, H.; Jiang, X. Hyperactive mutant of a wheat plasma membrane Na^+/H^+ antiporter improves the growth and salt tolerance of transgenic tobacco. *Plant Sci.* **2016**, *253*, 176–186. [CrossRef]
79. Feki, K.; Quintero, F.J.; Khoudi, H.; Leidi, E.O.; Masmoudi, K.; Pardo, J.M.; Brini, F. A constitutively active form of a durum wheat Na^+/H^+ antiporter SOS1 confers high salt tolerance to transgenic Arabidopsis. *Plant Cell Rep.* **2014**, *33*, 277–288. [CrossRef]
80. Jarvis, D.E.; Ryu, C.-H.; Beilstein, M.A.; Schumaker, K.S. Distinct Roles for SOS1 in the Convergent Evolution of Salt Tolerance in Eutrema salsugineum and Schrenkiella parvula. *Mol. Biol. Evol.* **2014**, *31*, 2094–2107. [CrossRef]
81. Fu, L.; Shen, Q.; Kuang, L.; Yu, J.; Wu, D.; Zhang, G. Metabolite profiling and gene expression of Na/K transporter analyses reveal mechanisms of the difference in salt tolerance between barley and rice. *Plant Physiol. Biochem.* **2018**, *130*, 248–257. [CrossRef]
82. Tiwari, B.K.; Aquib, A.; Anand, R. Analysis of physiological traits and expression of NHX and SOS3 genes in bread wheat (Triticum aestivum L.) under salinity stress. *J. Pharmacogn. Phytochem* **2020**, *9*, 362–366. [CrossRef]
83. Yarra, R.; Kirti, P.B. Expressing class I wheat NHX (TaNHX2) gene in eggplant (*Solanum melongena* L.) improves plant performance under saline condition. *Funct. Integr. Genom.* **2019**, *19*, 541–554. [CrossRef]
84. Mushke, R.; Yarra, R.; Kirti, P.B. Improved salinity tolerance and growth performance in transgenic sunflower plants via ectopic expression of a wheat antiporter gene (TaNHX2). *Mol. Biol. Rep.* **2019**, *46*, 5941–5953. [CrossRef] [PubMed]
85. Bayat, F.; Shiran, B.; Belyaev, D.V. Overexpression of HvNHX2, a Vacuolar Na^+/H^+ Antiporter Gene from Barley, Improves Salt Tolerance in "Arabidopsis thaliana". *Aust. J. Crop Sci.* **2011**, *5*, 428–432.
86. Chen, H.; An, R.; Tang, J.-H.; Cui, X.-H.; Hao, F.-S.; Chen, J.; Wang, X.-C. Over-expression of a vacuolar Na^+/H^+ antiporter gene improves salt tolerance in an upland rice. *Mol. Breed.* **2007**, *19*, 215–225. [CrossRef]
87. Ali, A.; Maggio, A.; Bressan, R.; Yun, D.-J. Role and Functional Differences of HKT1-Type Transporters in Plants under Salt Stress. *Int. J. Mol. Sci.* **2019**, *20*, 1059. [CrossRef]
88. Wangsawang, T.; Chuamnakthong, S.; Kohnishi, E.; Sripichitt, P.; Sreewongchai, T.; Ueda, A. A salinity-tolerant japonica cultivar has Na^+ exclusion mechanism at leaf sheaths through the function of a Na^+ transporter OsHKT1;4 under salinity stress. *J. Agron. Crop Sci.* **2018**, *204*, 274–284. [CrossRef]
89. Kobayashi, N.I.; Yamaji, N.; Yamamoto, H.; Okubo, K.; Ueno, H.; Costa, A.; Tanoi, K.; Matsumura, H.; Fujii-Kashino, M.; Horiuchi, T.; et al. OsHKT1;5 mediates Na^+ exclusion in the vasculature to protect leaf blades and reproductive tissues from salt toxicity in rice. *Plant J.* **2017**, *91*, 657–670. [CrossRef]
90. Rizk, M.S.; Mekawy, A.M.M.; Assaha, D.V.M.; Chuamnakthong, S.; Shalaby, N.E.; Ueda, A. Regulation of Na^+ and K^+ Transport and Oxidative Stress Mitigation Reveal Differential Salt Tolerance of Two Egyptian Maize (*Zea mays* L.) Hybrids at the Seedling Stage. *J. Plant Growth Regul.* **2021**, *40*, 1629–1639. [CrossRef]
91. Hao, L.; Wen, Y.; Zhao, Y.; Lu, W.; Xiao, K. Wheat mitogen-activated protein kinase gene TaMPK4 improves plant tolerance to multiple stresses through modifying root growth, ROS metabolism, and nutrient acquisitions. *Plant Cell Rep.* **2015**, *34*, 2081–2097. [CrossRef]
92. Gu, L.; Liu, Y.; Zong, X.; Liu, L.; Li, D.-P.; Li, D.-Q. Overexpression of maize mitogen-activated protein kinase gene, ZmSIMK1 in Arabidopsis increases tolerance to salt stress. *Mol. Biol. Rep.* **2010**, *37*, 4067–4073. [CrossRef]
93. Zaidi, I.; Ebel, C.; Belgaroui, N.; Ghorbel, M.; Amara, I.; Hanin, M. The wheat MAP kinase phosphatase 1 alleviates salt stress and increases antioxidant activities in Arabidopsis. *J. Plant Physiol.* **2016**, *193*, 12–21. [CrossRef]
94. Lee, S.-K.; Kim, B.-G.; Kwon, T.-R.; Jeong, M.-J.; Park, S.-R.; Lee, J.-W.; Byun, M.-O.; Kwon, H.-B.; Matthews, B.F.; Hong, C.-B.; et al. Overexpression of the mitogen-activated protein kinase gene OsMAPK33 enhances sensitivity to salt stress in rice (*Oryza sativa* L.). *J. Biosci.* **2011**, *36*, 139–151. [CrossRef] [PubMed]
95. Wei, M.-Y.; Liu, J.-Y.; Li, H.; Hu, W.-J.; Shen, Z.-J.; Qiao, F.; Zhu, C.-Q.; Chen, J.; Liu, X.; Zheng, H.-L. Proteomic analysis reveals the protective role of exogenous hydrogen sulfide against salt stress in rice seedlings. *Nitric Oxide* **2021**, *111*, 14–30. [CrossRef] [PubMed]

96. Ding, H.; Ma, D.; Huang, X.; Hou, J.; Wang, C.; Xie, Y.; Wang, Y.; Qin, H.; Guo, T. Exogenous hydrogen sulfide alleviates salt stress by improving antioxidant defenses and the salt overly sensitive pathway in wheat seedlings. *Acta Physiol. Plant* **2019**, *41*, 123. [CrossRef]
97. Chen, J.; Wang, W.-H.; Wu, F.-H.; He, E.-M.; Liu, X.; Shangguan, Z.-P.; Zheng, H.-L. Hydrogen sulfide enhances salt tolerance through nitric oxide-mediated maintenance of ion homeostasis in barley seedling roots. *Sci. Rep.* **2015**, *5*, 12516. [CrossRef]
98. Rizvi, A.; Zaidi, A.; Ameen, F.; Ahmed, B.; AlKahtani, M.D.F.; Khan, M.S. Heavy metal induced stress on wheat: Phytotoxicity and microbiological management. *RSC Adv.* **2020**, *10*, 38379–38403. [CrossRef]
99. Adrees, M.; Ali, S.; Rizwan, M.; Ibrahim, M.; Abbas, F.; Farid, M.; Zia-ur-Rehman, M.; Irshad, M.K.; Bharwana, S.A. The effect of excess copper on growth and physiology of important food crops: A review. *Environ. Sci. Pollut. Res.* **2015**, *22*, 8148–8162. [CrossRef]
100. Singh, A.; Prasad, S.M. Remediation of heavy metal contaminated ecosystem: An overview on technology advancement. *Int. J. Environ. Sci. Technol.* **2015**, *12*, 353–366. [CrossRef]
101. Mahmood, T.; Gupta, K.J.; Kaiser, W.M. Cadmium stress stimulates nitric oxide production by wheat roots. *Pak. J. Bot.* **2009**, *41*, 1285–1290.
102. Aslam, M.; Aslam, A.; Sheraz, M.; Ali, B.; Ulhassan, Z.; Najeeb, U.; Zhou, W.; Gill, R.A. Lead Toxicity in Cereals: Mechanistic Insight Into Toxicity, Mode of Action, and Management. *Front. Plant Sci.* **2021**, *11*, 587785. [CrossRef]
103. Jamla, M.; Khare, T.; Joshi, S.; Patil, S.; Penna, S.; Kumar, V. Omics approaches for understanding heavy metal responses and tolerance in plants. *Curr. Plant Biol.* **2021**, *27*, 100213. [CrossRef]
104. Fang, H.; Jing, T.; Liu, Z.; Zhang, L.; Jin, Z.; Pei, Y. Hydrogen sulfide interacts with calcium signaling to enhance the chromium tolerance in Setaria italica. *Cell Calcium* **2014**, *56*, 472–481. [CrossRef] [PubMed]
105. AbdElgawad, H.; Zinta, G.; Hamed, B.A.; Selim, S.; Beemster, G.; Hozzein, W.N.; Wadaan, M.A.M.; Asard, H.; Abuelsoud, W. Maize roots and shoots show distinct profiles of oxidative stress and antioxidant defense under heavy metal toxicity. *Environ. Pollut.* **2020**, *258*, 113705. [CrossRef]
106. Singh, I.; Shah, K. Exogenous application of methyl jasmonate lowers the effect of cadmium-induced oxidative injury in rice seedlings. *Phytochemistry* **2014**, *108*, 57–66. [CrossRef] [PubMed]
107. Chakrabarty, D.; Trivedi, P.K.; Misra, P.; Tiwari, M.; Shri, M.; Shukla, D.; Kumar, S.; Rai, A.; Pandey, A.; Nigam, D.; et al. Comparative transcriptome analysis of arsenate and arsenite stresses in rice seedlings. *Chemosphere* **2009**, *74*, 688–702. [CrossRef]
108. Raghuram, B.; Sheikh, A.H.; Sinha, A.K. Regulation of MAP kinase signaling cascade by microRNAs in *Oryza sativa*. *Plant Signal. Behav.* **2014**, *9*, e972130. [CrossRef] [PubMed]
109. Awasthi, S.; Chauhan, R.; Srivastava, S.; Tripathi, R.D. The Journey of Arsenic from Soil to Grain in Rice. *Front. Plant Sci.* **2017**, *8*, 1007. [CrossRef]
110. Rao, K.P.; Vani, G.; Kumar, K.; Wankhede, D.P.; Misra, M.; Gupta, M.; Sinha, A.K. Arsenic stress activates MAP kinase in rice roots and leaves. *Arch. Biochem. Biophys.* **2011**, *506*, 73–82. [CrossRef] [PubMed]
111. Huang, J.; Zhang, Y.; Peng, J.-S.; Zhong, C.; Yi, H.-Y.; Ow, D.W.; Gong, J.-M. Fission Yeast HMT1 Lowers Seed Cadmium through Phytochelatin-Dependent Vacuolar Sequestration in Arabidopsis. *Plant Physiol.* **2012**, *158*, 1779–1788. [CrossRef]
112. Ma, C.; Burd, S.; Lers, A. *miR408* is involved in abiotic stress responses in Arabidopsis. *Plant J.* **2015**, *84*, 169–187. [CrossRef]
113. Sharma, D.; Tiwari, M.; Lakhwani, D.; Tripathi, R.D.; Trivedi, P.K. Differential expression of microRNAs by arsenate and arsenite stress in natural accessions of rice. *Metallomics* **2015**, *7*, 174–187. [CrossRef]
114. Liu, Y.; Zhang, C.; Wang, D.; Su, W.; Liu, L.; Wang, M.; Li, J. EBS7 is a plant-specific component of a highly conserved endoplasmic reticulum-associated degradation system in *Arabidopsis*. *Proc. Natl. Acad. Sci. USA* **2015**, *112*, 12205–12210. [CrossRef] [PubMed]
115. Agrawal, G.K.; Iwahashi, H.; Rakwal, R. Rice MAPKs. *Biochem. Biophys. Res. Commun.* **2003**, *302*, 171–180. [CrossRef]
116. Yeh, C.-M.; Chien, P.-S.; Huang, H.-J. Distinct signalling pathways for induction of MAP kinase activities by cadmium and copper in rice roots. *J. Exp. Bot.* **2006**, *58*, 659–671. [CrossRef] [PubMed]
117. Zhao, F.Y.; Wang, K.; Zhang, S.Y.; Ren, J.; Liu, T.; Wang, X. Crosstalk between ABA, auxin, MAPK signaling, and the cell cycle in cadmium-stressed rice seedlings. *Acta Physiol. Plant* **2014**, *36*, 1879–1892. [CrossRef]
118. Tang, M.; Mao, D.; Xu, L.; Li, D.; Song, S.; Chen, C. Integrated analysis of miRNA and mRNA expression profiles in response to Cd exposure in rice seedlings. *BMC Genom.* **2014**, *15*, 835. [CrossRef]
119. Ogawa, I.; Nakanishi, H.; Mori, S.; Nishizawa, N.K. Time course analysis of gene regulation under cadmium stress in rice. *Plant Soil* **2009**, *325*, 97–108. [CrossRef]
120. Hu, S.; Yu, Y.; Chen, Q.; Mu, G.; Shen, Z.; Zheng, L. OsMYB45 plays an important role in rice resistance to cadmium stress. *Plant Sci.* **2017**, *264*, 1–8. [CrossRef]
121. Wang, F.-Z.; Chen, M.-X.; Yu, L.-J.; Xie, L.-J.; Yuan, L.-B.; Qi, H.; Xiao, M.; Guo, W.; Chen, Z.; Yi, K.; et al. OsARM1, an R2R3 MYB Transcription Factor, Is Involved in Regulation of the Response to Arsenic Stress in Rice. *Front. Plant Sci.* **2017**, *8*, 1868. [CrossRef]
122. Vitorello, V.A.; Capaldi, F.R.; Stefanuto, V.A. Recent advances in aluminum toxicity and resistance in higher plants. *Braz. J. Plant Physiol.* **2005**, *17*, 129–143. [CrossRef]
123. Hodson, M.J.; Evans, D.E. Aluminium–silicon interactions in higher plants: An update. *J. Exp. Bot.* **2020**, *71*, 6719–6729. [CrossRef]
124. Osawa, H.; Matsumoto, H. Possible Involvement of Protein Phosphorylation in Aluminum-Responsive Malate Efflux from Wheat Root Apex. *Plant Physiol.* **2001**, *126*, 411–420. [CrossRef] [PubMed]

125. Jones, D.L.; Kochian, L.V. Aluminum interaction with plasma membrane lipids and enzyme metal binding sites and its potential role in Al cytotoxicity. *FEBS Lett.* **1997**, *400*, 51–57. [CrossRef]
126. Tian, Q.; Zhang, X.; Ramesh, S.; Gilliham, M.; Tyerman, S.D.; Zhang, W.-H. Ethylene negatively regulates aluminium-induced malate efflux from wheat roots and tobacco cells transformed with TaALMT1. *J. Exp. Bot.* **2014**, *65*, 2415–2426. [CrossRef] [PubMed]
127. Lima, J.C.; Arenhart, R.A.; Margis-Pinheiro, M.; Margis, R. Aluminum triggers broad changes in microRNA expression in rice roots. *Genet. Mol. Res.* **2011**, *10*, 2817–2832. [CrossRef]
128. Kong, X.; Zhang, M.; Xu, X.; Li, X.; Li, C.; Ding, Z. System analysis of microRNAs in the development and aluminium stress responses of the maize root system. *Plant Biotechnol. J.* **2014**, *12*, 1108–1121. [CrossRef]
129. Jalmi, S.K.; Bhagat, P.K.; Verma, D.; Noryang, S.; Tayyeba, S.; Singh, K.; Sharma, D.; Sinha, A.K. Traversing the Links between Heavy Metal Stress and Plant Signaling. *Front. Plant Sci.* **2018**, *9*, 12. [CrossRef]
130. Khanam, R.; Kumar, A.; Nayak, A.K.; Shahid, M.; Tripathi, R.; Vijayakumar, S.; Bhaduri, D.; Kumar, U.; Mohanty, S.; Panneerselvam, P.; et al. Metal(loid)s (As, Hg, Se, Pb and Cd) in paddy soil: Bioavailability and potential risk to human health. *Sci. Total Environ.* **2020**, *699*, 134330. [CrossRef]
131. Chen, Y.-A.; Chi, W.-C.; Trinh, N.N.; Huang, L.-Y.; Chen, Y.-C.; Cheng, K.-T.; Huang, T.-L.; Lin, C.-Y.; Huang, H.-J. Transcriptome Profiling and Physiological Studies Reveal a Major Role for Aromatic Amino Acids in Mercury Stress Tolerance in Rice Seedlings. *PLoS ONE* **2014**, *9*, e95163. [CrossRef]
132. Wojas, S.; Ruszczyńska, A.; Bulska, E.; Wojciechowski, M.; Antosiewicz, D.M. Ca2+-dependent plant response to Pb2+ is regulated by LCT1. *Environ. Pollut.* **2007**, *147*, 584–592. [CrossRef]
133. Huang, T.-L.; Huang, H.-J. ROS and CDPK-like kinase-mediated activation of MAP kinase in rice roots exposed to lead. *Chemosphere* **2008**, *71*, 1377–1385. [CrossRef]
134. Hasan, M.K.; Ahammed, G.J.; Yin, L.; Shi, K.; Xia, X.; Zhou, Y.; Yu, J.; Zhou, J. Melatonin mitigates cadmium phytotoxicity through modulation of phytochelatins biosynthesis, vacuolar sequestration, and antioxidant potential in *Solanum lycopersicum* L. *Front. Plant Sci.* **2015**, *6*, 601. [CrossRef] [PubMed]
135. Huang, T.-L.; Nguyen, Q.T.T.; Fu, S.-F.; Lin, C.-Y.; Chen, Y.-C.; Huang, H.-J. Transcriptomic changes and signalling pathways induced by arsenic stress in rice roots. *Plant Mol. Biol.* **2012**, *80*, 587–608. [CrossRef] [PubMed]
136. Steffens, B. The role of ethylene and ROS in salinity, heavy metal, and flooding responses in rice. *Front. Plant Sci.* **2014**, *5*, 685. [CrossRef] [PubMed]
137. Trinh, N.-N.; Huang, T.-L.; Chi, W.-C.; Fu, S.-F.; Chen, C.-C.; Huang, H.-J. Chromium stress response effect on signal transduction and expression of signaling genes in rice: Chromium stress response effect on signal transduction. *Physiol. Plant.* **2014**, *150*, 205–224. [CrossRef] [PubMed]
138. Aprile, A.; Sabella, E.; Vergine, M.; Genga, A.; Siciliano, M.; Nutricati, E.; Rampino, P.; De Pascali, M.; Luvisi, A.; Miceli, A.; et al. Activation of a gene network in durum wheat roots exposed to cadmium. *BMC Plant Biol.* **2018**, *18*, 238. [CrossRef] [PubMed]
139. Yang, Z.; Chu, C. Towards Understanding Plant Response to Heavy Metal Stress. In *Abiotic Stress in Plants-Mechanisms and Adaptations*; Shanker, A., Ed.; InTech: London, UK, 2011; ISBN 978-953-307-394-1.
140. Hasan, M.K.; Cheng, Y.; Kanwar, M.K.; Chu, X.-Y.; Ahammed, G.J.; Qi, Z.-Y. Responses of Plant Proteins to Heavy Metal Stress—A Review. *Front. Plant Sci.* **2017**, *8*, 1492. [CrossRef] [PubMed]
141. Ansarypour, Z.; Shahpiri, A. Heterologous expression of a rice metallothionein isoform (OsMTI-1b) in Saccharomyces cerevisiae enhances cadmium, hydrogen peroxide and ethanol tolerance. *Braz. J. Microbiol.* **2017**, *48*, 537–543. [CrossRef] [PubMed]
142. Liu, Q.; Hu, H.; Zhu, L.; Li, R.; Feng, Y.; Zhang, L.; Yang, Y.; Liu, X.; Zhang, H. Involvement of miR528 in the Regulation of Arsenite Tolerance in Rice (*Oryza sativa* L.). *J. Agric. Food Chem.* **2015**, *63*, 8849–8861. [CrossRef]
143. Jones, J.D.G.; Dangl, J.L. The plant immune system. *Nature* **2006**, *444*, 323–329. [CrossRef]
144. Basu, S.; Varsani, S.; Louis, J. Altering Plant Defenses: Herbivore-Associated Molecular Patterns and Effector Arsenal of Chewing Herbivores. *MPMI* **2018**, *31*, 13–21. [CrossRef]
145. Woźniak, A.; Formela, M.; Bilman, P.; Grześkiewicz, K.; Bednarski, W.; Marczak, Ł.; Narożna, D.; Dancewicz, K.; Mai, V.; Borowiak-Sobkowiak, B.; et al. The Dynamics of the Defense Strategy of Pea Induced by Exogenous Nitric Oxide in Response to Aphid Infestation. *Int. J. Mol. Sci.* **2017**, *18*, 329. [CrossRef] [PubMed]
146. Vos, I.A.; Pieterse, C.M.J.; van Wees, S.C.M. Costs and benefits of hormone-regulated plant defences. *Plant Pathol.* **2013**, *62*, 43–55. [CrossRef]
147. Howe, G.A.; Jander, G. Plant Immunity to Insect Herbivores. *Annu. Rev. Plant Biol.* **2008**, *59*, 41–66. [CrossRef] [PubMed]
148. Backer, R.; Naidoo, S.; van den Berg, N. The NONEXPRESSOR OF PATHOGENESIS-RELATED GENES 1 (NPR1) and Related Family: Mechanistic Insights in Plant Disease Resistance. *Front. Plant Sci.* **2019**, *10*, 102. [CrossRef]
149. Mai, V.C.; Drzewiecka, K.; Jeleń, H.; Narożna, D.; Rucińska-Sobkowiak, R.; Kęsy, J.; Floryszak-Wieczorek, J.; Gabryś, B.; Morkunas, I. Differential induction of Pisum sativum defense signaling molecules in response to pea aphid infestation. *Plant Sci.* **2014**, *221–222*, 1–12. [CrossRef] [PubMed]
150. Formela-Luboińska, M.; Chadzinikolau, T.; Drzewiecka, K.; Jeleń, H.; Bocianowski, J.; Kęsy, J.; Labudda, M.; Jeandet, P.; Morkunas, I. The Role of Sugars in the Regulation of the Level of Endogenous Signaling Molecules during Defense Response of Yellow Lupine to Fusarium oxysporum. *Int. J. Mol. Sci.* **2020**, *21*, 4133. [CrossRef] [PubMed]

151. Wei, X.; Shan, T.; Hong, Y.; Xu, H.; Liu, X.; Zhang, Z. TaPIMP2, a pathogen-induced MYB protein in wheat, contributes to host resistance to common root rot caused by Bipolaris sorokiniana. *Sci. Rep.* **2017**, *7*, 1754. [CrossRef]
152. Yang, J.; Duan, G.; Li, C.; Liu, L.; Han, G.; Zhang, Y.; Wang, C. The Crosstalks Between Jasmonic Acid and Other Plant Hormone Signaling Highlight the Involvement of Jasmonic Acid as a Core Component in Plant Response to Biotic and Abiotic Stresses. *Front. Plant Sci.* **2019**, *10*, 1349. [CrossRef]
153. Al-Zahrani, W.; Bafeel, S.O.; El-Zohri, M. Jasmonates mediate plant defense responses to *Spodoptera exigua* herbivory in tomato and maize foliage. *Plant Signal. Behav.* **2020**, *15*, 1746898. [CrossRef]
154. Verhage, A. Rewiring of the jasmonate signaling pathway in Arabidopsis during insect herbivory. *Front. Plant Sci.* **2011**, *2*, 47. [CrossRef]
155. Li, N.; Han, X.; Feng, D.; Yuan, D.; Huang, L.-J. Signaling Crosstalk between Salicylic Acid and Ethylene/Jasmonate in Plant Defense: Do We Understand What They Are Whispering? *Int. J. Mol. Sci.* **2019**, *20*, 671. [CrossRef] [PubMed]
156. Dong, N.; Liu, X.; Lu, Y.; Du, L.; Xu, H.; Liu, H.; Xin, Z.; Zhang, Z. Overexpression of TaPIEP1, a pathogen-induced ERF gene of wheat, confers host-enhanced resistance to fungal pathogen Bipolaris sorokiniana. *Funct. Integr. Genom.* **2010**, *10*, 215–226. [CrossRef] [PubMed]
157. Esmail, S.M.; Omara, R.I.; Abdelaal, K.A.A.; Hafez, Y.M. Histological and biochemical aspects of compatible and incompatible wheat- Puccinia striiformis interactions. *Physiol. Mol. Plant Pathol.* **2019**, *106*, 120–128. [CrossRef]
158. Jisha, V.; Dampanaboina, L.; Vadassery, J.; Mithöfer, A.; Kappara, S.; Ramanan, R. Overexpression of an AP2/ERF Type Transcription Factor OsEREBP1 Confers Biotic and Abiotic Stress Tolerance in Rice. *PLoS ONE* **2015**, *10*, e0127831. [CrossRef] [PubMed]
159. Otuka, A. Migration of rice planthoppers and their vectored re-emerging and novel rice viruses in East Asia. *Front. Microbiol.* **2013**, *4*, 309. [CrossRef] [PubMed]
160. Zhan, Y.; Sun, X.; Rong, G.; Hou, C.; Huang, Y.; Jiang, D.; Weng, X. Identification of two transcription factors activating the expression of OsXIP in rice defence response. *BMC Biotechnol.* **2017**, *17*, 26. [CrossRef]
161. Gottwald, S.; Samans, B.; Lück, S.; Friedt, W. Jasmonate and ethylene dependent defence gene expression and suppression of fungal virulence factors: Two essential mechanisms of Fusarium head blight resistance in wheat? *BMC Genom.* **2012**, *13*, 369. [CrossRef]
162. Labudda, M.; Różańska, E.; Prabucka, B.; Muszyńska, E.; Marecka, D.; Kozak, M.; Dababat, A.A.; Sobczak, M. Activity profiling of barley vacuolar processing enzymes provides new insights into the plant and cyst nematode interaction. *Mol. Plant Pathol.* **2020**, *21*, 38–52. [CrossRef]
163. Darino, M.; Chia, K.; Marques, J.; Aleksza, D.; Soto-Jiménez, L.M.; Saado, I.; Uhse, S.; Borg, M.; Betz, R.; Bindics, J.; et al. *Ustilago maydis* effector Jsi1 interacts with Topless corepressor, hijacking plant jasmonate/ethylene signaling. *New Phytol.* **2021**, *229*, 3393–3407. [CrossRef]
164. El-Wakeil, N.E.; Volkmar, C.; Sallam, A.A. Jasmonic acid induces resistance to economically important insect pests in winter wheat: Jasmonic acid induces resistance to insect pests in wheat. *Pest. Manag. Sci.* **2010**, *66*, 549–554. [CrossRef]
165. Kazan, K.; Manners, J.M. MYC2: The Master in Action. *Mol. Plant* **2013**, *6*, 686–703. [CrossRef] [PubMed]
166. Lorenzo, O.; Chico, J.M.; Saénchez-Serrano, J.J.; Solano, R. *JASMONATE-INSENSITIVE1* Encodes a MYC Transcription Factor Essential to Discriminate between Different Jasmonate-Regulated Defense Responses in Arabidopsis[W]. *Plant Cell* **2004**, *16*, 1938–1950. [CrossRef] [PubMed]
167. Kraus, E.C.; Stout, M.J. Seed treatment using methyl jasmonate induces resistance to rice water weevil but reduces plant growth in rice. *PLoS ONE* **2019**, *14*, e0222800. [CrossRef] [PubMed]
168. Bhavanam, S.; Stout, M. Seed Treatment With Jasmonic Acid and Methyl Jasmonate Induces Resistance to Insects but Reduces Plant Growth and Yield in Rice, Oryza sativa. *Front. Plant Sci.* **2021**, *12*, 691768. [CrossRef] [PubMed]
169. Lozano-Durán, R.; Zipfel, C. Trade-off between growth and immunity: Role of brassinosteroids. *Trends Plant Sci.* **2015**, *20*, 12–19. [CrossRef]
170. Denancé, N.; Sánchez-Vallet, A.; Goffner, D.; Molina, A. Disease resistance or growth: The role of plant hormones in balancing immune responses and fitness costs. *Front. Plant Sci.* **2013**, *4*, 155. [CrossRef]
171. Huot, B.; Yao, J.; Montgomery, B.L.; He, S.Y. Growth–Defense Tradeoffs in Plants: A Balancing Act to Optimize Fitness. *Mol. Plant* **2014**, *7*, 1267–1287. [CrossRef]
172. Ku, Y.-S.; Sintaha, M.; Cheung, M.-Y.; Lam, H.-M. Plant Hormone Signaling Crosstalks between Biotic and Abiotic Stress Responses. *Int. J. Mol. Sci.* **2018**, *19*, 3206. [CrossRef]
173. Pieterse, C.M.J.; Van der Does, D.; Zamioudis, C.; Leon-Reyes, A.; Van Wees, S.C.M. Hormonal Modulation of Plant Immunity. *Annu. Rev. Cell Dev. Biol.* **2012**, *28*, 489–521. [CrossRef]
174. Koornneef, A.; Leon-Reyes, A.; Ritsema, T.; Verhage, A.; Den Otter, F.C.; Van Loon, L.C.; Pieterse, C.M.J. Kinetics of Salicylate-Mediated Suppression of Jasmonate Signaling Reveal a Role for Redox Modulation. *Plant Physiol.* **2008**, *147*, 1358–1368. [CrossRef]
175. Luna, E.; Bruce, T.J.A.; Roberts, M.R.; Flors, V.; Ton, J. Next-Generation Systemic Acquired Resistance. *Plant Physiol.* **2012**, *158*, 844–853. [CrossRef] [PubMed]
176. Spoel, S.H.; Johnson, J.S.; Dong, X. Regulation of tradeoffs between plant defenses against pathogens with different lifestyles. *Proc. Natl. Acad. Sci. USA* **2007**, *104*, 18842–18847. [CrossRef] [PubMed]

177. Leon-Reyes, A.; Spoel, S.H.; De Lange, E.S.; Abe, H.; Kobayashi, M.; Tsuda, S.; Millenaar, F.F.; Welschen, R.A.M.; Ritsema, T.; Pieterse, C.M.J. Ethylene Modulates the Role of NONEXPRESSOR OF PATHOGENESIS-RELATED GENES1 in Cross Talk between Salicylate and Jasmonate Signaling. *Plant Physiol.* **2009**, *149*, 1797–1809. [CrossRef] [PubMed]
178. Cui, J.; Jander, G.; Racki, L.R.; Kim, P.D.; Pierce, N.E.; Ausubel, F.M. Signals Involved in Arabidopsis Resistance to *Trichoplusia ni* Caterpillars Induced by Virulent and Avirulent Strains of the Phytopathogen *Pseudomonas syringae. Plant Physiol.* **2002**, *129*, 551–564. [CrossRef] [PubMed]
179. Anderson, J.P.; Badruzsaufari, E.; Schenk, P.M.; Manners, J.M.; Desmond, O.J.; Ehlert, C.; Maclean, D.J.; Ebert, P.R.; Kazan, K. Antagonistic Interaction between Abscisic Acid and Jasmonate-Ethylene Signaling Pathways Modulates Defense Gene Expression and Disease Resistance in Arabidopsis. *Plant Cell* **2004**, *16*, 3460–3479. [CrossRef]
180. Nickstadt, A.; Thomma, B.P.H.J.; Feussner, I.; Kangasjarvi, J.; Zeier, J.; Loeffler, C.; Scheel, D.; Berger, S. The jasmonate-insensitive mutant jin1 shows increased resistance to biotrophic as well as necrotrophic pathogens. *Mol. Plant Pathol.* **2004**, *5*, 425–434. [CrossRef]
181. Adie, B.A.T.; Pérez-Pérez, J.; Pérez-Pérez, M.M.; Godoy, M.; Sánchez-Serrano, J.-J.; Schmelz, E.A.; Solano, R. ABA Is an Essential Signal for Plant Resistance to Pathogens Affecting JA Biosynthesis and the Activation of Defenses in *Arabidopsis. Plant Cell* **2007**, *19*, 1665–1681. [CrossRef]
182. Sánchez-Vallet, A.; López, G.; Ramos, B.; Delgado-Cerezo, M.; Riviere, M.-P.; Llorente, F.; Fernández, P.V.; Miedes, E.; Estevez, J.M.; Grant, M.; et al. Disruption of Abscisic Acid Signaling Constitutively Activates Arabidopsis Resistance to the Necrotrophic Fungus *Plectosphaerella cucumerina. Plant Physiol.* **2012**, *160*, 2109–2124. [CrossRef]
183. Aerts, N.; Pereira Mendes, M.; Van Wees, S.C.M. Multiple levels of crosstalk in hormone networks regulating plant defense. *Plant J.* **2021**, *105*, 489–504. [CrossRef]
184. Heil, M. Fitness costs of induced resistance: Emerging experimental support for a slippery concept. *Trends Plant Sci.* **2002**, *7*, 61–67. [CrossRef]
185. van Hulten, M.; Pelser, M.; van Loon, L.C.; Pieterse, C.M.J.; Ton, J. Costs and benefits of priming for defense in Arabidopsis. *Proc. Natl. Acad. Sci. USA* **2006**, *103*, 5602–5607. [CrossRef] [PubMed]
186. Walters, D.; Heil, M. Costs and trade-offs associated with induced resistance. *Physiol. Mol. Plant Pathol.* **2007**, *71*, 3–17. [CrossRef]
187. Pieterse, C.M.J.; Dicke, M. Plant interactions with microbes and insects: From molecular mechanisms to ecology. *Trends Plant Sci.* **2007**, *12*, 564–569. [CrossRef] [PubMed]
188. Thaler, J.S.; Humphrey, P.T.; Whiteman, N.K. Evolution of jasmonate and salicylate signal crosstalk. *Trends Plant Sci.* **2012**, *17*, 260–270. [CrossRef]
189. Vos, I.A.; Moritz, L.; Pieterse, C.M.J.; Van Wees, S.C.M. Impact of hormonal crosstalk on plant resistance and fitness under multi-attacker conditions. *Front. Plant Sci.* **2015**, *6*, 639. [CrossRef]
190. Zhang, H.; Sonnewald, U. Differences and commonalities of plant responses to single and combined stresses. *Plant J.* **2017**, *90*, 839–855. [CrossRef]
191. Labudda, M.; Tokarz, K.; Tokarz, B.; Muszyńska, E.; Gietler, M.; Górecka, M.; Różańska, E.; Rybarczyk-Płońska, A.; Fidler, J.; Prabucka, B.; et al. Reactive oxygen species metabolism and photosynthetic performance in leaves of Hordeum vulgare plants co-infested with Heterodera filipjevi and Aceria tosichella. *Plant Cell Rep.* **2020**, *39*, 1719–1741. [CrossRef]
192. Labudda, M.; Muszyńska, E.; Gietler, M.; Różańska, E.; Rybarczyk-Płońska, A.; Fidler, J.; Prabucka, B.; Dababat, A.A. Efficient antioxidant defence systems of spring barley in response to stress induced jointly by the cyst nematode parasitism and cadmium exposure. *Plant Soil* **2020**, *456*, 189–206. [CrossRef]
193. Sharma, R.; De Vleesschauwer, D.; Sharma, M.K.; Ronald, P.C. Recent Advances in Dissecting Stress-Regulatory Crosstalk in Rice. *Plant* **2013**, *6*, 250–260. [CrossRef]
194. Kissoudis, C.; van de Wiel, C.; Visser, R.G.F.; van der Linden, G. Enhancing crop resilience to combined abiotic and biotic stress through the dissection of physiological and molecular crosstalk. *Front. Plant Sci.* **2014**, *5*, 207. [CrossRef]
195. Rohila, J.S.; Yang, Y. Rice Mitogen-activated Protein Kinase Gene Family and Its Role in Biotic and Abiotic Stress Response. *J. Integr. Plant Biol.* **2007**, *49*, 751–759. [CrossRef]
196. Lee, S.C.; Luan, S. ABA signal transduction at the crossroad of biotic and abiotic stress responses: ABA in drought and pathogen responses. *Plant Cell Environ.* **2012**, *35*, 53–60. [CrossRef] [PubMed]
197. Jensen, M.K.; Hagedorn, P.H.; de Torres-Zabala, M.; Grant, M.R.; Rung, J.H.; Collinge, D.B.; Lyngkjaer, M.F. Transcriptional regulation by an NAC (NAM-ATAF1,2-CUC2) transcription factor attenuates ABA signalling for efficient basal defence towards *Blumeria graminis* f. sp. *hordei* in Arabidopsis. *Plant J.* **2008**, *56*, 867–880. [CrossRef] [PubMed]
198. Li, A.; Wang, X.; Leseberg, C.H.; Jia, J.; Mao, L. Biotic and abiotic stress responses through calcium-dependent protein kinase (CDPK) signaling in wheat (*Triticum aestivum* L.). *Plant Signal. Behav.* **2008**, *3*, 654–656. [CrossRef]
199. Asano, T.; Hayashi, N.; Kobayashi, M.; Aoki, N.; Miyao, A.; Mitsuhara, I.; Ichikawa, H.; Komatsu, S.; Hirochika, H.; Kikuchi, S.; et al. A rice calcium-dependent protein kinase OsCPK12 oppositely modulates salt-stress tolerance and blast disease resistance: OsCPK12 modulates abiotic and biotic stress responses. *Plant J.* **2012**, *69*, 26–36. [CrossRef]
200. Xiao, J.; Cheng, H.; Li, X.; Xiao, J.; Xu, C.; Wang, S. Rice WRKY13 Regulates Cross Talk between Abiotic and Biotic Stress Signaling Pathways by Selective Binding to Different cis-Elements. *Plant Physiol.* **2013**, *163*, 1868–1882. [CrossRef]

201. Yokotani, N.; Sato, Y.; Tanabe, S.; Chujo, T.; Shimizu, T.; Okada, K.; Yamane, H.; Shimono, M.; Sugano, S.; Takatsuji, H.; et al. WRKY76 is a rice transcriptional repressor playing opposite roles in blast disease resistance and cold stress tolerance. *J. Exp. Bot.* **2013**, *64*, 5085–5097. [CrossRef]
202. Peng, X.; Tang, X.; Zhou, P.; Hu, Y.; Deng, X.; He, Y.; Wang, H. Isolation and Expression Patterns of Rice WRKY82 Transcription Factor Gene Responsive to Both Biotic and Abiotic Stresses. *Agric. Sci. China* **2011**, *10*, 893–901. [CrossRef]
203. Qiu, Y.; Yu, D. Over-expression of the stress-induced OsWRKY45 enhances disease resistance and drought tolerance in Arabidopsis. *Environ. Exp. Bot.* **2009**, *65*, 35–47. [CrossRef]
204. Liu, H.; Zhou, X.; Dong, N.; Liu, X.; Zhang, H.; Zhang, Z. Expression of a wheat MYB gene in transgenic tobacco enhances resistance to Ralstonia solanacearum, and to drought and salt stresses. *Funct. Integr. Genom.* **2011**, *11*, 431–443. [CrossRef]
205. Yokotani, N.; Ichikawa, T.; Kondou, Y.; Iwabuchi, M.; Matsui, M.; Hirochika, H.; Oda, K. Role of the rice transcription factor JAmyb in abiotic stress response. *J. Plant Res.* **2013**, *126*, 131–139. [CrossRef] [PubMed]
206. Yang, Z.; Huang, Y.; Yang, J.; Yao, S.; Zhao, K.; Wang, D.; Qin, Q.; Bian, Z.; Li, Y.; Lan, Y.; et al. Jasmonate Signaling Enhances RNA Silencing and Antiviral Defense in Rice. *Cell Host Microbe* **2020**, *28*, 89–103.e8. [CrossRef] [PubMed]
207. Nakashima, K.; Tran, L.-S.P.; Van Nguyen, D.; Fujita, M.; Maruyama, K.; Todaka, D.; Ito, Y.; Hayashi, N.; Shinozaki, K.; Yamaguchi-Shinozaki, K. Functional analysis of a NAC-type transcription factor OsNAC6 involved in abiotic and biotic stress-responsive gene expression in rice: Rice OsNAC6 functions in stress responses. *Plant J.* **2007**, *51*, 617–630. [CrossRef] [PubMed]
208. Xia, N.; Zhang, G.; Liu, X.-Y.; Deng, L.; Cai, G.-L.; Zhang, Y.; Wang, X.-J.; Zhao, J.; Huang, L.-L.; Kang, Z.-S. Characterization of a novel wheat NAC transcription factor gene involved in defense response against stripe rust pathogen infection and abiotic stresses. *Mol. Biol. Rep.* **2010**, *37*, 3703–3712. [CrossRef]
209. Winfield, M.O.; Lu, C.; Wilson, I.D.; Coghill, J.A.; Edwards, K.J. Plant responses to cold: Transcriptome analysis of wheat: Plant responses to cold. *Plant Biotechnol. J.* **2010**, *8*, 749–771. [CrossRef]
210. Talanova, V.V.; Titov, A.F.; Repkina, N.S.; Topchieva, L.V. Cold-responsive COR/LEA genes participate in the response of wheat plants to heavy metals stress. *Dokl. Biol. Sci.* **2013**, *448*, 28–31. [CrossRef]

plants

MDPI

Review

Transcriptional and Post-Translational Regulation of Plant bHLH Transcription Factors during the Response to Environmental Stresses

Yasmina Radani [1,†], Rongxue Li [1,†], Harriet Mateko Korboe [1], Hongyu Ma [2,*] and Liming Yang [1,*]

1 State Key Laboratory of Tree Genetics and Breeding, Co-Innovation Center for Sustainable Forestry in Southern China, College of Biology and the Environment, Nanjing Forestry University, Nanjing 210037, China; radani.yasmina@gmail.com (Y.R.); lirongxue@njfu.edu.cn (R.L.); harrietmatekokorboe27@gmail.com (H.M.K.)
2 College of Plant Protection, Nanjing Agricultural University, Nanjing 210095, China
* Correspondence: mahongyu@njau.edu.cn (H.M.); yangliming@njfu.edu.cn (L.Y.)
† These authors contributed equally to this work.

Abstract: Over the past decades, extensive research has been conducted to identify and characterize various plant transcription factors involved in abiotic stress responses. Therefore, numerous efforts have been made to improve plant stress tolerance by engineering these transcription factor genes. The plant basic Helix–Loop–Helix (bHLH) transcription factor family represents one of the most prominent gene families and contains a bHLH motif that is highly conserved in eukaryotic organisms. By binding to specific positions in promoters, they activate or repress the transcription of specific response genes and thus affect multiple variables in plant physiology such as the response to abiotic stresses, which include drought, climatic variations, mineral deficiencies, excessive salinity, and water stress. The regulation of bHLH transcription factors is crucial to better control their activity. On the one hand, they are regulated at the transcriptional level by other upstream components; on the other hand, they undergo various modifications such as ubiquitination, phosphorylation, and glycosylation at the post-translational level. Modified bHLH transcription factors can form a complex regulatory network to regulate the expression of stress response genes and thus determine the activation of physiological and metabolic reactions. This review article focuses on the structural characteristics, classification, function, and regulatory mechanism of bHLH transcription factor expression at the transcriptional and post-translational levels during their responses to various abiotic stress conditions.

Keywords: bHLH transcription factors; abiotic stress; transcriptional regulation; post-translational regulation

Citation: Radani, Y.; Li, R.; Korboe, H.M.; Ma, H.; Yang, L. Transcriptional and Post-Translational Regulation of Plant bHLH Transcription Factors during the Response to Environmental Stresses. *Plants* **2023**, *12*, 2113. https://doi.org/10.3390/plants12112113

Academic Editors: Małgorzata Nykiel, Mateusz Labudda, Beata Prabucka, Marta Gietler and Justyna Fidler

Received: 19 April 2023
Revised: 16 May 2023
Accepted: 19 May 2023
Published: 26 May 2023

1. Introduction

The basic Helix–Loop–Helix (bHLH) Transcription Factors (TF) are a large and diverse family of transcription factors characterized by a highly conserved bHLH domain [1]. The bHLH transcription factor conservatively contains two connected subregions, namely the basic region (b), which is an essential DNA-binding region, and the HLH region, which consists of 40–50 amino acid residues and participates in homodimerization or heterodimerization [2] (Figure 1A,B). The amino acid sequences outside the bHLH region are divergent, even in closely related proteins from the same species. These short conserved amino acid motifs are commonly present in related plant bHLH proteins and are generally conserved within each subfamily [3,4].

The bHLH motif was first discovered in the murine transcription factors E12 and E4 [5]. Since then, several *bHLH* genes have been discovered, providing an initial classification of animals that divided bHLH transcription factors into six subgroups (A to F) based on their protein sequences, differences in bHLH domains, and comparisons of the functions of different family members [6,7]. Furthermore, outside of mammals, multiple classifications of 15 to 25 subfamilies based on bHLH and non-bHLH motifs have been obtained [3,4,8]. As more species were included, the most recent classification allows them to be divided into 26 subgroups [1], reflecting a deep evolutionary relationship in plants. In addition, phylogenetic analysis of several atypical bHLH proteins extended the number of subfamilies to 32 [8].

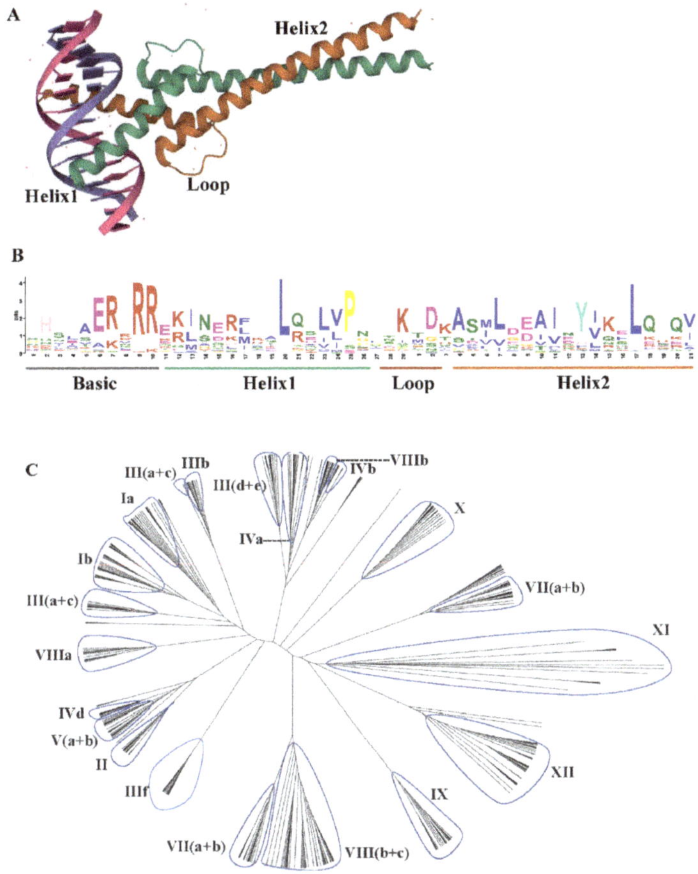

Figure 1. (**A**) The crystal structure of bHLH domain heterodimer bound to the DNA. (**B**) The sequence logo of the bHLH domain. The overall height of each stack represents the conservation of the sequence at that position. Each color of the letters represents a type of amino acid residue. (**C**) The phylogenetic tree analysis of *bHLH* gene family in *Oryza sativa* [monocot]; *Arabidopsis thaliana* [eudicot]; *Selaginella moellendorffii* [lycophyte]; *Physcomitrella patens* [moss]; *Volvox carteri; Chlamydomonas reinhardtii; Chlorella vulgaris; Ostreococcus tauri;* and *Cyanidioschyzon merolae* was constructed using MEGA7.0 with the JTT method and 1000 replicates. Then, the trees were visualized using Figtree. Group names were marked outside the circle. The protein sequences were downloaded from the report in 2010 [1].

Maize (*Zea mays* L.) was the first plant species in which the bHLH superfamily was first identified. Subsequently, 162 *bHLH* genes were identified in the model plant *Arabidopsis thaliana* [9], 167 in rice (*Oryza sativa* L.) [4] and 202 *bHLH* genes in *Poplar* [10]. The number of characterized and identified *bHLH* genes in plants has increased, showing their extensive and diverse functional involvement; 100 *PmbHLH* genes were identified in *Prunus mume* [11], 102 *bHLH* genes in walnut (*Juglans regia* L.) [12], 212 *MibHLH* genes in mango

(*Mangifera indica* L.) [13], 37 *SsbHLH* genes in sugarcane (*Saccharum spontaneum*) [14], 110 *IbbHLHs* in sweetpotato (*Ipomoea batatas* (L.) Lam.) [15], 118 *bHLH* genes were identified in melon [16], 107 *CabHLHs* were identified in *Capsicum annuum* [17], and 85 bHLH proteins (GbbHLH) were obtained from *Ginkgo biloba* [18]. *bHLH* gene families are widespread in plants and have demonstrated their essential roles in various biological processes involved in normal plant growth and development [1], flowering [19], and metabolic biosynthesis, including anthocyanin [20]. Therefore, many of them have a role and regulatory function in signal transduction [21,22] and gene expression in response to abiotic stresses such as salinity, drought, low temperature, and nutrient deficiency [23,24].

Many research studies have paid close attention to regulatory genes, including bHLH TFs, which play essential roles in multiple abiotic stress responses by regulating the expression of a wide range of downstream stress-responsive genes by interacting with the specific cis-elements in their promoter region [25]. Therefore, genetically modifying the expression of TFs can strongly affect plant stress tolerance as it mimics or enhances stress signals to simultaneously regulate many downstream stress-responsive genes. In recent years, several mechanisms of regulation and tolerance of bHLH TF to abiotic stress in model and non-model plants have been revealed, providing a better and more detailed explanation of their intervention under specific stress conditions. *bHLH*s are involved in various functional gene approaches to significantly enhance stress tolerance in plants. They can be activated under multiple stresses and play an essential role in abiotic stress responses by regulating a wide range of downstream stress-responsive genes [26]. Meanwhile, bHLH TFs themselves undergo various modifications at the post-translational level, such as ubiquitination [27], phosphorylation [28], and sumoylation [29], thereby forming a regulatory complex network to modulate the expression of *bHLH* stress-responsive genes.

Both transcriptional and post-translational modifications contribute to this plant stress response regulatory function. In this review, we focused on the structure and gene classification of bHLH TFs, as well as recent studies on their expression and the different mechanisms of transcriptional and post-translational regulation under different abiotic stress conditions.

2. Structure and Gene Family of bHLH TFs

The bHLH TF family, known for its conserved bHLH domain [30], is considered the second largest gene family after the MYB family gene and is found in the majority of eukaryotes. With a total of ~60 amino acids, the bHLH transcription factor is divided into two functionally distinct regions. On the one hand, there is the primary region, which is located in the N-terminal domain and contains a total of 15 to 20 amino acids, including basic amino acid residues, and is responsible for DNA binding, with certain conserved amino acids responsible for recognition of the E-box, a hexanucleotide consensus sequence in DNA (5-CANNTG-3). Other residues, on the contrary, are specific for a different E-box region in DNA (e.g., the G-box [5-CACGTG-3]) in target genes [25,31]. According to Toledo-Ortiz et al., *bHLH* genes can be divided into E-box or non-E-box binders and DNA-binding or non-DNA-binding proteins. The Helix–Loop–Helix (HLH) region, on the other hand, is found in the C-terminal domain and is composed of hydrophobic residues rearranged into two amphipathic regions connected to a loop region forming a hydrophobic ring (Figure 1B). They are involved in homo- or heterodimerization, as shown in Figure 1A, and thus participate in protein–protein interaction and gene expression control [2]. The dimer structure is stabilized by the hydrophobic amino acids Isoleucine (I), Leucine (L), and Valine (V) in conserved positions in the bHLH domain [30].

Outside of the bHLH domain, bHLH proteins show little to nonexistent conservation [3]. However, some groups of bHLH show some conserved domains outside of bHLH, most of which have been previously characterized in animals. For example, there is the highly conserved Leucine Zipper (LZ) motif adjacent to the second helix of the bHLH domain, which is predicted to adopt a coiled-coil structure, allowing protein dimerization. The PAS domain, the orange domain, the WRPW motif, and the COE domain are other domains also found in bHLH proteins [6,32]. The bHLH motif was first identified by Massari and Murre in 1989. Then, with more identified bHLH proteins, the first classification of different family members of animal bHLH TFs was performed using only the bHLH motif. This classification led to the selection of four distinct groups based on amino-acid patterns and E-box-binding specificity [7]. This classification separated bHLH proteins into classes A, B, C, and D [7].

Using the entire protein sequence and not just the bHLH domain allowed the inclusion of additional domains associated with many other proteins (e.g., Zip, Orange, and PAS) that may play important roles in protein dimerization and DNA binding. Therefore, the classification was extended to include two additional groups (E and F) [6,33]. The characterized bHLH proteins have been restricted to animals. It has been suggested that the ancestral plant bHLH sequence was a group B protein present in early eukaryote evolution, from which bHLHs of different lineages independently evolved [6,7] and function in transcriptional regulation associated with various diverse functions in plants including anthocyanin biosynthesis, phytochrome signaling, globulin expression, fruit dehiscence, and carpel and epidermal development [3]. Furthermore, some studies have aimed to divide the large bHLH family into smaller subgroups based on their sequence homology [1,3,8]. In 2003, Heim et al. compared the genetic structure, number of introns, and conservative motifs of 133 *bHLH* isolates from *Arabidopsis* and rearranged them into 12 subgroups. Since then, Pires and Dolan have divided bHLH into 26 subgroups by phylogenetic analysis, using a total of 544 *bHLH* genes from 9 land plants and algae [1] (Figure 1C). Meanwhile, Carretero-Paulet et al. added 29 to the original 133 *Arabidopsis bHLH* and classified them into 32 subfamilies by phylogenetic analysis, intron pattern, and conservative protein motif structure using 9 plant species [8].

Mainly, members of the same bHLH subfamily generally share similar gene structure and the same non-bHLH conserved motifs, such as a Leucine Zipper (LZ) domain (shared between IVb subfamilies and between IVc subfamilies) [7]. There is also an ACT domain, which is a ligand-binding regulatory domain found in a diverse group of proteins, mainly metabolic enzymes. The occurrence of the ACT domain in some proteins from different bHLH subfamilies suggests that the ACT bHLH association occurred multiple times, possibly through domain shuffling processes, suggesting that domain shuffling processes may play a small role in the evolution of *bHLH* genes in plants [1]. Members of the same plant bHLH subfamily are also often involved in the same biological processes, such as subgroup Va, which is involved in brassinosteroid signaling [34], and the Active Phytochrome Binding (APB) motif, encoded in several proteins of subfamily VII (a + b), which mediate the binding of multiple *A. thaliana* bHLH proteins to phytochrome B [1,35].

Plant bHLH diversity was already present in the early land plants before the moss and vascular plant lineages separated over 440 million years ago. Many current bHLH interactions also occurred in early land plants and were essentially conserved in the major plant groups [1]. The expansion of this family is closely related to plant evolution and diversity, not only in higher plants but also in lower plants such as algae, lichens, and mosses or non-plants such as mycobacteria [1]. The availability of a large number of land plants and algal genomes allowed further analysis of the conserved motifs of the *bHLH* gene family throughout the photosynthetic plant spectrum (Figure 2). Different numbers of members of the *bHLH* gene family have been identified in different species by searching for proteins containing the bHLH domain. The *bHLH* gene family was found to originate from Rhodophyta. The conservation of the bHLH conservative domain in different species was further predicted by MEME (Figure 2). According to the conservative motif, the bHLH domain will evolve to be more complete as species evolve. The evolution of the *bHLH* gene family also provides insights regarding the evolution of green algae into flowering plants through their adaptation to environmental changes [8]. Specifically, the genome-wide analysis of *bHLH* gene families from different species will help us to better understand the biological function, evolutionary origin, and expansion outcome of *bHLH* genes.

Lineage	Species	bHLH domain
Algae	*Chondrus crispus*	
Algae	*Chara braunii*	
Moss	*Marchantia polymorpha*	
Moss	*Physcomitrella patens*	
Fern	*Selaginella moellendorffii*	
Fern	*Anthoceros formosae*	
Fern	*Salvinia cucullata*	
Gymnospermae	*Picea abies*	
Gymnospermae	*Gnetum montanum*	
Angiosperm	*Liriodendron chinense*	
Angiosperm	*Aristolochia fimbriata*	
Eudicots	*Arabidopsis thaliana*	
Eudicots	*Malus domestica*	
Eudicots	*Populus trichocarpa*	
Eudicots	*Vitis vinifera*	
Eudicots	*Cucumis sativus*	
Eudicots	*Clonorchis sinensis*	
Eudicots	*Glycine max*	
Eudicots	*Nicotiana tabacum*	
Monocots	*Oryza sative*	
Monocots	*Zea mays*	
Monocots	*Triticum aestivum*	

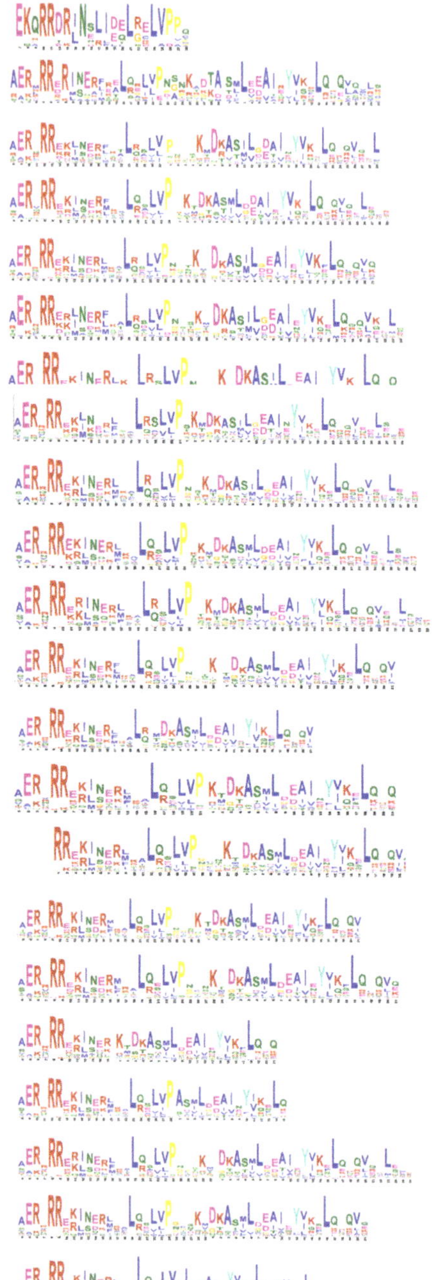

Figure 2. The bHLH domain conservative motif analysis expressed in different plants. Conservative domain motifs are used in MEME online web analytics. The bHLH protein sequences of these species were downloaded from the JGI and NCBI databases.

3. Regulation of bHLH TFs in Plant Stress Response

Previous research has revealed that the bHLH TF family can participate in the regulation of abiotic stresses such as drought, high salinity, low temperature, and nutrient deficiencies through transcriptional and post-translational modifications [36–41]. Transcriptome is commonly used in abiotic stress studies. Transcriptome technology can be used to identify key nodes or genes for stress resistance at the transcriptional regulatory level for subsequent research. In order to reveal the potential roles of bHLH members in various stress responses in different species, most functional signaling pathways of bHLH TF regulations were identified (Table 1 and Figure 3).

Table 1. bHLH transcription factors involved in plant response to abiotic stress.

Original Plant	Nomenclature	Stress Response	Regulation Type	Refs.
Ipomoea batatas (L.) *Lam.*	IbbHLH79	Cold	Positive regulation	[15]
Arabidopsis thaliana L.	rd22BP1/AtMYC2/ AtbHLH006	Drought	Positive regulation	[21]
Arachis hypogaea L.	AhbHLH112	Drought/salt	Positive regulation	[42]
Oryza sativa	OsWIH2/OsbHLH130	Drought	Positive regulation	[43]
Fagopyrum tataricum	FtbHLH3	Drought/oxidative	Unknown	[44]
Solanum lycopersicum	SlbHLH22	Drought	Positive regulation	[45]
Zea mays L.	ZmPIF1	Drought	Positive regulation	[46]
Malus × domestica Borkh.	MdCIB1	Drought	Positive regulation	[47]
Arabidopsis thaliana L.	AtbHLH68	Drought	Unknown	[48]
Populus euphratica	PebHLH35	Drought	Positive regulation	[49]
Arabidopsis thaliana L.	AtbHLH122	Drought/salt	Positive regulation	[50]
Oryza sativa	OsbHLH148	Drought	Positive regulation	[51]
Oriza rufipogon	OrbHLH001	Salt/cold	Positive regulation	[52]
Oryza sativa	BEAR1/OsbHLH014	Salt	Positive regulation	[53]
Oryza sativa	OsbHLH062	Salt	Unknown	[54]
Eleusine coracana L.	EcbHLH57	Salt/oxidative/drought	Positive regulation	[55]
Beta vulgaris L.	BvbHLH93	Salt	Positive regulation	[56]
Selaginella lepidophylla	SlbHLHopt	Salt	Positive regulation	[57]
Arabidopsis thaliana L.	AtbHLH112	Drought/salt	Positive regulation	[58]
Oryza sativa	OsbHLH035	Salt	Unknown	[59]
Oriza rufipogon	OrbHLH2	Salt/osmotic	Positive regulation	[60]
Panax ginseng C.A. Meyer	PgbHLH102	Salt	Unknown	[61]
Arabidopsis thaliana L.	AtICE1/AtbHLH116	Cold	Positive regulation	[62]
Arabidopsis thaliana L.	ICE2	Cold	Positive regulation	[63]
Malus × domestica Borkh.	MdCIbHLH1	Cold	Positive regulation	[64]
Pyrus ussuriensis	PuICE1	Cold	Positive regulation	[65]
Oryza sativa	OsbHLH1	Cold	Positive regulation	[66]
Nicotiana tabacum L.	NtbHLH123	Cold/salt	Positive regulation	[67]
Zoysia japonica	ZjbHLH76/ZjICE1	Cold	Positive regulation	[68,69]
Vitis amurensis	VaICE1/VaICE2	Cold	Positive regulation	[70]
Dimocarpus longan Lour.	DlICE1	Cold	Positive regulation	[71]
Citrus sinensis	CsbHLH18	Cold/salt	Positive regulation	[72]
Poncirus trifoliate	PtrbHLH	Cold/oxidative	Positive regulation	[73,74]
Rosa multiflora	RmICE1	Cold/salt	Positive regulation	[75]
Prunus avium L.	PavbHLHs	Cold	Unknown	[76]
Arabidopsis thaliana L.	FIT/AtbHLH29	Fe deficiency	Unknown	[77]
Arabidopsis thaliana L.	AtbHLH38	Fe deficiency	Positive regulation	[78]
Arabidopsis thaliana L.	AtbHLH39	Fe deficiency	Positive regulation	[78]
Arabidopsis thaliana L.	AtbHLH100	Fe deficiency	Positive regulation	[79]
Arabidopsis thaliana L.	AtbHLH101	Fe deficiency	Positive regulation	[79]
Arabidopsis thaliana L.	PYE/AtbHLH47	Fe deficiency	Unknown	[80]

Table 1. *Cont.*

Original Plant	Nomenclature	Stress Response	Regulation Type	Refs.
Arabidopsis thaliana L.	ILR3/AtbHLH105	Fe deficiency	Positive regulation/Negative regulation	[81]
Arabidopsis thaliana L.	AtbHLH104	Fe deficiency	Positive regulation	[82]
Arabidopsis thaliana L.	AtbHLH18	Fe deficiency	Negative regulation	[83]
Arabidopsis thaliana L.	AtbHLH19	Fe deficiency	Negative regulation	[83]
Arabidopsis thaliana L.	AtbHLH20	Fe deficiency	Negative regulation	[83]
Arabidopsis thaliana L.	AtbHLH25	Fe deficiency	Negative regulation	[83]
Chrysanthemum morifolium	CmbHLH1	Fe deficiency	Unknown	[84]
Oryza sativa	OsIRO3/OsbHLH2	Fe deficiency	Positive regulation	[85]
Oryza sativa	OsIRO2/OsbHLH056	Fe deficiency	Positive regulation	[86]
Oryza sativa	OsbHLH156	Fe deficiency	Positive regulation	[87]
Triticum aestivum L.	TabHLH1	Pi and N deficiency	Positive regulation	[88]
Chlamydomonas reinhardtii	NRI1	N starvation	Positive regulation	[89]
Oryza sativa	OsPTF1	Pi starvation	Positive regulation	[90]
Nicotiana tabacum L.	NtbHLH1	Fe deficiency	Positive regulation	[91]
Zea mays L.	ZmICE1	Cold	Positive regulation	[92]
Oryza sativa	OsPRI2/OsbHLH058 OsPRI3/OsbHLH059	Fe deficiency	Positive regulation	[93]
Solanum lycopersicum	SlICE1	Cold/salinity	Positive regulation	[94]
Cucumis sativus L.	CsbHLH041	Salt	Positive regulation	[95]
Nicotiana tabacum L.	NtbHLH86	Drought	Positive regulation	[96]
Malus × *domestica Borkh.*	MdbHLH3	Cold	Positive regulation	[97]
Malus × *domestica Borkh.*	MdbHLH106L	Salt	Positive regulation	[98]
Malus × *domestica Borkh.*	MdbHLH130	Salt	Positive regulation	[99]
Pyrus ussuriensis	PubHLH1	Cold	Positive regulation	[100]
Arabidopsis thaliana L.	AtbHLH92	Salt/osmotic stress	Unknown	[101]
Arabidopsis thaliana L.	AtAIB/AtbHLH17	Drought/salt	Positive regulation	[102]
Fagopyrum tataricum	FtbHLH2	Cold	Positive regulation	[103]
Solanum tuberosum	StbHLH45	Drought	Unknown	[104]
Eucalyptus camaldulensis	EcaICE1	Cold	Positive regulation	[105]
Triticum aestivum L.	TabHLH39	Osmotic	Unknown	[106]
Arabidopsis thaliana L.	AtNIG1/AtbHLH028	Salt	Positive regulation	[107]
Glycine Max (L.) *Merrill*	GmbHLH57	Fe deficiency	Unknown	[108]
Glycine Max (L.) *Merrill*	GmbHLH300	Fe deficiency	Unknown	[108]
Zea mays L.	ZmPIF3	Drought	Positive regulation	[109]
Oryza sativa	OsbHLH133	Fe deficiency	Negative regulation	[110]
Oryza sativa	OsPRI1/OsbHLH115	Fe deficiency	Positive regulation	[111]
Arabidopsis thaliana L.	AtbHLH121	Fe deficiency	Unknown	[112]
Arabidopsis thaliana L.	AtbHLH106	Salt	Positive regulation/Negative regulation	[113]
Arabidopsis thaliana L.	AtbHLH11	Fe deficiency	Negative regulation	[114]
Oryza sativa	OsbHLH006	Drought	Unknown	[115]
Oryza sativa	OsbHLH068	Salt	Positive regulation	[116]
Vitis vinifera	VvbHLH1	Drought/salt/cold	Positive regulation	[117]

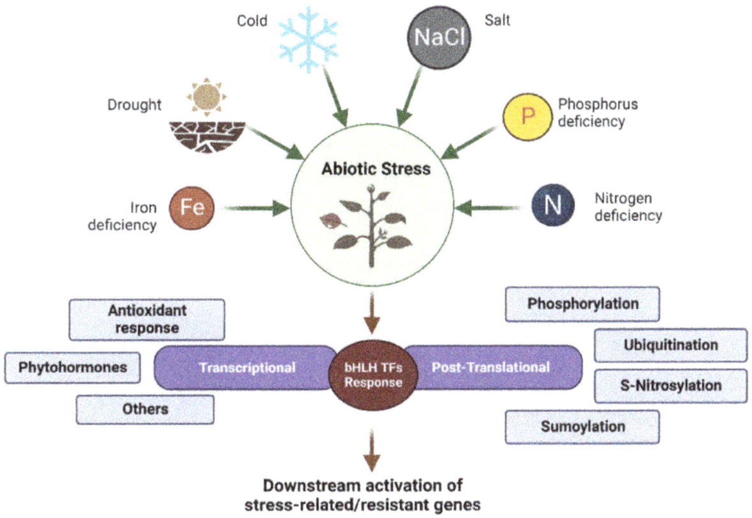

Figure 3. Summary of the different types of bHLH TFs responses to abiotic stresses in plants.

4. Drought Stress Response

One of the major environmental issues that reduces plant growth and development is drought. Drought-induced yield declines have been reported for many crop species, which depend on the severity and duration of the stress period and are associated with a decline in crop yield productivity [118]. Abscisic acid (ABA) is an important phytohormone that plays an important role in abiotic stress tolerance, whose main function appears to be the regulation of plant water balance and osmotic stress tolerance. Plant bHLH transcription factors generally respond to drought stress through both ABA-dependent and ABA-independent regulatory systems [119–121].

In peanuts (*Arachis hypogaea*), AhbHLH112 responds to drought stress by regulating antioxidant gene-mediated reactive oxidative species (ROS) scavenging or ABA-dependent pathways as, under drought stress, ABA accumulates, and blocks Ca2+ influx and membrane proton pumps, resulting in plasma membrane depolarization and stomatal closure, which is required to reduce water loss and maintain high water potential. Under drought stress, AhbHLH112 is induced and could activate antioxidant genes and promote ROS scavenging (POD) [42]. *MfbHLH38* also acts positively on plant defenses via the ABA-dependent signaling pathway and enhances the ROS scavenging system, thereby reducing oxidative damage under stress [40]. Drought stress can also induce the accumulation of OsbHLH130, which activates *OsWIH2* expression, and the latter improves rice drought tolerance by reducing the water loss rate and ROS accumulation [43]. *FtbHLH3* responds to drought stress by increasing photosynthetic efficiency and upregulating the expression of critical genes in the ABA signaling pathway, the proline biosynthetic pathway, the ROS scavenging system, and the drought-responsive pathway [44]. *SlbHLH22* also improved tomato plant stress resistance by inducing the expression of genes involved in flavonoid biosynthesis. Flavonoids can enhance plant tolerance to drought and salt stress due to their ability to scavenge superoxides, peroxides, and free radicals generated during stress through ABA biosynthesis and the ROS scavenging pathway, leading to stomatal closure, increased proline content, and enhanced CAT, POD, and SOD activities with reduced ROS accumulation, resulting in an improved tolerance under abiotic stress [45]. ZmPIF1 is a positive regulator of root development, ABA synthesis, signaling pathways, and drought tolerance. It was found that ZmPIF1 binds to the G-box element in the promoters of NCED, CBF4, ATAF2/NAC081, NAC30, and other transcription factors and positively regulates their expression [46]. Cryptochrome-interacting bHLH1 (MdCIB1), through ABA-dependent and ABA-independent signaling pathways, regulates ABA signal transduction, antioxidant system, osmotic balance, and expression levels of stress-related genes [47]. AtbHLH68 may positively respond to drought stress through ABA signaling and by regulating ABA homeostasis in *Arabidopsis* [48]. *PebHLH35* in transgenic *Arabidopsis* confers drought tolerance by reducing stomatal density, opening, transpiration rate, and water loss, and increasing the chlorophyll content and photosynthesis rate [49]. AtMYC2 (AtbHLH006) is a transcriptional activator in ABA-inducible gene expression under drought stress in plants as it regulates the expression of *rd22* [21]. The *rd22* gene is a dehydration-responsive gene that is activated in *Arabidopsis* plants by exogenous ABA [122]. *AtbHLH122* responds to drought stress via an ABA-independent pathway as it represses CYP707A3, an important ABA 8′-hydroxylase [123], increasing ABA content, and then expressing ABA-inducible genes. It functions as a positive regulator in drought, salt and osmotic signaling [50]. OsbHLH148 also responds to drought stress via a jasmonate signaling pathway. Jasmonates (JAs) are a class of oxygen-containing lipid derivatives (oxylipins) that are considered plant hormones necessary for plant growth and environmental adaptation [124]. In the absence of stress and JA, OsbHLH148 is repressed through its interaction with OsJAZ. Upon exposure to drought stress, JA increases, leading to the degradation of OsJAZ proteins. The released OsbHLH148 activates drought tolerance genes, including *OsDREB1s*, leading to drought tolerance [51] (Figure 4A).

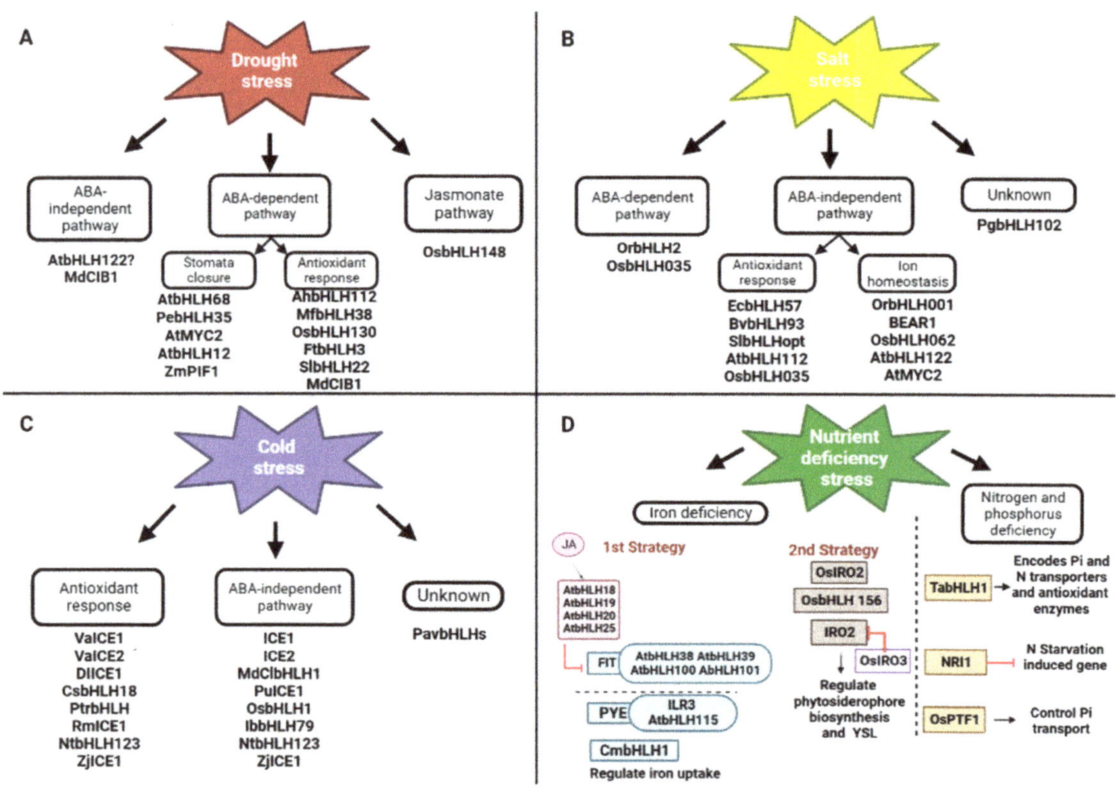

Figure 4. Transcriptional regulations of bHLH TF under abiotic stresses. (**A**) For drought stress; (**B**) for salt stress; (**C**) for cold stress; and (**D**) for nutrient deficiency stress.

Under drought and salt stress, the overexpression of CgbHLH001 can confer stress tolerance. Regulated by phosphorylation by a protein kinase such as calcium-dependent protein kinase (CDPK), CgbHLH001 acts as a positive regulator in controlling downstream relevant genes, thereby reducing ROS production and enhancing ROS scavenging ability and improving physiological performance [18] (Figure 5D).

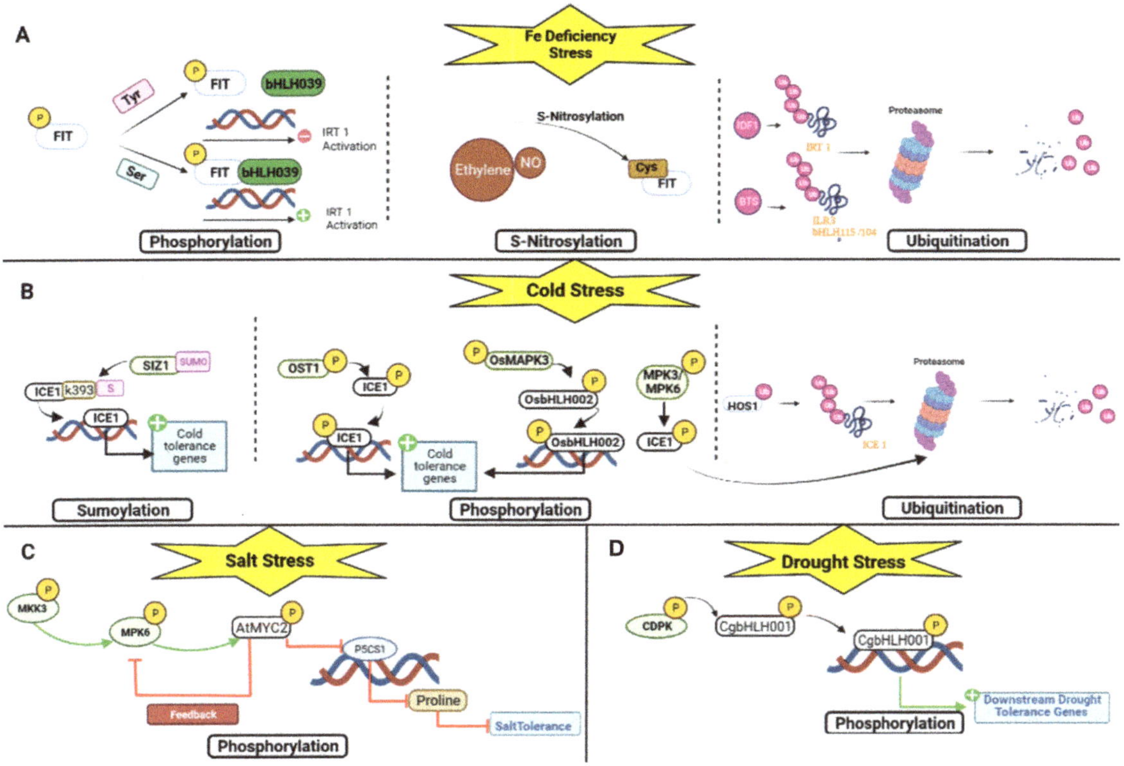

Figure 5. Post-translational regulations of bHLH TF under abiotic stresses. (**A**) For drought stress; (**B**) for salt stress; (**C**) for cold stress; and (**D**) for nutrient deficiency stress.

5. Salt Stress Response

Salinity has two main effects on plants. It either induces osmotic stress or induces the instability of ions. Several bHLH transcription factors have been characterized in rice. There is *OrbHLH001*, a homologue of *ICE1* (*AtbHLH116*), which activates *OsAKT1*, which enhances Na+ efflux and K+ uptake, thus controlling the Na+/K+ ratio in salt stress to confer salt stress tolerance [52]. BEAR1 (OsbHLH014), a bHLH transcription factor, plays a vital role in rice salt stress response. After receiving the salt stress signal, the BEAR1 protein as a transcriptional activator induces two pathways. On the one hand, it controls the gene expression level of the salt response signaling cascade. On the other hand, it regulates the Na+ /K+ balance and membrane stabilization genes and thus induces salt stress tolerance [53]. OsbHLH062 is involved in a jasmonate-dependent transcription regulatory module complex: Osbhlh062-JAZ9-OsNINJA. OsJAZ9 represents a transcriptional regulator and an essential element that modulates salt stress tolerance by binding OsbHLH062, a transcriptional activator, to OsNINJA, an essential repressor of basal jasmonate (JA) signaling. Under salt stress, JA levels increase. The JAZ protein OsJAZ9 is recruited by SCFCOI1 for ubiquitination by the 26S proteasome, leading to its degradation, releasing OsNINJA, and allowing OsbHLH062 to bind to the E-box and activate target genes. OsbHLH062 can bind to the E-box cis-element and the promoter of OsHAK21, including some of the ion transporters, thereby conferring salt stress tolerance [54]. *AtbHLH122* inhibits *CYP707A3* gene expression under NaCl stress [50]. In addition, Krishnamurthy et al. (2019) identified AtMYC2 and *AtbHLH122* as upstream regulators of ABA-mediated *AtNHX1* and *AtNHX6*, both Na+/H+ exchangers. In addition, the overexpression of EcbHLH57 enhanced tobacco resistance to salt stress by increasing the expression of stress-responsive genes such as *LEA14*, *rd29A*, *rd29B*, *SOD*, *APX*, *ADH1*, *HSP70*, and *PP2C* [55]. *BvbHLH93* regulates salt stress tolerance by enhancing antioxidant activity and reducing ROS production, thereby maintaining ion homeostasis, but it needs further investigation at the transcriptional level [56]. SlbHLHopt is an essential regulator of water deficit and NaCl stress in transgenic *Arabidopsis* since it increases flavonoid content [57]. In *Arabidopsis*, *AtbHLH112* regulates the expression of genes involved in abiotic stress tolerance by increasing proline levels and reducing ROS accumulation and water loss through the expression of *Proline Dehydrogenase 1* (*ProDH*) [58]. OsbHLH035 responds to salt stress through an ABA-independent pathway by indirectly mediating the expression of *OsHKT1s* such as *OsHKT1;3* and *OsHKT1;5*, which are sodium transporters that transfer Na+ from the xylem into xylem parenchyma cells, thus preventing the accumulation of Na+ in aerial tissues, since a higher concentration of Na+ in the roots causes osmotic stress [59]. *OrbHLH2* in transgenic *Arabidopsis* improved salt tolerance through an ABA-independent pathway by increasing expression levels of *DREB1A/CBF* to confer improved plant tolerance to salt stress; however, more research is needed to determine its mode of action [60]. *PgbHLH102* could respond to salt stress and higher salt concentrations, but its mode of action needs further investigation [61] (Figure 4B). Through an MKK3-MPK6-MYC2 cascade, AtMYC2 (AtbHLH006) also functions as a negative regulator of salt tolerance in *A. thaliana*. In brief, the MKK3-MPK6 module phosphorylates and hence activates AtMYC2 in response to salt stress. AtMYC2 binds to *P5CS1*, which is the main contributor to proline accumulation induced by stress [125], and regulates salt tolerance in *Arabidopsis thaliana* by inhibiting the synthesis of *P5CS1* and proline [126] (Figure 5C).

6. Cold Stress Response

Cold stress, including chilling, adversely affects plant growth and development, limits the geographical distribution of plant species, and reduces crop yields worldwide [127]. For acclimation to cold stress, cold tolerance requires a downstream cascade of transcriptional regulation of target genes. Cold-responsive genes contain DRE/CRT cis-elements with a core sequence CCGAC [25,128]. C-repeat Binding Factors (*CBFs*) (*CBF1*, *CBF2*, and *CBF3*; also known as *DREB1b*, *DREB1c*, and *DREB1a*, respectively) are known genes that bind to these elements, triggering transcription of cold-responsive genes [129]. Since *CBFs* are only induced after 15 min of cold exposure, the intervention of another TF present in the cell is required [130]. In *Arabidopsis*, the most well-defined pathway is the *ICE 1* (*AtbHLH116*) transcription factor, an ABA-independent pathway that binds to the CBF promoter and activates transcription of *CBF1/3*, which in turn induces the *CRT/BRE* genes responsible for cold tolerance [62]. *ICE2* (*BAC42644*), a homologue of *ICE1* (*AtbHLH116*), mediates the same ABA-independent pathway in cold tolerance [63]. *MdCIbHLH1* also interferes with the *CBF* pathway by activating the transcription of *CBF 1* and *2* [64]. *PuICE1* demonstrates a *CBF/DREB* pathway that requires cold-induced protein-protein interactions with the Hepta-helical protein 1 (PuHHP1), which acts as a positive regulator of ABA-mediated stomatal closure in response to cold stress, thus increasing the levels of PuDREBa transcripts and positively regulating cold stress; however, more research is needed to experimentally clarify the physiological mechanism between PuHHP1 and *PuICE1*. [65,131]. In rice, *OsbHLH1* expression is significant for cold tolerance as it might also regulate *CBF/DREB1* gene expression [66]. The IbbHLH79 protein, an *ICE1*-like gene, can activate the *CBF* pathway and IbbHLH79-overexpressing transgenic plants show enhanced cold tolerance [15]. Using the *NtCBFs* signaling pathways, cold-activated *NtbHLH123* and *ZjbHLH76/ZjICE1* regulate the expression of their target genes or regulate stress-responsive genes such as CBFs associated with ROS scavenging, resulting in improved tolerance to cold stress [67–69]. *VaICE1* and *VaICE2* *ICE*-like TF play a positive role in freezing tolerance and influence cold stress-related factors such as electrolyte leakage and proline and malondialdehyde (MDA) levels, thereby reducing ROS damage and improving osmotic protection [70]. The overexpression of *DlICE1* in transgenic *Arabidopsis thaliana* enhances cold tolerance by increasing proline content, reducing ion leakage and accumulation of MDA and ROS [71]. *CsbHLH18* mediates a cold response by regulating antioxidant genes by binding to and activating the *CsPOD* promoter, thereby inducing ROS scavenging [72]. *PtrbHLH* also responds to cold stress by modulating POD and CAT [73,74]. *RmICE1* also responds to cold stress by regulating antioxidant genes [75]. Several *PavbHLH* also demonstrated a cold response regulatory mechanism, but the molecular details require further studies [76] (Figure 4C). HOS1 encodes a ring finger protein with E3 ligase activity that reduces *ICE1* (*AtbHLH116*) activity via the ubiquitination/proteasome pathway [132,133]. In addition, OST1 phosphorylates *ICE1* (*AtbHLH116*) and inhibits HOS1-mediated degradation of *ICE1*, since the OST1 protein competes with HOS1 for binding to *ICE1*, thereby releasing *ICE1* from the HOS1–*ICE1* complex. The dual role of OST1 helps enhance *ICE1* stability to increase *CBF* expression and freeze tolerance [134]. MKK2 is also activated by cold stress in plants and consequently activates MPK4 and MPK6 [135]. In brief, MPK3/6-mediated phosphorylation promotes *ICE1* (*AtbHLH116*) degradation, with Ser94, Thr366, and Ser403 being important sites for phosphorylation-dependent degradation [28]. Subsequently, when *ICE1* (*AtbHLH116*) levels accumulate, it is phosphorylated by MPK3/MPK6 and consequently cleared via the 26S proteasome pathway, which inhibits *CBF*-dependent cold signaling [136]. Another positive regulation is found in SUMO conjugation, which enhances *ICE1* (*AtbHLH116*) activity. SIZ1 (SAP and Miz) encodes a SUMO-E3 ligase that is necessary for freezing tolerance, facilitates sumoylation of *ICE1* (*AtbHLH116*), and recognizes K393 as the SUMO conjugation residue in the protein. SIZ-mediated SUMO1 conjugation to K393 affects the activity of *ICE1* (*AtbHLH116*) to control *CBF3/DREB1A* expression and prevents access to the ubiquitination complex. A K393R mutation blocks *ICE1* (*AtbHLH116*) sumoylation, represses expression of *CBF3/DREB1A* and its regulon genes, and reduces cold tolerance [29]. Under cold stress, active OsMAPK3 phosphorylates OsbHLH002, accumulating phospho-OsbHLH002, promoting trehalose-6-phosphate phosphatase 1 (OsTPP1) expression, and increasing trehalose content and resistance to chilling damage [137] (Figure 5B).

7. Iron Deficiency Response

Iron (Fe) is an essential element for plant survival and development as it is involved in several vital processes: photosynthesis, DNA synthesis, respiration, and chlorophyll synthesis. In *Arabidopsis*, FIT (AtbHLH029) is a well-known TF that plays a crucial role in Fe uptake via the first strategy [77], which consists of the acidification of the external environment mediated by membrane proton pumps, the *Autoinhibited Plasma Membrane H+/ATPases* (*AHA2*) [138], to release and solubilize the iron. It is then reduced from ferric iron to ferrous iron by *Ferric Reduction Oxidase 2* (*FRO2*) [139] and transported through specific channels within the *Iron-Regulated Transporter 1* (*IRT1*) in the plant [140]. It binds to the fer uptake genes *FRO2* and *IRT1*. Since it cannot function alone, it forms heterodimers with the four bHLH TFs AtbHLH38, AtbHLH39, AtbHLH100, and AtbHLH101 [78,79,141,142]. There is also another protein, POPEYE (PYE) (AtbHLH47), that responds to Fe deficiency by using a network independent of FIT [80]. The dimers of ILR3 (AtbHLH105) and AtbHLH104 or ILR3 (AtbHLH105) and AtbHLH34 regulate PYE (AtbHLH47), which interacts with two bHLH transcription factors, ILR3 (AtbHLH105) and AtbHLH115 [81,82,143]. Four other bHLH partners of FIT (AtbHLH029) (AtbHLH18, AtbHLH19, AtbHLH20, and AtbHLH25) promote its degradation in response to JA induction, thereby antagonizing the activity of AtbHLH38, AtbHLH39, AtbHLH100, and AtbHLH101 and limiting Fe uptake [83]. At the transcriptional level, the active and inactive states are distinguished by specific covalent modifications. Suppose the transition from the inactive state to the active state is bottlenecked. In that case, this could be achieved by limiting the enzymes that can confer or remove covalent modifications to "activate" FIT (AtbHLH029). FIT (AtbHLH029) could be in a negative feedback loop, limiting its abundance. In the context of Fe regulation, the rapid switching off of FIT (AtbHLH029) could be to prevent Fe toxicity symptoms [39]. A study by Zhao et al. showed that *chrysanthemum* CmbHLH1 promoted iron absorption through H+-ATPase-mediated rhizosphere acidification. The second strategy is not based on the reduction of iron [84]. Nonetheless, it is chelated by phytosiderophores (PS) [144] and then transported by a specific transporter, the oligopeptide transporter YS1. This strategy led to the discovery of the transcription factors IDEF1 and IDEF2 in rice, which control the expression of phytosiderophore biosynthesis and YSL2 [145,146]. *IRO2* in rice is a close homologue of bHLH39 and positively regulates phytosiderophore biosynthesis and YSL15 [41]. In rice, the iron-related transcription factor 3 (*OsIRO3*) (*OsbHLH063*), which belongs to the bHLH gene family, plays a critical role in maintaining iron homeostasis in an iron-deficient environment, as it negatively controls transcript levels of *OsIRO2* (*OsbHLH056*) [85–87]. Ogo et al. (2007) found that the overexpression of *OsIRO2/OsbHLH056* promotes iron uptake in rice following the 2nd strategy; however, its mode of action still needs further studies. Wang et al. studied the role of a new rice bHLH-type transcription factor, OsbHLH156, in iron homeostasis and found that OsbHLH156 greatly increased iron deficiency. They concluded that OsbHLH156 is required for a strategy II uptake mechanism in rice (Figure 4D). Ethylene (ET) and Nitric Oxide (NO) are required for complete upregulation of Fe-deficiency gene expression and FIT (AtbHLH029) protein abundance as they increase the accumulation of FIT by inhibiting proteasomal FIT (AtbHLH029) and thereby prevent its degradation. Therefore, they stabilize FIT (AtbHLH029) expression by nitrosylation of Cys residues present in FIT [147]. Therefore, this finding requires further elucidation. To limit Fe accumulation in plants, a RING-type E3 ubiquitin ligase called Degradation Factor 1 (IDF1) (AT4G30370) is directly involved in the degradation of *IRT1* through ubiquitination [148,149]. To limit the excess of Fe, which can be toxic to plants, there is BRUTUS (BTS) (AT3G18290), a RING-type E3 ubiquitin ligase that plays a role as a negative regulator since it is induced in Fe deficiency and has been shown to target the bHLH transcription factors ILR3 (AtbHLH105) and bHLH115/104 for degradation via the 26S proteasome by its RING domain (E3 ligase), thus inhibiting the formation of heterodimer complexes of PYE/PYEL (AtbHLH47). This includes disrupting the activation of Fe-deficiency genes regulated by PYE (AtbHLH47). BTS (AT3G18290) contains a hemerythrin-like domain that can bind Fe. Fe binding to the hemerythrin-like domain of BTS is involved in its destabilization and subsequent degradation [27,150]. On the one hand, it has been shown that the phosphorylation of Ser 221 and Ser 272 in FIT (AtbHLH029) positively regulates FIT accumulation in the nucleus, enabling its dimerization to AtbHLH039 and its activation of *IRT1* promoters [39]. On the other hand, phosphorylation of Tyr238 and Tyr278 reduces and inhibits heterodimerization of FIT (AtbHLH029) with AtbHLH039 and expression of *IRT1* promoters, rendering FIT (AtbHLH029) inactive [41] (Figure 5A).

8. Phosphorus and Nitrogen Deprivation Stress Response

Phosphorus (P) and Nitrogen (N) are both indispensable macronutrients for plant growth and crop productivity. *TabHLH1* is sensitive to external Pi and N deficiency stress. It confers enhanced tolerance to both Pi and N deficiency through the transcriptional modulation of a panel of genes encoding *Phosphate Transporter* (*PT*), *Nitrate Transporter* (*NRT*), and antioxidant enzymes [88]. *NRI1*, previously named *Nitrogen Starvation-Induced Gene 17* (*NSG17*) (BAD44756.1), a basic Helix–Loop–Helix (bHLH) TF, represses the transcriptional activators of N starvation-induced genes to regulate N starvation-specific responses when sufficient N is supplied to *C. reinhardtii* [89]. *OsPTF1* (AAO73566), responsible for tolerance to Pi starvation in rice, controls the Pi transport system in plants. *OsPTF1* is an essential element involved in higher root growth and consequently higher uptake rates of Pi, and it is also involved in the efficient use of Pi in plants. Therefore, *OsPTF1* has potential in engineering plants with higher Pi utilization efficiency [90] (Figure 4D).

9. Conclusions and Future Perspectives

To withstand environmental stresses, plants have evolved interdependent regulatory pathways that allow them to respond and adapt to the environment in a timely manner. Abiotic stress conditions affect many aspects of plant physiology and cause widespread changes in cellular processes. Research into plant defense mechanisms after abiotic stress is of great importance for subsequent breeding. As one of the most abundant transcription factor families in eukaryotes, members of the bHLH family have complex structures, large numbers, and diverse functions. Many studies have shown that bHLHs can regulate plant resistance to a variety of abiotic stresses. This review summarized the roles of the bHLH transcription factor family in the corresponding abiotic stresses in plants from two aspects. First, bHLH transcription factors show evolutionary differences in the conservative domain. Previous studies have shown that members of the same subfamily may regulate the synthesis of the same plant hormones [151]. Thus, there is reasonable speculation that members of the same subfamily may have evolved through genetic replication from the same ancestor. However, these assumptions still need to be verified. Second, we discussed how the bHLH transcription factor specifically regulates plant tolerance to abiotic stresses through transcriptional and post-translational modification pathways. Furthermore, with the development of transcriptome sequencing, transcriptome technology has been widely used in the study of plant stress tolerance. RNA-seq can quickly screen important functional genes by looking for genes with significant differences through samples, in order to find key nodes or genes for stress resistance at the transcriptional level for subsequent research. This naturally raises an interesting question: can bHLHs regulate plant tolerance to stress by controlling the synthesis of plant hormones or compounds via transcriptional or post-translational modification pathways?

Author Contributions: Project administration, L.Y. and H.M.; writing—original draft preparation, Y.R. and R.L.; visualization, R.L.; writing—review and editing, H.M.K., L.Y. and H.M.; funding acquisition, L.Y. All authors have read and agreed to the published version of the manuscript.

Funding: This research was funded by the National Natural Science Foundation of China (No. 31971682) and the Research Startup Fund for High-Level and High-Educated Talents of Nanjing Forestry University.

Data Availability Statement: Not applicable.

Conflicts of Interest: The authors declare no conflict of interest.

Abbreviations

ACT domain, small regulatory domains initially found in AK, chorismate mutase, and TyrA (prephenate dehydrogenase); TFs, transcription factors; ACE, ACTIVATOR FOR CELL ELONGATION; AIF, ATBS1 (ACTIVATION-TAGGED BRI1 SUPPRESSOR 1)-INTERACTING FACTOR; ARF, AUXIN RESPONSE FACTOR; BEE, BR ENHANCED EXPRESSION; BEH, BES1-HOMOLOGUE; BES1, BRI1 EMS-SUPRESSOR 1; bHLH, basic helix–loop–helix; BIM, BES1-INTERACTING MYC-LIKE; BR, brassinosteroid; BRZ, brassinazole; CES, CESTA; CIB, CRYPTOCHROME INTERACTING bHLH; CIL, CIB1-LIKE PROTEIN; HOS1, high expression of osmotically responsive gene 1; NCED, 9-cis-epoxycarotenoid dioxygenase; CBF4, C-repeat-binding factor; GA, gibberellin; HBI1, HOMOLOGUE OF BEE2 INTERACTING WITH IBH1; HFR1, LONG HYPOCOTYL IN FAR-RED 1; IBH1, ILI1 BINDING BHLH 1; IBL1, IBH1-LIKE 1; KDR, KIDARI; PAR, PHYTOCHROME RAPIDLY REGULATED; phy, phytochrome; PIF, PHYTOCHROME INTERACTING FACTOR; PIL1, PIF3-LIKE 1; PRE, PACLOBUTRAZOL RESISTANCE; YSL, an iron-phytosiderophore transporter; ABA, abscisic acid; JA, jasmonic acid; P5CS1, pyrroline-5-carboxylate synthetase 1; FIT, FER-LIKE IRON DEFICIENCY-INDUCED TRANSCRIPTION FACTOR.

References

1. Pires, N.; Dolan, L. Origin and Diversification of Basic-Helix-Loop-Helix Proteins in Plants. *Mol. Biol. Evol.* **2010**, *27*, 862–874. [CrossRef]
2. Massari, M.E.; Murre, C. Helix-Loop-Helix Proteins: Regulators of Transcription in Eucaryotic Organisms. *Mol. Cell. Biol.* **2000**, *20*, 429–440. [CrossRef] [PubMed]
3. Heim, M.A.; Jakoby, M.; Werber, M.; Martin, C.; Weisshaar, B.; Bailey, P.C. The Basic Helix-Loop-Helix Transcription Factor Family in Plants: A Genome-Wide Study of Protein Structure and Functional Diversity. *Mol. Biol. Evol.* **2003**, *20*, 735–747. [CrossRef] [PubMed]
4. Li, X.; Duan, X.; Jiang, H.; Sun, Y.; Tang, Y.; Yuan, Z.; Guo, J.; Liang, W.; Chen, L.; Yin, J.; et al. Genome-Wide Analysis of Basic/Helix-Loop-Helix Transcription Factor Family in Rice and Arabidopsis. *Plant Physiol.* **2006**, *141*, 1167–1184. [CrossRef] [PubMed]
5. Murre, C.; McCaw, P.S.; Baltimore, D. A new DNA binding and dimerization motif in immunoglobulin enhancer binding, daughterless, MyoD, and myc proteins. *Cell* **1989**, *56*, 777–783. [CrossRef]
6. Ledent, V.; Vervoort, M. The Basic Helix-Loop-Helix Protein Family: Comparative Genomics and Phylogenetic Analysis. *Genome Res.* **2001**, *11*, 754–770. [CrossRef]
7. Atchley, W.R.; Fitch, W.M. A natural classification of the basic helix–loop–helix class of transcription factors. *Proc. Natl. Acad. Sci. USA* **1997**, *94*, 5172–5176. [CrossRef]
8. Carretero-Paulet, L.; Galstyan, A.; Roig-Villanova, I.; Martínez-García, J.F.; Bilbao-Castro, J.R.; Robertson, D.L. Genome-wide classification and evolutionary analysis of the bHLH family of transcription factors in Arabidopsis, poplar, rice, moss, and algae. *Plant Physiol.* **2010**, *153*, 1398–1412. [CrossRef]
9. Bailey, P.C.; Martin, C.; Toledo-Ortiz, G.; Quail, P.H.; Huq, E.; Heim, M.A.; Jakoby, M.; Werber, M.; Weisshaar, B. Update on the Basic Helix-Loop-Helix Transcription Factor Gene Family in *Arabidopsis thaliana*. *Plant Cell* **2003**, *15*, 2497–2502. [CrossRef]
10. Zhao, K.; Li, S.; Yao, W.; Zhou, B.; Li, R.; Jiang, T. Characterization of the basic helix–loop–helix gene family and its tissue-differential expression in response to salt stress in poplar. *PeerJ* **2018**, *6*, e4502. [CrossRef]
11. Wu, Y.; Wu, S.; Wang, X.; Mao, T.; Bao, M.; Zhang, J. Genome-wide identification and characterization of the bHLH gene family in an ornamental woody plant Prunus mume. *Hortic. Plant J.* **2022**, *8*, 531–544. [CrossRef]
12. Zhao, W.; Liu, Y.; Li, L.; Meng, H.; Yang, Y.; Dong, Z.; Wang, L.; Wu, G. Genome-Wide Identification and Characterization of bHLH Transcription Factors Related to Anthocyanin Biosynthesis in Red Walnut (*Juglans regia* L.). *Front. Genet.* **2021**, *12*, 632509. [CrossRef] [PubMed]
13. Salih, H.; Tan, L.; Htet, N.N.W. Genome-Wide Identification, Characterization of bHLH Transcription Factors in Mango. *Trop. Plant Biol.* **2021**, *14*, 72–81. [CrossRef]
14. Ali, A.; Javed, T.; Zaheer, U.; Zhou, J.-R.; Huang, M.-T.; Fu, H.-Y.; Gao, S.-J. Genome-Wide Identification and Expression Profiling of the bHLH Transcription Factor Gene Family in Saccharum spontaneum Under Bacterial Pathogen Stimuli. *Trop. Plant Biol.* **2021**, *14*, 283–294. [CrossRef]
15. Jin, R.; Kim, H.S.; Yu, T.; Zhang, A.; Yang, Y.; Liu, M.; Yu, W.; Zhao, P.; Zhang, Q.; Cao, Q.; et al. Identification and function analysis of bHLH genes in response to cold stress in sweetpotato. *Plant Physiol. Biochem.* **2021**, *169*, 224–235. [CrossRef]
16. Tan, C.; Qiao, H.; Ma, M.; Wang, X.; Tian, Y.; Bai, S.; Hasi, A. Genome-Wide Identification and Characterization of Melon bHLH Tran-scription Factors in Regulation of Fruit Development. *Plants* **2021**, *10*, 2721. Available online: https://pubmed.ncbi.nlm.nih.gov/34961193/ (accessed on 10 December 2021). [CrossRef] [PubMed]
17. Liu, R.; Song, J.; Liu, S.; Chen, C.; Zhang, S.; Wang, J.; Xiao, Y.; Cao, B.; Lei, J.; Zhu, Z. Genome-wide identification of the Capsicum bHLH transcription factor family: Discovery of a candidate regulator involved in the regulation of species-specific bioactive metabolites. *BMC Plant Biol.* **2021**, *21*, 262. Available online: https://pubmed.ncbi.nlm.nih.gov/34098881/ (accessed on 7 June 2021). [CrossRef]
18. Zhou, X.; Liao, Y.; Kim, S.-U.; Chen, Z.; Nie, G.; Cheng, S.; Ye, J.; Xu, F. Genome-wide identification and characterization of *bHLH* family genes from *Ginkgo biloba*. *Sci. Rep.* **2020**, *10*, 13723. [CrossRef]
19. Castelain, M.; Le Hir, R.; Bellini, C. The non-DNA-binding bHLH transcription factor PRE3/bHLH135/ATBS1/TMO7 is involved in the regulation of light signaling pathway in Arabidopsis. *Physiol. Plant.* **2012**, *145*, 450–460. [CrossRef]
20. Qi, Y.; Zhou, L.; Han, L.; Zou, H.; Miao, K.; Wang, Y. *PsbHLH1*, a novel transcription factor involved in regulating anthocyanin biosynthesis in tree peony (*Paeonia suffruticosa*). *Plant Physiol. Biochem.* **2020**, *154*, 396–408. [CrossRef]
21. Abe, H.; Urao, T.; Ito, T.; Seki, M.; Shinozaki, K.; Yamaguchi-Shinozaki, K. Arabidopsis AtMYC2 (bHLH) and AtMYB2 (MYB) Function as Transcriptional Activators in Abscisic Acid Signaling. *Plant Cell* **2002**, *15*, 63–78. [CrossRef] [PubMed]
22. Zhang, L.Y.; Bai, M.Y.; Wu, J.; Zhu, J.Y.; Wang, H.; Zhang, Z.; Wang, W.; Sun, Y.; Zhao, J.; Sun, X.; et al. Antagonistic HLH/bHLH transcription factors mediate brassino-steroid regulation of cell elongation and plant development in rice and Arabidopsis. *Plant Cell* **2009**, *21*, 3767–3780. [CrossRef] [PubMed]
23. Bernhardt, C.; Lee, M.M.; Gonzalez, A.; Zhang, F.; Lloyd, A.; Schiefelbein, J. Faculty Opinions recommendation of the bHLH genes GLABRA3 (GL3) and ENHANCER OF GLABRA3 (EGL3) specify epidermal cell fate in the Arabidopsis root. *Development* **2003**, *130*, 6431–6439. [CrossRef]
24. Morohashi, K.; Zhao, M.; Yang, M.; Read, B.; Lloyd, A.; Lamb, R.; Grotewold, E. Participation of the Arabidopsis bHLH Factor GL3 in Trichome Initiation Regulatory Events. *Plant Physiol.* **2007**, *145*, 736–746. [CrossRef] [PubMed]

25. Yamaguchi-Shinozaki, K.; Shinozaki, K. A novel cis-acting element in an Arabidopsis gene is involved in responsiveness to drought, low-temperature, or high-salt stress. *Plant Cell* **1994**, *6*, 251–264. [CrossRef] [PubMed]
26. Krishnamurthy, P.; Vishal, B.; Khoo, K.; Rajappa, S.; Loh, C.-S.; Kumar, P.P. Expression of AoNHX1 increases salt tolerance of rice and Arabidopsis, and bHLH transcription factors regulate AtNHX1 and AtNHX6 in Arabidopsis. *Plant Cell Rep.* **2019**, *38*, 1299–1315. [CrossRef] [PubMed]
27. Selote, D.; Samira, R.; Matthiadis, A.; Gillikin, J.W.; Long, T.A. Iron-binding e3 ligase mediates iron response in plants by targeting basic helix-loop-helix transcription factors. *Plant Physiol.* **2015**, *167*, 273–286. [CrossRef]
28. Zhao, C.; Wang, P.; Si, T.; Hsu, C.-C.; Wang, L.; Zayed, O.; Yu, Z.; Zhu, Y.; Dong, J.; Tao, W.A.; et al. MAP Kinase Cascades Regulate the Cold Response by Modulating ICE1 Protein Stability. *Dev. Cell* **2017**, *43*, 618–629.e5. [CrossRef]
29. Miura, K.; Jin, J.B.; Lee, J.; Yoo, C.Y.; Stirm, V.; Miura, T.; Ashworth, E.N.; Bressan, R.A.; Yun, D.J.; Hasegawa, P.M. SIZ1-mediated sumoylation of ICE1 controls CBF3/DREB1A ex-pression and freezing tolerance in Arabidopsis. *Plant Cell.* **2007**, *19*, 1403–1414. [CrossRef]
30. Ferre-D'Amare, A.R.; Prendergast, G.C.; Ziff, E.B.; Burley, S.K. Recognition by Max of its cognate DNA through a dimeric b/HLH/Z domain. *Nature* **1993**, *363*, 38–45. [CrossRef]
31. Toledo-Ortiz, G.; Huq, E.; Quail, P.H. The Arabidopsis Basic/Helix-Loop-Helix Transcription Factor Family. *Plant Cell* **2003**, *15*, 1749–1770. [CrossRef] [PubMed]
32. Stevens, J.D.; Roalson, E.; Skinner, M.K. Phylogenetic and expression analysis of the basic helix-loop-helix transcription factor gene family: Genomic approach to cellular differentiation. *Differentiation* **2008**, *76*, 1006–1042. [CrossRef]
33. Jones, S. An overview of the basic helix-loop-helix proteins. *Genome Biol.* **2004**, *5*, 226. [CrossRef]
34. Yin, Y.; Vafeados, D.; Tao, Y.; Yoshida, S.; Asami, T.; Chory, J. A new class of transcription factors mediates brassinosteroid-regulated gene expression in Arabidopsis. *Cell* **2005**, *120*, 249–259. [CrossRef]
35. Khanna, R.; Huq, E.; Kikis, E.A.; Al-Sady, B.; Lanzatella, C.; Quail, P.H. A Novel Molecular Recognition Motif Necessary for Targeting Photoactivated Phytochrome Signaling to Specific Basic Helix-Loop-Helix Transcription Factors. *Plant Cell* **2004**, *16*, 3033–3044. [CrossRef]
36. Verma, S.; Nizam, S.; Verma, P.K. Biotic and Abiotic Stress Signaling in Plants. *Stress Signal. Plants Genom. Proteom. Perspect.* **2013**, *1*, 25–49. [CrossRef]
37. Suzuki, N.; Rivero, R.M.; Shulaev, V.; Blumwald, E.; Mittler, R. Abiotic and biotic stress combinations. *New Phytol.* **2014**, *203*, 32–43. [CrossRef]
38. Wu, H.; Ye, H.; Yao, R.; Zhang, T.; Xiong, L. OsJAZ9 acts as a transcriptional regulator in jasmonate signaling and modulates salt stress tolerance in rice. *Plant Sci.* **2015**, *232*, 136. [CrossRef]
39. Gratz, R.; Manishankar, P.; Ivanov, R.; Köster, P.; Mohr, I.; Trofimov, K.; Steinhorst, L.; Meiser, J.; Mai, H.-J.; Drerup, M.; et al. CIPK11-Dependent Phosphorylation Modulates FIT Activity to Promote Arabidopsis Iron Acquisition in Response to Calcium Signaling. *Dev. Cell* **2019**, *48*, 726–740.e10. [CrossRef]
40. Qiu, J.R.; Huang, Z.; Xiang, X.Y.; Xu, W.X.; Wang, J.T.; Chen, J.; Song, L.; Xiao, Y.; Li, X.; Ma, J.; et al. MfbHLH38, a Myrothamnus flabellifolia bHLH transcription factor, confers tolerance to drought and salinity stresses in Arabidopsis. *BMC Plant Biol.* **2020**, *20*, 542. [CrossRef] [PubMed]
41. Wang, S.; Li, L.; Ying, Y.; Wang, J.; Shao, J.F.; Yamaji, N.; Whelan, J.; Ma, J.F.; Shou, H. A transcription factor OsbHLH156 regulates Strategy II iron acquisition through localising IRO2 to the nucleus in rice. *New Phytol.* **2019**, *225*, 1247–1260. [CrossRef] [PubMed]
42. Li, C.; Yan, C.; Sun, Q.; Wang, J.; Yuan, C.; Mou, Y.; Shan, S.; Zhao, X. The bHLH transcription factor *AhbHLH112* improves the drought tolerance of peanut. *BMC Plant Biol.* **2021**, *21*, 540. [CrossRef] [PubMed]
43. Gu, X.; Gao, S.; Li, J.; Song, P.; Zhang, Q.; Guo, J.; Wang, X.; Han, X.; Wang, X.; Zhu, Y.; et al. The bHLH transcription factor regulated gene OsWIH2 is a positive regulator of drought tolerance in rice. *Plant Physiol. Biochem.* **2021**, *169*, 269–279. [CrossRef] [PubMed]
44. Yao, P.-F.; Li, C.-L.; Zhao, X.-R.; Li, M.-F.; Zhao, H.-X.; Guo, J.-Y.; Cai, Y.; Chen, H.; Wu, Q. Overexpression of a Tartary Buckwheat Gene, FtbHLH3, Enhances Drought/Oxidative Stress Tolerance in Transgenic Arabidopsis. *Front. Plant Sci.* **2017**, *8*, 625. [CrossRef]
45. Waseem, M.; Rong, X.; Li, Z. Dissecting the Role of a Basic Helix-Loop-Helix Transcription Factor, SlbHLH22, Under Salt and Drought Stresses in Transgenic *Solanum lycopersicum* L. *Front. Plant Sci.* **2019**, *10*, 734. [CrossRef]
46. Gao, Y.; Wu, M.; Zhang, M.; Jiang, W.; Ren, X.; Liang, E.; Zhang, D.; Zhang, C.; Xiao, N.; Li, Y.; et al. A maize phytochrome-interacting factors protein ZmPIF1 enhances drought tolerance by inducing stomatal closure and improves grain yield in *Oryza sativa*. *Plant Biotechnol. J.* **2018**, *16*, 1375–1387. [CrossRef] [PubMed]
47. Ren, Y.-R.; Yang, Y.-Y.; Zhao, Q.; Zhang, T.-E.; Wang, C.-K.; Hao, Y.-J.; You, C.-X. MdCIB1, an apple bHLH transcription factor, plays a positive regulator in response to drought stress. *Environ. Exp. Bot.* **2021**, *188*, 104523. [CrossRef]
48. Le Hir, R.; Castelain, M.; Chakraborti, D.; Moritz, T.; Dinant, S.; Bellini, C. At*bHLH68* transcription factor contributes to the regulation of ABA homeostasis and drought stress tolerance in *Arabidopsis thaliana*. *Physiol. Plant.* **2017**, *160*, 312–327. [CrossRef] [PubMed]
49. Dong, Y.; Wang, C.; Han, X.; Tang, S.; Liu, S.; Xia, X.; Yin, W. A novel bHLH transcription factor PebHLH35 from Populus euphratica confers drought tolerance through regulating stomatal development, photosynthesis and growth in *Arabidopsis*. *Biochem. Biophys. Res. Commun.* **2014**, *450*, 453–458. [CrossRef]

50. Liu, W.; Tai, H.; Li, S.; Gao, W.; Zhao, M.; Xie, C.; Li, W.X. bHLH122 is important for drought and osmotic stress resistance in Ara-bidopsis and in the repression of ABA catabolism. *New Phytol.* **2014**, *201*, 1192–1204. [CrossRef]
51. Seo, J.-S.; Joo, J.; Kim, M.-J.; Kim, Y.-K.; Nahm, B.H.; Song, S.I.; Cheong, J.-J.; Lee, J.S.; Kim, J.-K.; Choi, Y.D. OsbHLH148, a basic helix-loop-helix protein, interacts with OsJAZ proteins in a jasmonate signaling pathway leading to drought tolerance in rice. *Plant J.* **2011**, *65*, 907–921. [CrossRef]
52. Chen, Y.; Li, F.; Ma, Y.; Chong, K.; Xu, Y. Overexpression of OrbHLH001, a putative helix–loop–helix transcription factor, causes increased expression of AKT1 and maintains ionic balance under salt stress in rice. *J. Plant Physiol.* **2013**, *170*, 93–100. [CrossRef] [PubMed]
53. Teng, Y.; Lv, M.; Zhang, X.; Cai, M.; Chen, T. BEAR1, a bHLH Transcription Factor, Controls Salt Response Genes to Regulate Rice Salt Response. *J. Plant Biol.* **2022**, *65*, 217–230. [CrossRef]
54. Singh, A.P.; Pandey, B.K.; Mehra, P.; Heitz, T.; Giri, J. OsJAZ9 overexpression modulates jasmonic acid biosynthesis and potassium deficiency responses in rice. *Plant Mol. Biol.* **2020**, *104*, 397–410. [CrossRef] [PubMed]
55. Babitha, K.C.; Vemanna, R.S.; Nataraja, K.N.; Udayakumar, M. Overexpression of EcbHLH57 Transcription Factor from Eleusine coracana L. in Tobacco Confers Tolerance to Salt, Oxidative and Drought Stress. *PLoS ONE* **2015**, *10*, e0137098. [CrossRef]
56. Wang, Y.; Wang, S.; Tian, Y.; Wang, Q.; Chen, S.; Li, H.; Ma, C.; Li, H. Functional Characterization of a Sugar Beet *BvbHLH93* Transcription Factor in Salt Stress Tolerance. *Int. J. Mol. Sci.* **2021**, *22*, 3669. [CrossRef]
57. Ariyarathne, M.A.; Wone, B.W. Overexpression of the Selaginella lepidophylla bHLH transcription factor enhances water-use efficiency, growth, and development in Arabidopsis. *Plant Sci.* **2022**, *315*, 111129. [CrossRef]
58. Liu, Y.; Ji, X.; Nie, X.; Qu, M.; Zheng, L.; Tan, Z.; Zhao, H.; Huo, L.; Liu, S.; Zhang, B.; et al. Arabidopsis AtbHLH112 regulates the expression of genes involved in abiotic stress tolerance by binding to their E-box and GCG-box motifs. *New Phytol.* **2015**, *207*, 692–709. [CrossRef]
59. Chen, H.-C.; Cheng, W.-H.; Hong, C.-Y.; Chang, Y.-S.; Chang, M.-C. The transcription factor OsbHLH035 mediates seed germination and enables seedling recovery from salt stress through ABA-dependent and ABA-independent pathways, respectively. *Rice* **2018**, *11*, 50. [CrossRef]
60. Zhou, J.; Li, F.; Wang, J.-L.; Ma, Y.; Chong, K.; Xu, Y.-Y. Basic helix-loop-helix transcription factor from wild rice (OrbHLH2) improves tolerance to salt- and osmotic stress in *Arabidopsis*. *J. Plant Physiol.* **2009**, *166*, 1296–1306. [CrossRef]
61. Zhu, L.; Zhao, M.; Chen, M.; Li, L.; Jiang, Y.; Liu, S.; Jiang, Y.; Wang, K.; Wang, Y.; Sun, C.; et al. The bHLH gene family and its response to saline stress in Jilin ginseng, Panax ginseng C.A. Meyer. *Mol. Genet. Genom.* **2020**, *295*, 877–890. [CrossRef] [PubMed]
62. Chinnusamy, V.; Ohta, M.; Kanrar, S.; Lee, B.-H.; Hong, X.; Agarwal, M.; Zhu, J.-K. ICE1: A regulator of cold-induced transcriptome and freezing tolerance in *Arabidopsis*. *Genes Dev.* **2003**, *17*, 1043–1054. [CrossRef]
63. Fursova, O.V.; Pogorelko, G.V.; Tarasov, V.A. Identification of ICE2, a gene involved in cold acclimation which determines freezing tolerance in Arabidopsis thaliana. *Gene* **2009**, *429*, 98–103. [CrossRef] [PubMed]
64. Feng, X.-M.; Zhao, Q.; Zhao, L.-L.; Qiao, Y.; Xie, X.-B.; Li, H.-F.; Yao, Y.-X.; You, C.-X.; Hao, Y.-J. The cold-induced basic helix-loop-helix transcription factor gene MdCIbHLH1 encodes an ICE-like protein in apple. *BMC Plant Biol.* **2012**, *12*, 22. [CrossRef] [PubMed]
65. Huang, X.; Li, K.; Jin, C.; Zhang, S. ICE1 of Pyrus ussuriensis functions in cold tolerance by enhancing PuDREBa transcriptional levels through interacting with PuHHP1. *Sci. Rep.* **2015**, *5*, 17620. [CrossRef]
66. Wang, Y.-J.; Zhang, Z.-G.; He, X.-J.; Zhou, H.-L.; Wen, Y.-X.; Dai, J.-X.; Zhang, J.-S.; Chen, S.-Y. A rice transcription factor OsbHLH1 is involved in cold stress response. *Theor. Appl. Genet.* **2003**, *107*, 1402–1409. [CrossRef]
67. Zhao, Q.; Xiang, X.; Liu, D.; Yang, A.; Wang, Y. Tobacco transcription factor NtbHLH123 confers tolerance to cold stress by regulating the NtCBF pathway and reactive oxygen species homeostasis. *Front. Plant Sci.* **2018**, *9*, 381. [CrossRef]
68. Zuo, Z.F.; Sun, H.J.; Lee, H.Y.; Kang, H.G. Identification of bHLH genes through genome-wide association study and antisense expression of ZjbHLH076/ZjICE1 influence tolerance to low temperature and salinity in Zoysia japonica. *Plant Sci.* **2021**, *313*, 111088. [CrossRef]
69. Zuo, Z.-F.; Kang, H.-G.; Park, M.-Y.; Jeong, H.; Sun, H.-J.; Song, P.-S.; Lee, H.-Y. Zoysia japonica MYC type transcription factor ZjICE1 regulates cold tolerance in transgenic Arabidopsis. *Plant Sci.* **2019**, *289*, 110254. [CrossRef]
70. Xu, W.; Jiao, Y.; Li, R.; Zhang, N.; Xiao, D.; Ding, X.; Wang, Z. Chinese Wild-Growing Vitis amurensis ICE1 and ICE2 Encode MYC-Type bHLH Transcription Activators that Regulate Cold Tolerance in Arabidopsis. *PLoS ONE* **2014**, *9*, e102303. [CrossRef]
71. Yang, X.; Wang, R.; Hu, Q.; Li, S.; Mao, X.; Jing, H.; Zhao, J.; Hu, G.; Fu, J.; Liu, C. DlICE1, a stress-responsive gene from Dimocarpus longan, enhances cold tolerance in transgenic Arabidopsis. *Plant Physiol. Biochem.* **2019**, *142*, 490–499. [CrossRef] [PubMed]
72. Geng, J.; Liu, J.-H. The transcription factor CsbHLH18 of sweet orange functions in modulation of cold tolerance and homeostasis of reactive oxygen species by regulating the antioxidant gene. *J. Exp. Bot.* **2018**, *69*, 2677–2692. [CrossRef] [PubMed]
73. Huang, X.S.; Wang, W.; Zhang, Q.; Liu, J.H. A basic helix-loop-helix transcription factor, PtrbHLH, of Poncirus trifoliata confers cold tolerance and modulates peroxidase-mediated scavenging of Hydrogen Peroxide. *Plant Physiol.* **2013**, *162*, 1178–1194. [CrossRef]
74. Geng, J.; Wei, T.-L.; Wang, Y.; Huang, X.; Liu, J.-H. Overexpression of PtrbHLH, a basic helix-loop-helix transcription factor from Poncirus trifoliata, confers enhanced cold tolerance in pummelo (Citrus grandis) by modulation of H_2O_2 level via regulating a CAT gene. *Tree Physiol.* **2019**, *39*, 2045–2054. [CrossRef]

75. Luo, P.; Li, Z.; Chen, W.; Xing, W.; Yang, J.; Cui, Y. Overexpression of RmICE1, a bHLH transcription factor from Rosa multiflora, enhances cold tolerance via modulating ROS levels and activating the expression of stress-responsive genes. *Environ. Exp. Bot.* **2020**, *178*, 104160. [CrossRef]
76. Shen, T.; Wen, X.; Wen, Z.; Qiu, Z.; Hou, Q.; Li, Z.; Mei, L.; Yu, H.; Qiao, G. Genome-wide identification and expression analysis of bHLH transcription factor family in response to cold stress in sweet cherry (*Prunus avium* L.). *Sci. Hortic.* **2021**, *279*, 109905. [CrossRef]
77. Colangelo, E.P.; Guerinot, M.L. The essential basic helix-loop-helix protein FIT1 is required for the iron deficiency response. *Plant Cell* **2004**, *16*, 3400–3412. [CrossRef] [PubMed]
78. Yuan, Y.; Wu, H.; Wang, N.; Li, J.; Zhao, W.; Du, J.; Wang, D.; Ling, H.-Q. FIT interacts with AtbHLH38 and AtbHLH39 in regulating iron uptake gene expression for iron homeostasis in Arabidopsis. *Cell Res.* **2008**, *18*, 385–397. [CrossRef]
79. Sivitz, A.B.; Hermand, V.; Curie, C.; Vert, G. Arabidopsis bHLH100 and bHLH101 Control Iron Homeostasis via a FIT-Independent Pathway. *PLoS ONE* **2012**, *7*, e44843. [CrossRef]
80. Long, T.A.; Tsukagoshi, H.; Busch, W.; Lahner, B.; Salt, D.E.; Benfey, P.N. The bHLH Transcription Factor POPEYE Regulates Response to Iron Deficiency in *Arabidopsis* Roots. *Plant Cell* **2010**, *22*, 2219–2236. [CrossRef]
81. Samira, R.; Li, B.; Kliebenstein, D.; Li, C.; Davis, E.; Gillikin, J.W.; Long, T.A. The bHLH transcription factor ILR3 modulates multiple stress responses in Arabidopsis. *Plant Mol. Biol.* **2018**, *97*, 297–309. [CrossRef] [PubMed]
82. Li, X.; Zhang, H.; Ai, Q.; Liang, G.; Yu, D. Two bHLH Transcription Factors, bHLH34 and bHLH104, Regulate Iron Homeostasis in *Arabidopsis thaliana*. *Plant Physiol.* **2016**, *170*, 2478–2493. [CrossRef] [PubMed]
83. Cui, Y.; Chen, C.-L.; Cui, M.; Zhou, W.-J.; Wu, H.-L.; Ling, H.-Q. Four IVa bHLH Transcription Factors Are Novel Interactors of FIT and Mediate JA Inhibition of Iron Uptake in *Arabidopsis*. *Mol. Plant* **2018**, *11*, 1166–1183. [CrossRef] [PubMed]
84. Zhao, M.; Song, A.; Li, P.; Chen, S.; Jiang, J.; Chen, F. A bHLH transcription factor regulates iron intake under Fe deficiency in chrysanthemum. *Sci. Rep.* **2014**, *4*, 1–6. [CrossRef] [PubMed]
85. Wang, W.; Ye, J.; Ma, Y.; Wang, T.; Shou, H.; Zheng, L. OsIRO3 Plays an Essential Role in Iron Deficiency Responses and Regulates Iron Homeostasis in Rice. *Plants* **2020**, *9*, 1095. [CrossRef]
86. Ogo, Y.; Itai, R.N.; Nakanishi, H.; Kobayashi, T.; Takahashi, M.; Mori, S.; Nishizawa, N.K. The rice bHLH protein OsIRO2 is an essential regulator of the genes involved in Fe uptake under Fe-deficient conditions. *Plant J.* **2007**, *51*, 366–377. [CrossRef] [PubMed]
87. Liang, G.; Zhang, H.; Li, Y.; Pu, M.; Yang, Y.; Li, C.; Lu, C.; Xu, P.; Yu, D. *Oryza sativa* FER-LIKE FE DEFICIENCY-INDUCED TRANSCRIPTION FACTOR (OsFIT/OsbHLH156) interacts with OsIRO2 to regulate iron homeostasis. *J. Integr. Plant Biol.* **2020**, *62*, 668–689. [CrossRef]
88. Yang, T.; Hao, L.; Yao, S.; Zhao, Y.; Lu, W.; Xiao, K. TabHLH1, a bHLH-type transcription factor gene in wheat, improves plant tolerance to Pi and N deprivation via regulation of nutrient transporter gene transcription and ROS homeostasis. *Plant Physiol. Biochem.* **2016**, *104*, 99–113. [CrossRef]
89. Jia, M.; Munz, J.; Lee, J.; Shelley, N.; Xiong, Y.; Joo, S.; Jin, E.; Lee, J.H. The bHLH family NITROGEN-REPLETION INSENSITIVE1 represses nitrogen starvation-induced responses in Chlamydomonas reinhardtii. *Plant J.* **2022**, *110*, 337–357. Available online: https://onlinelibrary.wiley.com/doi/full/10.1111/tpj.15673. (accessed on 18 January 2022). [CrossRef]
90. Yi, K.; Wu, Z.; Zhou, J.; Du, L.; Guo, L.; Wu, Y.; Wu, P. OsPTF1, a novel transcription factor involved in tolerance to phosphate starvation in rice. *Plant Physiol.* **2005**, *138*, 2087–2096. [CrossRef]
91. Li, Y.Y.; Sui, X.Y.; Yang, J.S.; Xiang, X.H.; Li, Z.Q.; Wang, Y.Y.; Zhou, Z.C.; Hu, R.S.; Liu, D. A novel bHLH transcription factor, NtbHLH1, modulates iron homeostasis in tobacco (*Nicotiana tabacum* L.). *Biochem. Biophys. Res. Commun.* **2020**, *522*, 233–239. [CrossRef] [PubMed]
92. Lu, X.; Yang, L.; Yu, M.; Lai, J.; Wang, C.; McNeil, D.; Zhou, M.; Yang, C. A novel Zea mays ssp. mexicana L. MYC-type ICE-like transcription factor gene ZmICE1, enhances freezing tolerance in transgenic Arabidopsis thaliana. *Plant Physiol. Biochem.* **2017**, *113*, 78–88. [CrossRef] [PubMed]
93. Zhang, H.; Li, Y.; Pu, M.; Xu, P.; Liang, G.; Yu, D. *Oryza sativa* POSITIVE REGULATOR OF IRON DEFICIENCY RESPONSE 2 (OsPRI2) and OsPRI3 are involved in the maintenance of Fe homeostasis. *Plant Cell Environ.* **2020**, *43*, 261–274. [CrossRef] [PubMed]
94. Miura, K.; Shiba, H.; Ohta, M.; Kang, S.W.; Sato, A.; Yuasa, T.; Iwaya-Inoue, M.; Kamada, H.; Ezura, H. SlICE1 encoding a MYC-type transcription factor controls cold tolerance in tomato, *Solanum lycopersicum*. *Plant Biotechnol.* **2012**, *29*, 253–260. [CrossRef]
95. Li, J.; Wang, T.; Han, J.; Ren, Z. Genome-wide identification and characterization of cucumber bHLH family genes and the functional characterization of CsbHLH041 in NaCl and ABA tolerance in Arabidopsis and cucumber. *BMC Plant Biol.* **2020**, *20*, 272.
96. Bai, G.; Yang, D.H.; Chao, P.; Yao, H.; Fei, M.; Zhang, Y.; Chen, X.; Xiao, B.; Li, F.; Wang, Z.Y.; et al. Genome-wide identification and expression analysis of NtbHLH gene family in tobacco (*Nicotiana tabacum* L.) and the role of NtbHLH86 in drought adaptation. *Plant Divers.* **2021**, *43*, 510–522. [CrossRef]
97. Xie, X.B.; Li, S.; Zhang, R.F.; Zhao, J.; Chen, Y.C.; Zhao, Q.; Yao, Y.X.; You, C.X.; Zhang, X.S.; Hao, Y.J. The bHLH transcription factor MdbHLH3 promotes anthocyanin accumulation and fruit colouration in response to low temperature in apples. *Plant Cell Environ.* **2012**, *35*, 1884–1897. [CrossRef]

98. Zou, Q.; Xu, H.; Yang, G.; Yu, L.; Jiang, H.; Mao, Z.; Hu, J.; Zhang, Z.; Wang, N.; Chen, X. MdbHLH106-like transcription factor enhances apple salt tolerance by upregulating MdNHX1 expression. *Plant Cell Tissue Organ Cult.* **2021**, *145*, 333–345. [CrossRef]
99. Zhao, Q.; Fan, Z.; Qiu, L.; Che, Q.; Wang, T.; Li, Y.; Wang, Y. MdbHLH130, an Apple bHLH Transcription Factor, Confers Water Stress Resistance by Regulating Stomatal Closure and ROS Homeostasis in Transgenic Tobacco. *Front. Plant Sci.* **2020**, *11*, 543696. [CrossRef]
100. Jin, C.; Huang, X.-S.; Li, K.-Q.; Yin, H.; Li, L.-T.; Yao, Z.-H.; Zhang, S.-L. Overexpression of a bHLH1 Transcription Factor of Pyrus ussuriensis Confers Enhanced Cold Tolerance and Increases Expression of Stress-Responsive Genes. *Front. Plant Sci.* **2016**, *7*, 441. [CrossRef]
101. Jiang, Y.; Yang, B.; Deyholos, M.K. Functional characterization of the *Arabidopsis* bHLH92 transcription factor in abiotic stress. *Mol. Genet. Genom.* **2009**, *282*, 503–516. [CrossRef]
102. Babitha, K.C.; Ramu, S.V.; Pruthvi, V.; Mahesh, P.; Nataraja, K.N.; Udayakumar, M. Co-expression of *AtbHLH17* and *AtWRKY28* confers resistance to abiotic stress in *Arabidopsis*. *Transgenic Res.* **2013**, *22*, 327–341. [CrossRef] [PubMed]
103. Yao, P.; Sun, Z.; Li, C.; Zhao, X.; Li, M.; Deng, R.; Huang, Y.; Zhao, H.; Chen, H.; Wu, Q. Overexpression of Fagopyrum tataricum FtbHLH2 enhances tolerance to cold stress in transgenic *Arabidopsis*. *Plant Physiol. Biochem.* **2018**, *125*, 85–94. [CrossRef]
104. Wang, R.; Zhao, P.; Kong, N.; Lu, R.; Pei, Y.; Huang, C.; Ma, H.; Chen, Q. Genome-Wide Identification and Characterization of the Potato bHLH Transcription Factor Family. *Genes* **2018**, *9*, 54. [CrossRef] [PubMed]
105. Lin, Y.; Zheng, H.; Zhang, Q.; Liu, C.; Zhang, Z. Functional profiling of EcaICE1 transcription factor gene from Eucalyptus camaldulensis involved in cold response in tobacco plants. *J. Plant Biochem. Biotechnol.* **2014**, *23*, 141–150. [CrossRef]
106. Zhai, Y.; Zhang, L.; Xia, C.; Fu, S.; Zhao, G.; Jia, J.; Kong, X. The wheat transcription factor, TabHLH39, improves tolerance to multiple abiotic stressors in transgenic plants. *Biochem. Biophys. Res. Commun.* **2016**, *473*, 1321–1327. [CrossRef]
107. Kim, J.; Kim, H.-Y. Functional analysis of a calcium-binding transcription factor involved in plant salt stress signaling. *FEBS Lett.* **2006**, *580*, 5251–5256. [CrossRef]
108. Li, L.; Gao, W.; Peng, Q.; Zhou, B.; Kong, Q.; Ying, Y.; Shou, H. Two soybean bHLH factors regulate response to iron deficiency. *J. Integr. Plant Biol.* **2018**, *60*, 608–622. [CrossRef]
109. Gao, Y.; Jiang, W.; Dai, Y.; Xiaoyi, T.; Zhang, C.; Li, H.; Lu, Y.; Wu, M.; Tao, X.; Deng, D.; et al. A maize phytochrome-interacting factor 3 improves drought and salt stress tolerance in rice. *Plant Mol. Biol.* **2015**, *87*, 413–428. [CrossRef] [PubMed]
110. Wang, L.; Ying, Y.; Narsai, R.; Ye, L.; Zheng, L.; Tian, J.; Whelan, J.; Shou, H. Identification of OsbHLH133 as a regulator of iron distribution between roots and shoots in *Oryza sativa*. *Plant Cell Environ.* **2013**, *36*, 224–236. [CrossRef]
111. Zhang, H.; Li, Y.; Yao, X.; Liang, G.; Yu, D. Positive Regulator of Iron Homeostasis1, OsPRI1, Facilitates Iron Homeostasis. *Plant Physiol.* **2017**, *175*, 543–554. [CrossRef]
112. Lei, R.; Li, Y.; Cai, Y.; Li, C.; Pu, M.; Lu, C.; Yang, Y.; Liang, G. bHLH121 Functions as a Direct Link that Facilitates the Activation of FIT by bHLH IVc Transcription Factors for Maintaining Fe Homeostasis in *Arabidopsis*. *Mol. Plant* **2020**, *13*, 634–649. [CrossRef] [PubMed]
113. Ahmad, A.; Niwa, Y.; Goto, S.; Ogawa, T.; Shimizu, M.; Suzuki, A.; Kobayashi, K.; Kobayashi, H. bHLH106 Integrates Functions of Multiple Genes through Their G-Box to Confer Salt Tolerance on *Arabidopsis*. *PLoS ONE* **2015**, *10*, e0126872. [CrossRef] [PubMed]
114. Tanabe, N.; Noshi, M.; Mori, D.; Nozawa, K.; Tamoi, M.; Shigeoka, S. The basic helix-loop-helix transcription factor, bHLH11 functions in the iron-uptake system in *Arabidopsis thaliana*. *J. Plant Res.* **2019**, *132*, 93–105. [CrossRef]
115. Kiribuchi, K.; Jikumaru, Y.; Kaku, H.; Minami, E.; Hasegawa, M.; Kodama, O.; Seto, H.; Okada, K.; Nojiri, H.; Yamane, H. Involvement of the basic helix-loop-helix tran-scription factor RERJ1 in wounding and drought stress responses in rice plants. *Biosci. Biotechnol. Biochem.* **2005**, *69*, 1042–1044. [CrossRef] [PubMed]
116. Chen, H.C.; Hsieh-Feng, V.; Liao, P.C.; Cheng, W.H.; Liu, L.Y.; Yang, Y.W.; Lai, M.H.; Chang, M.C. The function of OsbHLH068 is partially redundant with its homolog, AtbHLH112, in the regulation of the salt stress response but has opposite functions to control flowering in Arabidopsis. *Plant Mol. Biol.* **2017**, *94*, 531–548. [CrossRef] [PubMed]
117. Wang, F.; Zhu, H.; Chen, D.; Li, Z.; Peng, R.; Yao, Q. A grape bHLH transcription factor gene, VvbHLH1, increases the accumulation of flavonoids and enhances salt and drought tolerance in transgenic Arabidopsis thaliana. *Plant Cell Tissue Organ Cult.* **2016**, *125*, 387–398. [CrossRef]
118. Farooq, M.; Wahid, A.; Kobayashi, N.; Fujita, D.; Basra, S.M.A. Plant drought stress: Effects, mechanisms and management. *Agron. Sustain. Dev.* **2011**, *29*, 185–212. [CrossRef]
119. Cutler, S.R.; Rodriguez, P.L.; Finkelstein, R.R.; Abrams, S.R. Abscisic acid: Emergence of a core signaling network. *Annu. Rev. Plant Biol.* **2010**, *61*, 651–679. [CrossRef]
120. Nakashima, K.; Yamaguchi-Shinozaki, K.; Shinozaki, K. The transcriptional regulatory network in the drought response and its crosstalk in abiotic stress responses including drought, cold, and heat. *Front. Plant Sci.* **2014**, *5*, 170. [CrossRef]
121. Takahashi, F.; Kuromori, T.; Sato, H.; Shinozaki, K. Regulatory Gene Networks in Drought Stress Responses and Resistance in Plants. In *Survival Strategies in Extreme Cold and Desiccation*; Springer: Berlin/Heidelberg, Germany, 2018; Volume 1081, pp. 189–214. [CrossRef]
122. Yamaguchi-Shinozaki, K.; Shinozaki, K. The plant hormone abscisic acid mediates the drought-induced expression but not the seed-specific expression of rd22, a gene responsive to dehydration stress in *Arabidopsis thaliana*. *MGG Mol. Gen. Genet.* **1993**, *238*, 17–25. [CrossRef] [PubMed]

123. Umezawa, T.; Okamoto, M.; Kushiro, T.; Nambara, E.; Oono, Y.; Seki, M.; Kobayashi, M.; Koshiba, T.; Kamiya, Y.; Shinozaki, K. CYP707A3, a major ABA 8′-hydroxylase involved in dehydration and rehydration response in Arabidopsis thaliana. *Plant J.* **2006**, *46*, 171–182. [CrossRef] [PubMed]
124. Kim, H.; Seomun, S.; Yoon, Y.; Jang, G. Jasmonic Acid in Plant Abiotic Stress Tolerance and Interaction with Abscisic Acid. *Agronomy* **2021**, *11*, 1886. [CrossRef]
125. Funck, D.; Baumgarten, L.; Stift, M.; von Wirén, N.; Schönemann, L. Differential Contribution of P5CS Isoforms to Stress Tolerance in *Arabidopsis*. *Front. Plant Sci.* **2020**, *11*, 565134. [CrossRef]
126. Verma, D.; Jalmi, S.; Bhagat, P.K.; Verma, N.; Sinha, A.K. A bHLH transcription factor, MYC2, imparts salt intolerance by regulating proline biosynthesis in *Arabidopsis*. *FEBS J.* **2020**, *287*, 2560–2576. [CrossRef]
127. Pearce, R.S. Plant Freezing and Damage. *Ann. Bot.* **2001**, *87*, 417–424. [CrossRef]
128. Stockinger, E.J.; Gilmour, S.J.; Thomashow, M.F. Arabidopsis thaliana CBF1 encodes an AP2 domain-containing transcriptional activator that binds to the C-repeat/DRE, a cis-acting DNA regulatory element that stimulates transcription in response to low temperature and water deficit. *Proc. Natl. Acad. Sci. USA* **1997**, *94*, 1035–1040. [CrossRef]
129. Gilmour, S.J.; Fowler, S.G.; Thomashow, M.F. Arabidopsis transcriptional activators CBF1, CBF2, and CBF3 have matching functional activities. *Plant Mol. Biol.* **2004**, *54*, 767–781. [CrossRef]
130. Gilmour, S.J.; Zarka, D.G.; Stockinger, E.J.; Salazar, M.P.; Houghton, J.M.; Thomashow, M.F. Low temperature regulation of the Arabidopsis CBF family of AP2 transcriptional activators as an early step in cold-induced COR gene expression. *Plant J.* **1998**, *16*, 433–442. [CrossRef]
131. Chen, C.-C.; Liang, C.-S.; Kao, A.-L.; Yang, C.-C. HHP1, a novel signalling component in the cross-talk between the cold and osmotic signalling pathways in Arabidopsis. *J. Exp. Bot.* **2010**, *61*, 3305–3320. [CrossRef]
132. Lee, H.; Xiong, L.; Gong, Z.; Ishitani, M.; Stevenson, B.; Zhu, J.-K. The *Arabidopsis HOS1* gene negatively regulates cold signal transduction and encodes a RING finger protein that displays cold-regulated nucleo–cytoplasmic partitioning. *Genes Dev.* **2001**, *15*, 912–924. [CrossRef]
133. Dong, C.-H.; Agarwal, M.; Zhang, Y.; Xie, Q.; Zhu, J.-K. The negative regulator of plant cold responses, HOS1, is a RING E3 ligase that mediates the ubiquitination and degradation of ICE1. *Proc. Natl. Acad. Sci. USA* **2006**, *103*, 8281–8286. [CrossRef] [PubMed]
134. Ding, Y.; Li, H.; Zhang, X.; Xie, Q.; Gong, Z.; Yang, S. OST1 kinase modulates freezing tolerance by enhancing ICE1 stability in arabidopsis. *Dev. Cell* **2015**, *32*, 278–289. [CrossRef] [PubMed]
135. Teige, M.; Scheikl, E.; Eulgem, T.; Doczi, R.; Ichimura, K.; Shinozaki, K.; Dangl, J.L.; Hirt, H. The MKK2 pathway mediates cold and salt stress signaling in Arabidopsis. *Mol. Cell* **2004**, *15*, 141–152. [CrossRef] [PubMed]
136. Li, H.; Ding, Y.; Shi, Y.; Zhang, X.; Zhang, S.; Gong, Z.; Yang, S. MPK3- and MPK6-Mediated ICE1 Phosphorylation Negatively Regulates ICE1 Stability and Freezing Tolerance in Arabidopsis. *Dev. Cell* **2017**, *43*, 630–642.e4. [CrossRef]
137. Zhang, Z.; Li, J.; Li, F.; Liu, H.; Yang, W.; Chong, K.; Xu, Y. OsMAPK3 Phosphorylates OsbHLH002/OsICE1 and Inhibits Its Ubiquitination to Activate OsTPP1 and Enhances Rice Chilling Tolerance. *Dev. Cell* **2017**, *43*, 731–743.e5. [CrossRef]
138. Santi, S.; Schmidt, W. Dissecting iron deficiency-induced proton extrusion in Arabidopsis roots. *New Phytol.* **2009**, *183*, 1072–1084. [CrossRef]
139. Robinson, N.J.; Procter, C.M.; Connolly, E.L.; Guerinot, M.L. A ferric-chelate reductase for iron uptake from soils. *Nature* **1999**, *397*, 694–697. [CrossRef]
140. Eide, D.; Broderius, M.; Fett, J.; Guerinot, M.L. A novel iron-regulated metal transporter from plants identified by functional expression in yeast. *Proc. Natl. Acad. Sci. USA* **1996**, *93*, 5624–5628. [CrossRef]
141. Wang, H.-Y.; Klatte, M.; Jakoby, M.; Bäumlein, H.; Weisshaar, B.; Bauer, P. Iron deficiency-mediated stress regulation of four subgroup Ib BHLH genes in Arabidopsis thaliana. *Planta* **2007**, *226*, 897–908. [CrossRef]
142. Wang, N.; Cui, Y.; Liu, Y.; Fan, H.; Du, J.; Huang, Z.; Yuan, Y.; Wu, H.; Ling, H.-Q. Requirement and Functional Redundancy of Ib Subgroup bHLH Proteins for Iron Deficiency Responses and Uptake in Arabidopsis thaliana. *Mol. Plant* **2013**, *6*, 503–513. [CrossRef] [PubMed]
143. Zhang, J.; Liu, B.; Li, M.; Feng, D.; Jin, H.; Wang, P.; Liu, J.; Xiong, F.; Wang, J.; Wang, H.-B. The bHLH Transcription Factor bHLH104 Interacts with IAA-LEUCINE RESISTANT3 and Modulates Iron Homeostasis in Arabidopsis. *Plant Cell* **2015**, *27*, 787–805. [CrossRef] [PubMed]
144. Higuchi, K.; Suzuki, K.; Nakanishi, H.; Yamaguchi, H.; Nishizawa, N.-K.; Mori, S. Cloning of Nicotianamine Synthase Genes, Novel Genes Involved in the Biosynthesis of Phytosiderophores. *Plant Physiol.* **1999**, *119*, 471–480. [CrossRef] [PubMed]
145. Kobayashi, T.; Ogo, Y.; Itai, R.N.; Nakanishi, H.; Takahashi, M.; Mori, S.; Nishizawa, N.K. The transcription factor IDEF1 regulates the response to and tolerance of iron deficiency in plants. *Proc. Natl. Acad. Sci. USA* **2007**, *104*, 19150–19155. [CrossRef]
146. Kobayashi, T.; Ogo, Y.; Aung, M.S.; Nozoye, T.; Itai, R.N.; Nakanishi, H.; Yamakawa, T.; Nishizawa, N.K. The spatial expression and regulation of transcription factors IDEF1 and IDEF2. *Ann. Bot.* **2010**, *105*, 1109–1117. [CrossRef]
147. Lindermayr, C.; Durner, J. S-Nitrosylation in plants: Pattern and function. *J. Proteom.* **2009**, *73*, 1–9. [CrossRef] [PubMed]
148. Barberon, M.; Zelazny, E.; Robert, S.; Conéjéro, G.; Curie, C.; Friml, J.; Vert, G. Monoubiquitin-dependent endocytosis of the Iron-Regulated Transporter 1 (IRT1) transporter controls iron uptake in plants. *Proc. Natl. Acad. Sci. USA* **2011**, *108*, E450–E458. [CrossRef]
149. Shin, L.J.; Lo, J.C.; Chen, G.H.; Callis, J.; Fu, H.; Yeh, K.C. IRT1 degradation factor1, a ring E3 Ubiquitin ligase, regulates the degradation of iron-regulated transporter1 in Arabidopsis. *Plant Cell* **2013**, *25*, 3039–3051. [CrossRef]

150. Matthiadis, A.; Long, T.A. Further insight into BRUTUS domain composition and functionality. *Plant Signal. Behav.* **2016**, *11*, e1204508. [CrossRef]
151. Qian, Y.C.; Zhang, T.Y.; Yu, Y.; Gou, L.P.; Yang, J.T.; Xu, J.; Pi, E.X. Regulatory Mechanisms of bHLH Transcription Factors in Plant Adaptive Responses to Various Abiotic Stresses. *Front. Plant Sci.* **2021**, *12*, 677611. [CrossRef]

plants

MDPI

Article

During Water Stress, Fertility Modulated by ROS Scavengers Abundant in Arabidopsis Pistils

Ya-Ying Wang [1], Donald J. Head [1] and Bernard A. Hauser [1,2,*]

[1] Department of Biology, University of Florida, Gainesville, FL 32611, USA; yaying.wang@gmail.com (Y.-Y.W.); djhead@me.com (D.J.H.)

[2] Plant Molecular and Cellular Biology Program, University of Florida, Gainesville, FL 32611, USA

* Correspondence: bahauser@ufl.edu; Tel.: +1-352-392-0009

Abstract: Hours after watering plants with 75 mM NaCl, the water potential of reproductive structures precipitously decreases. In flowers with mature gametes, this change in water potential did not alter the rate of fertilization but caused 37% of the fertilized ovules to abort. We hypothesize that the accumulation of reactive oxygen species (ROS) in ovules is an early physiological manifestation associated with seed failure. In this study, we characterize ROS scavengers that were differentially expressed in stressed ovules to determine whether any of these genes regulate ROS accumulation and/or associate with seed failure. Mutants in an iron-dependent superoxide dismutase (FSD2), ascorbate peroxidase (APX4), and three peroxidases (PER17, PER28, and PER29) were evaluated for changes in fertility. Fertility was unchanged in *apx4* mutants, but the other mutants grown under normal conditions averaged a 140% increase in seed failure. In pistils, *PER17* expression increases three-fold after stress, while the other genes decreased two-fold or more following stress; this change in expression accounts for differences in fertility between healthy and stressed conditions for different genotypes. In pistils, H_2O_2 levels rose in *per* mutants, but only in the triple mutant was there a significant increase, indicating that other ROS or their scavengers be involved in seed failure.

Keywords: ovules; plant stress; fertility; seed formation; reactive oxygen

Citation: Wang, Y.-Y.; Head, D.J.; Hauser, B.A. During Water Stress, Fertility Modulated by ROS Scavengers Abundant in Arabidopsis Pistils. *Plants* **2023**, *12*, 2182. https://doi.org/10.3390/plants12112182

Academic Editors: Małgorzata Nykiel, Mateusz Labudda, Beata Prabucka, Marta Gietler and Justyna Fidler

Received: 6 April 2023
Revised: 24 May 2023
Accepted: 26 May 2023
Published: 31 May 2023

1. Introduction

Reactive oxygen species (ROS), such as superoxide (O_2^-), hydroxyl radicals (•OH), and hydrogen peroxide (H_2O_2), are a group of highly active small molecules that can be naturally produced during cellular metabolism. In plants, ROS are mainly produced in chloroplasts, mitochondria, and peroxisomes during photosynthesis, respiration, or other metabolic processes [1,2]. Low to moderate amounts of ROS signal different responses and regulate many biological processes, e.g., development and defense responses [3,4]. Excessive ROS, however, change the redox homeostasis and leads to oxidative stress [5–7]. A classic example of redox signaling is the activation of lumen enzymes after illumination. Recently, mutants affecting thioredoxin and glutathione reduction changed auxin transport, which interfered with floral initiation, showed vasculature defects, and altered root morphology [8]. Furthermore, the specific isoforms that interact to form an active auxin receptor show differential responses to osmotic and salt stress and can lead to different ROS levels [9]. Among various ROS in plants, H_2O_2 is generally considered to be the main molecule that serves as a long-range signaling molecule because it is relatively stable and can diffuse relatively rapidly crossing membranes. ROS can damage proteins, break down fatty acids, damage DNA, and induce programmed cell death (PCD) [6,10,11]. Understanding how plants regulate ROS levels can permit hypotheses of mechanisms to prevent excessive ROS production.

In plant cells, there are several ROS-scavenging enzymes to maintain redox homeostasis. Superoxide dismutases (SOD) convert superoxide to H_2O_2 and oxygen [12]. Catalases

(CAT) break down H_2O_2 into water and oxygen [13]. Glutathione peroxidases (GPX) reduce H_2O_2 by oxidizing glutathione (GSH), and glutathione reductases (GR) regenerate GSH utilizing NAD(P)H [14]. Ascorbate peroxidases (APX), which are members of class I peroxidases, reduce H_2O_2 by oxidizing ascorbate [15]. Each of these ROS scavengers is found in different cell compartments and locations [1,5,16]. The class III heme-containing peroxidases (PER or PRX) are unique to plants and make up a large gene family. In Arabidopsis, there are 73 of these annotated peroxidases [17,18]. Many of these peroxidases are predicted to localize to the cell wall or vacuole and affect lignification, defense responses, hormone metabolism, salt tolerance, and development [19,20]. Horseradish peroxidase (HRP) is the best-known class III peroxidase. The radical products generated after this catalytic process trigger pathogen defense responses [21]. The functions of most of these class III peroxidases, however, remain unknown.

ROS are recognized as key modulators of PCD [22,23]. Many environmental stresses, such as heat shock, chilling, salt, high light, and pathogen attack, cause ROS accumulation [7,13,24–27]. ROS accumulation was observed in salt-stressed ovules and was hypothesized to cause high abortion rates in Arabidopsis [26,28]; fertility is reduced because seeds arise from ovules.

Research into stress effects on plant reproduction usually attributes decreases in seed set to pollen defects or pollen tube growth, but few studies evaluated the effects of stress on ovules or embryos. In natural populations, pollen limitation reduces fertility for 60% of plant species. Pollen becomes limiting when plant stress induces pollen abortion or the development of low-quality pollen where the pollen tube fails to transmit the sperm nuclei to the ovule. Pollen is particularly sensitive to stress, especially in monocots. In this work, plants were stressed after male and female gametophyte development was complete, and the effect on subsequent reproduction was evaluated. Pollination and fertilization rates were examined to determine the effect that the male gametophyte had on fertility.

Previous work showed that the expression of a group of ROS-scavenging genes was significantly altered in Arabidopsis ovules following salt stress [28]. Here we analyze ROS-scavenging mutants that were differentially expressed in stressed ovules at a critical stage of ovule development. We hypothesize that these ROS-scavenging genes may regulate the ROS levels during ovule abortion. In this study, we characterized the role of three *PER* genes in regulating ROS levels in Arabidopsis ovules, and we tested whether mutants of these loci affect ROS accumulation in ovules or seed failure.

2. Results

2.1. Salt Stress Lowers Floral Water Potential

After male and female gametophyte development was complete, plants were stressed to determine the downstream effects on reproduction. Since flowers and gametophytes develop synchronously, the gametophyte stage can be inferred from the floral phenotype [29]. Flowering plants were watered once with 75 mM NaCl, the solution was absorbed until the soil reached 100% soil moisture content, and the excess salt solution was discarded. To determine how this osmotic stress affected flowers, inflorescence water potential was measured. Within 6 h, the water potential in flowers dropped significantly and fell to maximal levels within 3 days (Figure 1). Plants were watered every other day, so the water potential data oscillated as the soil dried and rehydrated (Figure 1). These measurements showed that watering with salt increases the number of solutes in the soil and lowers inflorescence water potential for days. Water potential could drop as a result of increased ion transport and accumulation in flowers or a drop in cell turgor in this region. Previous analyses showed that salt stress caused sodium ions to increase by 25% in flowers, but this increase was offset by a decrease in potassium ions [26], resulting in little change in flower osmotic potential. In flowers, salt stress primarily affected water potential and cell turgor.

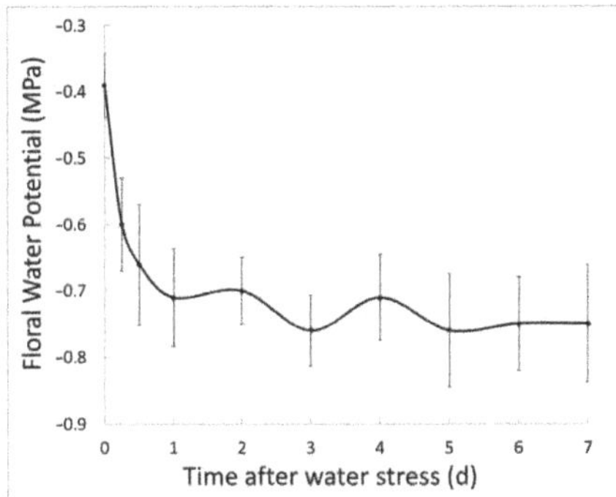

Figure 1. Using a pressure chamber, the water potential of inflorescences was measured periodically after plants were watered once with 75 mM NaCl. The initial point shows the water potential in healthy plants. Following stress, water potential decreases and plateaus after a day. Plants were watered on day 2, 4, and 6, so a slight rise in water potential was observed on those days. Five or more inflorescences were measured at each time point. The best-fit line through the average at each time is plotted. Average ±1 standard deviation shown.

2.2. Increased Ovule Abortion Rate in Peroxidase Mutants

Fitness was evaluated in five ROS-scavenging genes (*per17, 28, 29, apx4,* and *fsd2*). RT-PCR showed that no full-length transcripts were present in the RNA population of leaves and flowers (Figure 2). Previous work identified these genes as differential during ovule abortion [28]. Since the mutation lesion in each of these mutant lines occurs within or upstream of the peroxidase or SOD domain, these are all loss-of-function mutants (Figure 2). Previous research showed that stage 12 flowers are especially susceptible to environmental stress [26]. Therefore, stage 12 flowers were marked, and the fertility was measured after plants were either salt-stressed treated with 75 mM NaCl for 48 h or under healthy growth conditions. Under normal growth conditions, three class III peroxidase mutants (*per17, per28, per29*) and a superoxide dismutase mutant (*fsd2*) caused significant increases in the ovule abortion rates (Table 1). The *apx4* mutant did not significantly affect fertility. Except for *fsd2,* the stress treatment significantly altered the fertility of all genotypes. There was a highly significant interaction between the treatment and genotype when comparing the fertility of *fsd2* with wild-type plants so, according to Seltman [30], the p values for the main effects (genotype and treatment) are ignored and assumed to be statistically significant. The double and triple mutants of *PER* genes were created, and the ovule abortion rates of these mutants also rose significantly. Ovule abortion rates were also scored for Arabidopsis under 75 mM NaCl salt stress for 48 h. The ovule abortion rates increased significantly in *per17* mutant, *per17per28* and *per17per29* double mutants, and *per17per28per29* triple mutants (Table 1).

In order to compare the fertility and abortion rates in different experimental replicates, the fertility of each mutant was normalized with wild-type controls that were grown in the same trays as the experimental plants. For the wild-type controls, 95% of the ovules successfully set seed under healthy growth conditions; for the salt-stressed controls, 55% of the ovules set seed. When compared to reproduction rates in the respective single mutants, relative fertility decreased even more in salt-stressed *per17*, *per17per28* and *per17per29* double mutants, and the *per17per28per29* triple mutant (Table 1). Notably, the ovule abortion rates of these mutants significantly varied from those of wild-type fruits following salt stress (Table 1), indicating that their mutants responded more sensitively to salt stress. These results concur with the differential expression data reported by Sun et al. [28]: mutation of genes that were induced by environmental stress led to larger fertility effects when the plants were exposed to salt stress. Conversely, the fitness of mutants in genes that were most abundant in unstressed plants showed greater effects in healthy plants.

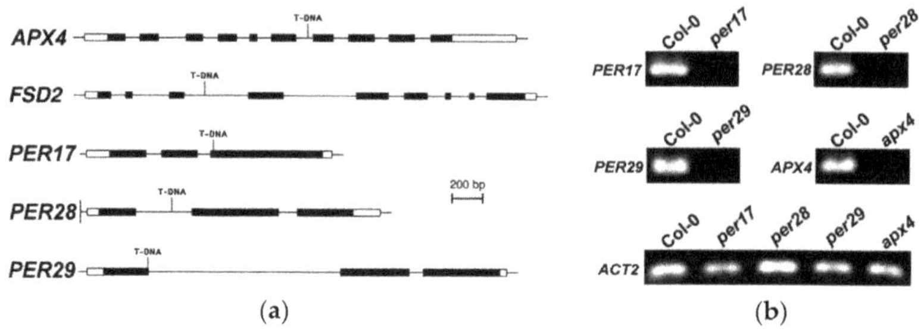

Figure 2. Analysis of ROS scavenger mutants. (**a**) Molecular models of *APX4, FSD2, PER17, PER28, and PER29,* including the T-DNA insertion sites, are shown. Genes are shown 5′ (left) to 3′ (right). Solid boxes denote exons, lines represent introns, and open boxes indicate untranslated regions. (**b**) RT-PCR analysis revealed that *per17, per28, per29,* and *apx4* homozygous mutant alleles contain no detectable full-length transcripts. *per17, per28, per29,* and *apx4* are null alleles. *ACTIN2* (*ACT2*) served as a positive control.

Table 1. Mutation of ROS-scavenging genes significantly reduced fertility. Prior to stress, stage 12 flowers were marked, and the subsequent ovule abortion rates for these fruits determined. Stressed plants were treated with 75 mM NaCl for 48 h. For each genotype, 30 pistils were scored. After arcsine transformation, ANOVA tests compared fertility between mutant genotypes and grouped controls: differences of less than 0.05 [a], 0.01 [b], 0.001 [c], and 0.0001 [e] indicated.

Genotype	Abortion Rate (%)	
	Healthy [1]	Stressed [1]
Wild type	12.9 ± 2.0	26.9 ± 4.6
per17	29.6 ± 5.8 [b]	44.6 ± 6.0 [a]
per28	23.2 ± 3.3 [b]	34.6 ± 4.4
fsd2	35.2 ± 6.8 [c]	29.0 ± 4.2
apx4	21.4 ± 4.2	23.2 ± 4.0
Wild type	18.0 ± 1.7	33.4 ± 3.3
per29	50.8 ± 3.5 [e]	41.6 ± 4.8
Wild type	14.9 ± 1.3	33.1 ± 2.7
per17 per28	39.7 ± 6.3 [c]	64.0 ± 6.6 [e]
Wild type	23.9 ± 3.1	49.5 ± 4.5
per17 per29	35.6 ± 6.0	67.7 ± 6.9 [b]
Wild type	21.6 ± 1.5	39.5 ± 3.6
per28 per29	33.7 ± 4.0 [b]	40.0 ± 2.9
Wild type	5.4 ± 0.7	25.4 ± 4.6
per17 per28 per29	31.5 ± 5.2 [e]	66.3 ± 5.6 [e]

[1] The ovule abortion rate is the average ± one standard error.

2.3. Fertilization Rates and Seed Failure

Images of aborting ovules are shown (Figure 3). After salt stress, Sun et al. [28] reported that ROS were first detected in the gametophyte. As ovule abortion progressed, ROS were found throughout the ovule [28]. To determine whether or not salt stress correlates with the formation of ROS in ovules, samples from stressed plants were stained with CH_2DCFDA, a ROS-sensitive dye. As opposed to traditional histochemical stains, nitroblue tetrazolium or cerium chloride that specifically measure superoxide or peroxide levels, CH_2DCFDA interacts with a wide variety of ROS molecules with different affinities so fluorescence of this dye yields a qualitative summation of ROS. In each *per* mutant genotype, ROS accumulation was evaluated in ovules 48 h after salt stress (Figure 4). We observed significantly higher levels of ROS accumulation in the ovules of *per17, 28,* and

29 single mutants than in wild-type controls (p <0.0001). Even greater ROS levels were detected in the double and triple mutants.

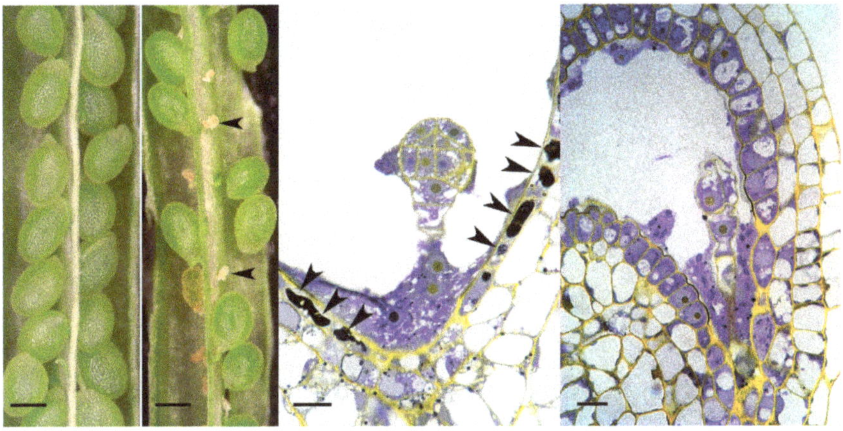

(a) (b) (c) (d)

Figure 3. Representative images of dissected pistils of (**a**) healthy and (**b**) salt-stressed plants. Two aborting/aborted ovules were indicated (arrowheads). Globular embryos from (**c**) stressed and (**d**) healthy ovules are shown. In the stressed ovule, most of the endothelium cells were necrotic and had degenerated (arrowheads). In the healthy control, endothelial cells were cytoplasmically dense, indicative of high metabolic activity associated with the movement of nutrients and metabolites from the maternal plant into the embryo sac. Size bars are 100 μm (**a**,**b**) and 10 μm (**c**,**d**).

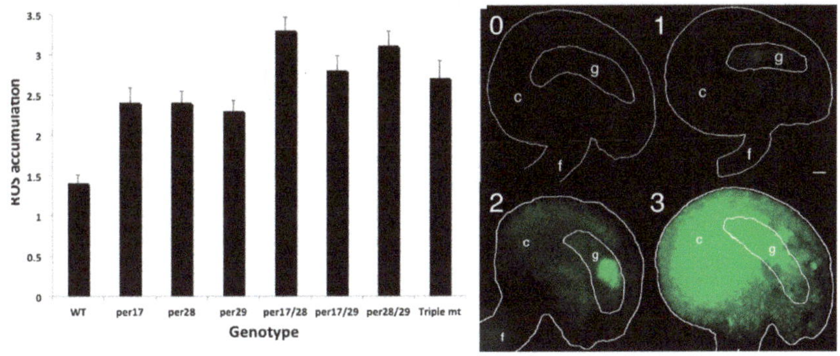

Figure 4. Following salt stress, peroxidase mutants accumulated ROS in ovules. Prior to stress, stage 12 flowers were marked. CH_2DCFDA fluorescence intensity in ovules was evaluated: level 0 had no detectable ROS; level 1 had detectable ROS in the embryo sac or gametophyte (g); level 2 had ROS accumulation in the chalaza (c); and level 3 had copious ROS accumulation throughout the ovule. All *per* mutants had significantly more ROS than controls ($p < 0.01$). ANOVA analyses were done using JMP8 (unequal variances and two-tailed distributions). Size bar is 10 μm.

2.4. H_2O_2 Levels in Pistils

Peroxidases neutralize H_2O_2 so the amount of this metabolite was measured in pistils. Pistils were used because this was the smallest segment of the plant that could be dissected from without producing wound-induced ROS. ROS levels increase in ovules (Figure 5), but low levels are detected in the carpel walls, stigma, and style. Measuring the level of H_2O_2 in *per* mutants will help determine whether or not the accumulation of ROS in ovules is a result of the production of H_2O_2 due to a reduction in peroxidase activity. A coupled enzyme assay using a fluorescent indicator was used to detect H_2O_2 levels in pistils. Results show H_2O_2 accumulated in the pistils of peroxidase triple mutants was significantly higher than in those of wild-type controls in both unstressed and salt-stressed conditions (Figure 5a). While H_2O_2 levels in *per* single and double mutants increased, they were not significantly higher than controls. Plant tissues contain many peroxidase isoforms, so the removal of one is likely to have an incremental effect.

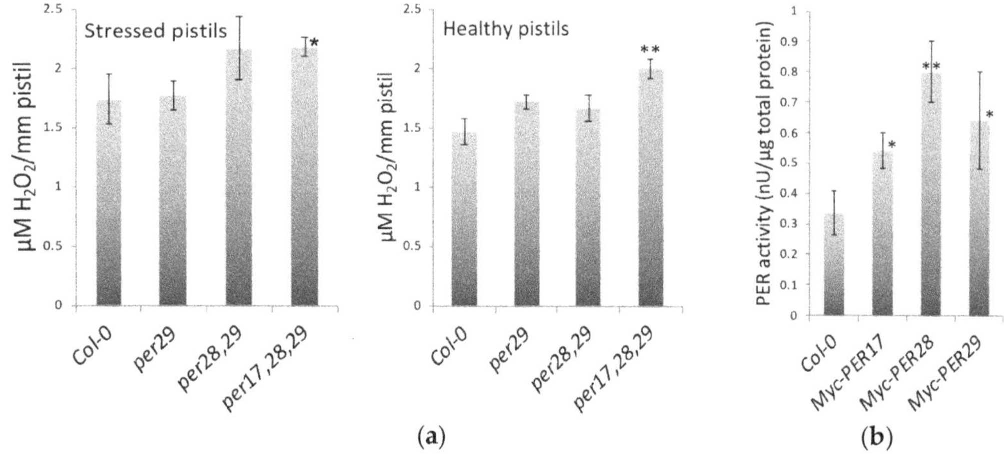

Figure 5. Peroxidase activity and H_2O_2 accumulation in *per* mutants. (**a**) The amount of hydrogen peroxide was compared between wild-type pistils and peroxidase mutant pistils. Healthy and stressed stage 12 Arabidopsis flowers were marked. Plants were treated with either 75 mM NaCl (stressed) or water (healthy). After 48 h, H_2O_2 was measured in five or more pistils from each genotype and treatment. Asterisks indicated significant differences between the mutant genotypes and the wild-type (Col-0) controls (* $p < 0.05$ and ** $p < 0.01$) (**b**) Peroxidase activity from myc-PER17, myc-PER28, and myc-PER29 transformants was measured. Each well of an ELISA plate was coated with a myc-tag antibody so that myc-tagged proteins could be fished from total protein extracts isolated from plants. Amplex Red substrate was used to determine the activity of these peroxidases. Peroxidase activity from these transformants was significantly greater than the wild-type controls (* $p < 0.05$ and ** $p < 0.01$). ANOVA analysis was done using JMP8 (unequal variances and two-tailed distributions).

2.5. *Myc-Tagged Peroxidases Exhibited Peroxidase Activity*

Peroxidases that have an N-terminal myc-tag were ectopically expressed in Arabidopsis. Myc-tagged PER17, 28, and 29 were fished from total protein extracts using an anti-myc antibody that coated the ELISA plate wells. The peroxidase activity of the tagged proteins was examined. Results showed that myc-tagged PER17, PER28, and PER29 showed significantly more peroxidase activity than the control that contained no myc-tagged protein (Figure 5). This indicates that myc-tagged proteins contain active peroxidases that neutralize H_2O_2.

2.6. *Expression and Sub-Cellular Localization of Peroxidases*

While *PER17*, *PER28*, and *PER29* mRNA were detected in many plant tissues, quantitative PCR results revealed that they were most abundant in pistils (Figure 6). The fact that these three *PER* mRNA levels are low in leaves may explain why there was no apparent mutant phenotype in vegetative tissues. To check whether other class III peroxidases might compensate for the loss of functions in mutant alleles, RNA levels for these genes abundant in stage 12 pistils were evaluated. The expression of each class III peroxidase was evaluated using the eFP browser [29]; *PER9*, *PER39*, and *PER68* transcripts are relatively abundant in stage 12 pistils, the organ evaluated in this study. In the stressed *per17*, *per28*, and *per29* single mutants, quantitative RT-PCR showed that the expression of the other five genes remained steady—none of the genes significantly differed from controls.

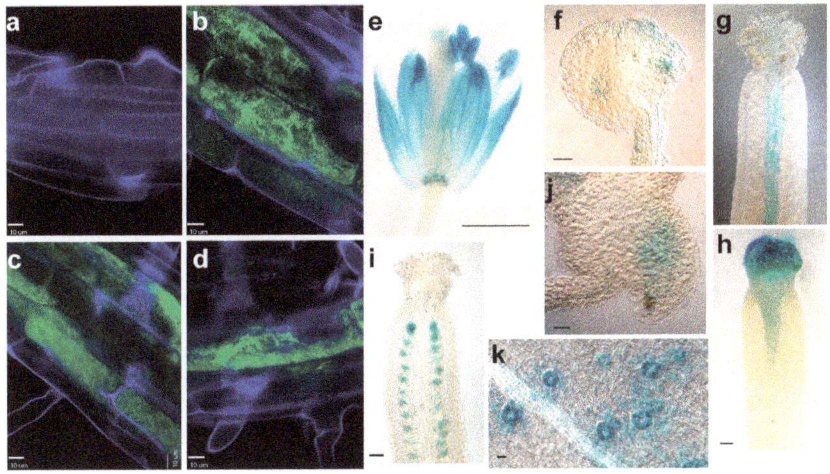

Figure 6. The expression of three PER proteins. Merged confocal images show GFP (green) and cell wall (blue) fluorescence in (**a**) wild type, (**b**) PER17-GFP, (**c**) PER28-GFP, and (**d**) PER29-GFP cells. GUS histochemical staining of *PER17::GUS* (**e**,**f**), *PER28::GUS* (**g**,**h**), and *PER29::GUS* (**i**–**k**) plants is shown. PER17 promoter was active in flowers (**e**) and the chalaza region of ovules (**f**) and leaves. *PER28* was initially expressed in the transmitting tract of stage 12 pistils (**g**) and increased in abundance in the stigma of stage 13 pistils (**h**). *PER29* promoter was active in ovules (**i**) and sepal guard cells (**k**). A higher magnification of panel i shows staining in the gametophyte (**j**). (**a**–**d**,**f**,**j**,**k**), bar = 10 μm; (**e**), bar = 1 mm; (**g**–**i**), bar =100 μm.

Many peroxidases are active in peroxisomes, so the PTS1 predictor [31] was used to evaluate the putative *PER17*, *PER28*, and *PER29* protein sequences. This bioinformatics analysis revealed a low probability that these proteins target peroxisomes. Constructs containing GFP translational fusion protein were created to infer the sub-cellular localization of each peroxidase. Transgenic plants containing peroxidase-GFP proteins were examined by confocal microscopy. Results showed that three peroxidase proteins were present in the cell wall or apoplast (Figure 6).

2.7. Promoter Activity of Peroxidases in Reproductive Tissues

To investigate the spatial expression of three PERs, Arabidopsis plants that were transformed with *PER::GUS* constructs were analyzed. Results revealed that the *PER17* promoter was active in sepals, stamen, nectaries, style, and the chalaza region of ovules (Figure 6e,f). In some ovules, GUS staining also appeared near the synergids and filiform apparatus of the *PER17::GUS* plants (Figure 6f). In the *PER28::GUS* plants, this gene was first expressed in the transmitting track of stage 12 flowers (Figure 6g). In stage 13 flowers, when the stigma is most receptive to pollination, *PER28* displayed maximal expression (Figure 6h). Plants that carried the *PER29::GUS* construct showed GUS staining in ovules, especially around the gametophyte (Figure 6i,j). In the siliques, *PER29::GUS* staining was only found in ovules but not developing seeds, indicating that the *PER29* promoter is active in gametophytes but not embryos. In addition, *PER29* was expressed in guard cells (Figure 6k). For each of these constructs, stems, rosette leaves, and cauline leaves were stained, but no signal was observed when floral parts were fully stained. Evaluation of eFP expression data for these constructs [32] indicates that these genes are expressed in other regions of the plant but at much lower levels, so the GUS signal appeared absent when the signal from reproductive tissues was observed.

3. Discussion

3.1. Peroxidase Mutants and Fertility

The genes in this study were chosen for analysis because they exhibited significant changes in expression at the initiation of ovule abortion. These ROS-scavenging genes were *PER17*, *PER28*, *PER29*, *APX4*, and *FSD2*. While *PER17* expression increased, the levels for the other four genes were significantly lower following salt stress [26]. Previous investigation correlated ROS accumulation in ovules with the rate of ovule abortion [33]. Results show that mature gametophytes or developing ovules tolerate up to 8 h of 75 mM salt stress with minimal disruption to reproduction, but treatment with 200 mM NaCl for 12 h causes more than 90% of the ovules to abort [26]. From a global microarray study [28], five ROS scavengers that exhibited significant changes in gene expression at the critical developmental stage were selected for further analysis.

Wide-scale mutational analysis revealed that organisms contain a small fraction of essential genes, but most genes were genetically redundant or exhibited distributed robustness [34,35]. Recent analysis indicates that the vast majority of genes serve a function, but these include chromatin effects, RNA interference, DNA modification, and components altering the genetic landscape (Encode Consortium, 2012). Partial redundancy and distributed robustness both lead to incremental phenotypes instead of lethality. While *APX4* transcripts are abundant in leaves and developing embryos [32], mutation of the *APX4* locus had no effect on ovule development or plant fertility (Table 1). The absence of an ovule or seed phenotype in *apx4* mutants may be due to genetic redundancy with another ROS scavenger. ROS scavengers that are expressed in developing seeds can limit oxidative stress during seed desiccation and germination, thereby reducing seed deterioration [36]. Mutation of the *APX4* locus primarily affects seedling growth and establishment [37].

3.2. Stress, Photosynthesis, and ROS Scavengers

The rate at which stress is applied to plants affects the physiological outcome. When water stress slowly increases, plants activate different regulatory networks, permitting acclimation and tolerance to water stress [38]. Once plants in this study were stressed, the rapid change in floral water potential allowed little time for this type of acclimation. Weeks after the imposition of stress, we observed a higher fraction of ovules formed seeds. This observation indicates plants acclimated to these soil conditions.

Myouga et al. [39] reported that FSD2 is active in chloroplasts. When plants were grown at five-fold higher fluence rates than those described here, *fsd2* leaves were chlorotic. At lower fluence rates, the other two Arabidopsis SOD loci were sufficient to protect chloroplasts from photo-oxidation. Reproductive analyses of *fsd2* mutants revealed a significant reduction in fertility for actively photosynthesizing plants but had minimal effect on stressed plants (Table 1). The rather small effect of this mutation on fertility after stress demonstrates the reduction in photosynthetic rates creates less superoxide production in the chloroplasts. The absence of a fertility phenotype indicates that the other two SODs work redundantly and scavenge the residual superoxide generated.

3.3. Peroxidases Affecting Fertility

In this study, we measured how three class III peroxidase loci affected ROS metabolism and ovule abortion. The nomenclature of class III peroxidases in *Arabidopsis* varies, where both PER or PRX have been used. In earlier studies, PER17 was reported to modulate lignification and pod shatter [19], and PER28 was hypothesized to affect pollen-pistil interactions [40]. We found that PER17, PER28, and PER29 affected H_2O_2 production in ovules. Mutation of individual loci showed modest increases in peroxide accumulation (Figure 5). The significant increase in ROS shown in Figure 4 indicates these pistil-specific peroxidases work together to cause this increase in activity.

The level of *PER17* mRNA increased after 24 h of stress, while the expression of *PER28* and *PER29* was repressed under these conditions (Figure 3C). Following salt stress, *per17*, *per17per28*, *per17per28*, and *per17per28per29* mutants exhibited significantly lower fertility (Figure 3B) than unstressed counterparts (Figure 3A). One simple explanation for this is that increased *PER17* activity after salt stress limits ovule failure. Conversely, *per28*, *per29*, and *per28per29* mutants showed significant ovule abortion rates in healthy plants, but not after environmental stress (Table 1), indicating that *PER28* and *PER29* scavenge peroxides in healthy plants but have negligible activity in those that are salt-stressed. These data suggest that under normal conditions, all three *PER* genes scavenge ROS in ovules or nearby areas to which this molecule can diffuse, thereby protecting ovules from abortion. When encountering salt stress, *PER17* remains an influential peroxide scavenger that removes ROS.

We observed ROS accumulation after salt stress in ovules of three peroxidase mutants (Figure 4), which indicated that either the rate of ROS removal decreased or its synthesis accelerated. ROS accumulation further increased in *per* double and triple mutants (Figure 4). However, the ovule abortion rates in some of the peroxidase mutants were not affected. We cannot distinguish which types of ROS accumulated in these tissues because CH_2DCFDA stains many ROS. One or more types of ROS may reach a critical threshold, signaling the described physiological changes. Data reported here show that the induction of ROS accumulation by the mutation of peroxidases was sufficient to increase the rate of seed abortion.

It is believed that H_2O_2 may be a better signaling molecule than other types of ROS because it can cross the plasma membrane and move into neighboring cells. In plants, class III peroxidases can reduce peroxides by using various donor molecules, such as auxin or secondary metabolites (e.g., lignin precursors and phenolic compounds) [19]. Some cell wall peroxidases modulate ROS levels, which affect plant defense or development, but this reaction is strongly pH-dependent [24,41–43]. Therefore, we quantified H_2O_2 levels in our peroxidase mutants. The H_2O_2 levels in peroxidase single and double mutants were higher than the wild-type controls but were not significantly higher as a result of the high variance. For given genotypes, this variance can be explained by variability in the activity in the family of 70+ peroxidases present in Arabidopsis, which might or might not be due to differences in ovule abortion rates among pistils. Purification of H_2O_2 is labor intensive, and samples cannot be stored frozen, so the sample size for each genotype was limited. The hydrogen peroxide levels were significantly higher in triple mutants under both unstressed ($p < 0.05$) and salt-stressed ($p < 0.01$) conditions (Figure 5). These data correlated with low fertility (Table 1 and Figure 3), indicating H_2O_2 may be the molecule that causes ovule abortion.

According to the expression patterns from GUS staining results, three Class III peroxidases may modulate ROS signaling and/or affect the fertilization process, nutrient transport, and/or pistil growth. *PER29::GUS* revealed that this gene is active in ovules, especially around the gametophyte (Figure 6j). Sun et al. [28] reported that after salt stress, ROS initially accumulated in the gametophyte and then spread throughout the ovule. Thus, *PER29* may be responsible for regulating ROS in the gametophyte prior to fertilization. Together with the fact that *PER29* has high levels of expression in healthy controls (Figure 3C), these results explain why the *per29* mutants exhibited the lowest relative fertility among the genotypes tested (Figure 3A).

The tip growth of pollen tubes requires ROS buildup and the presence of an antioxidant that alters pollen growth in vitro [44]. Before fertilization, a pollen grain adheres to the stigma, and the pollen tube emerges and grows along the transmitting tract, which contains nutrients and signaling molecules that affect pollen tube guidance. Therefore, the fertilization of *per17*, *28*, and *29* mutants was examined. Flowers were emasculated, and pollen tubes were stained 24 h after manual pollination [45]. After 24 h, pollen tubes entered 90% of the wild-type, *per28*, and *per29* ovules, but only 50% of the *per17* pollen tubes reached an ovule. Since *per17* had 70% fertility (Table 1), we conclude that pollen tube growth occurs more slowly or pollen tube guidance is disrupted in this genotype. Wild-type pollen fertilized significantly fewer *per17* ovules, which is consistent with a pollen guidance problem. Future experiments are necessary to discriminate between these possibilities.

PER28::GUS was expressed in the transmitting track (Figure 6g). When the stigma is most receptive to pollination, *PER28* displayed maximal expression in the stigma (Figure 6H). The expression of these genes is consistent with a role in pollen-stigma interaction or pollen tube growth/tracking. However, the pollination and pollen tube growth appeared normal in the *per28* mutant. During manual pollination of the pistil, it was noted that the pollen did not adhere as well to the stigma, which suggests the *PER28* locus may affect pollen adhesion to the stigma.

3.4. ROS, Fertility, and PCD

In a variety of plant cell types and plant species, PCD is induced with a pulse of ROS. Images in this paper show necrosis of the endothelium of stressed ovules, which coincides with ROS accumulation (Figures 3–6). Seed abortion is preceded by changes in mitochondrial potential [33]. Previous work shows that similar peroxidases can regulate apoplastic oxidative burst and then PCD in response to fungal infection [46]. When pollen nuclei are delivered to the synergids by the pollen tube, ROS induce PCD in these two cells [47]. This is mediated by the activity of a kinase MAP kinase cascade, which effects synergid PCD and immunity-associated PCD [48]. In tomatoes, disruption of fertilization leads to altered endothelium development and ectopic ROS spikes. This can produce parthenogenic seeds or altered developmental fates [49,50]. In wheat, drought stress increases ROS production and increases yield over a shortened fertility window [51]. In a number of different plants, ROS affect reproductive development by modulating PCD and changing cell fate.

4. Materials and Methods

4.1. Plant Growth and Fertility Measurements

Arabidopsis thaliana wild-type (Col-0) and ROS scavenger T-DNA insertion mutants (SALK_003180 for *per17*, SALK_076194 for *per28*, SALK_133065 for *per29*, SAIL_519_E04 for *apx4*, and SALK_080457 for *fsd2*) were identified from ABRC stocks (Columbus, OH). After backcrossing these mutant alleles with wild-type Arabidopsis (Col-0), homozygous mutants were isolated by a polymerase chain reaction (PCR). Two gene-specific primers from upstream and downstream coding sequences were used to verify the presence of wild-type alleles. One gene-specific primer and one primer from the T-DNA border sequence were used to confirm the presence of the mutant allele (Table S1). All mutants are null mutants since none encode a full-length transcript (Figure 2).

Plants were grown in 2 × 2 pots in a Percival plant growth chamber (Perry, IA) at 24 °C, 50% relative humidity, and continuous fluorescent light (100 μmol photon m^{-2} s^{-1}). To stress plants, the pots were soaked in irrigated water containing 75 mM NaCl for 6 h, drained, then not irrigated with plain water for an additional 42 h.

Healthy seeds and aborted ovules in each silique (developed from pistil) were recorded. From these data, ovule abortion rates and fertility with standard errors were calculated. Before performing ANOVA analysis using JMP8 software (Cary, NC, USA), ovule abortion rates were root arcsine transformed to generate a normal distribution. A normality test was done in JMP8 for all other ANOVA analyses.

4.2. RNA Extraction and PCR

Total RNA was isolated from 10-day-old seedlings and tissues from 30-day-old hydroponic plants using RNeasy mini kit (Qiagen, Valencia, CA, USA). Hydroponic propagation was described by Gibeaut et al. [52]. Complementary DNA (cDNA) was synthesized using SuperScript III reverse transcriptase (Invitrogen, Grand Island, NY, USA) from 1 μg of total RNA as suggested by the manufacturer. Quantitative PCR (qPCR) was performed in optical 96-well plates using an ABI StepOnePlus machine (Applied Biosystems, Grand Island, NY, USA). Each 10-μL reaction consisted of GoTaq qPCR Master Mix (Promega, Madison, WI, USA), 2 μL of 1:50 diluted cDNA, and 0.2 mM each of gene-specific primer pairs (see Table S1). The following thermal profile was used for qPCR: 95 °C for 2 min; 30 cycles of 95 °C for 15 s and 60 °C for 1 min; and a melt curve analysis at 95 °C for 15 s, 60 °C for 15 s, and 95 °C for 15 s. C_T and standard curve were extracted using ABI StepOne v2.2 software (Grand Island, NY, USA). The fold change of gene expression and standard deviations were calculated according to the guide from ABI (http://www.appliedbiosystems.com/absite/us/en/home/support/tutorials.html, accessed on 6 April 2023). Melt curve analyses were used to verify that a single product was produced from each qPCR.

4.3. Metabolite Assays

Pistils were dissected lengthwise and incubated in 10 μM 5-(and 6-) carboxy-2′, 7′-dichlorodihydrofluorescein diacetate (CH_2DCFDA, Molecular Probes, Grand Island, NY, USA) for 30 min, then samples were washed twice in 10 mM of phosphate buffer pH 7.0. In the presence of ROS, CH_2DCFDA was oxidized to form carboxy-dichloro-fluorescein (CDCF), which is highly fluorescent. The CDCF was visualized by excitation between 490 and 510 nm, and the fluorescent emissions were detected using a 520 nm long-pass filter. This filter set detects both CDCF emission (green) and plant cell auto-fluorescence (red).

Flowers were frozen in N_2(l). Using a steel block immersed in N_2(l) and frozen forceps, pistils were dissected from flowers, and pistil length was measured with a dissecting microscope. After the pistils were ground to powder, 0.1 mL of 0.2 M $HClO_4$ was added to stop all protein activity. To ensure that wound-induced H_2O_2 was not produced, samples were kept frozen until they were dissolved in $HClO_4$. Samples were kept in this buffer at 0 °C for 5 min and then centrifuged at 20,800× *g* for 10 min at 4 °C. The supernatant was neutralized with 110 μL of 0.2 M NH_4OH and was centrifuged again at 3000× *g*. The supernatant was passed through a column of AG resin (Bio-Rad, Hercules, CA, USA) and then eluted with 0.1 mL deionized water.

Amplex Red Peroxidase Assay Kit (Invitrogen, A22188) was used to quantify H_2O_2 levels in the extracts, as recommended by the manufacturer. Fluorescence was measured using Synergy HT fluorescence plate reader (Bio-Tek, Highland Park, IL, USA) with excitation/emission at 540/590 nm. Fluorescence intensity was measured every 10 min to ensure that fluorescence detection of the standards and unknown samples exhibited a linear relationship between accumulation and time. The concentration of H_2O_2 in the samples was calculated using a standard curve with HRP of known activity, which was included with the kit as a standard.

4.4. Constructs and Plant Transformants

The pEW201ML plasmid was created to contain the PER17 coding sequence (upstream) that is translationally fused to the GFP coding sequence (downstream) and is driven by the CAMV 35S promoter. Similarly, the pEW202ML and pEW203ML binary plasmids were created to contain PER28 and PER29 with translationally fused GFP, respectively. Plants were transformed with this *A. tumefaciens* containing the plasmid, according to Clough and Bent [53]. Transformants were selected by spraying 1000-fold diluted BASTA (Finale, AgrEvo, Pikeville, NC, USA).

To synthesize proteins for the assay, PER17, 28, and 29 genes were ectopically expressed using CaMV 35S promoter. The pEW401ML plasmid (*35S::myc-PER17*) contains the PER17 coding sequence and an upstream in-frame myc tag (MEQKLISEEDL). The pEW402ML and pEW403ML plasmids contain, respectively, PER28 and PER29, each with a myc tag. Transformants containing each construct were generated as described above. Myc-tagged protein expression was confirmed in these transformants (Figure S1). Total soluble proteins were extracted from 12-day-old seedlings. The extraction buffer contained 50 mM Tris-HCl pH 8.5, 1X protease inhibitor cocktail VI (RPI Corp), and 0.2 M β-mercaptoethanol. Prior to binding, each well of the FLUOTRAC 600 immunology plate (Greiner Bio-One, Monroe,

NC, USA) was coated with a 0.5 μg myc-tag antibody (Millipore, Billerica, MA) at 4 °C overnight as described by Pratt et al. [54]. After washing and blocking, myc-tagged proteins were fished from total protein extracts (1 mg of myc-PER17; 0.9 mg of myc-PER28; and 0.8 mg of myc-PER29). Following additional washes, Amplex Red was applied to determine the activity of myc-peroxidases as recommended by the manufacturer. Fluorescence was measured with excitation/emission at 540/590 nm. The relative fluorescence of each myc-peroxidase was normalized with wild-type controls in each experimental replicate.

To construct *PER17::GUS*, 1.9 kbp of *PER17* upstream sequence was PCR amplified and transcriptionally fused to the upstream of the *uidA* gene start codon [55]. *PER28::GUS* and *PER29::GUS* constructs were created similarly. GUS assays were done as previously described [56].

5. Conclusions

Here we report three loss-of-function peroxidase mutants (*per17*, *per28*, and *per29*) were found to accumulate more ROS and H_2O_2 in ovules, and the ovule abortion rates significantly increased in these mutants. A simple hypothesis explaining the results reported here is that PER17 and PER29 proteins, which are predominantly expressed in ovules, cause oxidative bursts in ovules, inducing ovule abortion.

Supplementary Materials: The following supporting information can be downloaded at: https://www.mdpi.com/article/10.3390/plants12112182/s1, Figure S1: SDS PAGE of Myc-PER fusion proteins; Table S1: Primers used for genotyping and cloning.

Author Contributions: Conceptualization, Y.-Y.W. and B.A.H.; methodology Y.-Y.W. and B.A.H.; investigation, Y.-Y.W., D.J.H.; original draft preparation, Y.-Y.W.; writing, review, and editing, all authors; supervision, project administration, and funding acquisition, B.A.H. All authors have read and agreed to the published version of the manuscript.

Funding: This work was supported by a grant from the USDA Cooperative State Research, Education, and Extension Service grant number 2008-35100-19244.

Data Availability Statement: Mutants are available from the Arabidopsis Biological Resource Center. Constructs are available from the corresponding authors by request.

Acknowledgments: We thank George Casella and Meng Xu for advice on statistical analyses and Logan Peoples for work done on protein expression experiments.

Conflicts of Interest: The authors declare no conflict of interest.

References

1. Mittler, R.; Vanderwera, S.; Golery, M.; Breusegem, F. Reactive oxygen gene network of plants. *Trends Plant Sci.* **2004**, *9*, 490–498. [CrossRef] [PubMed]
2. Moller, I.M. Plant mitochondria and oxidative stress: Electron transport, NADPH turnover, and metabolism of reactive oxygen species. *Annu. Rev. Plant Physiol. Plant Mol. Biol.* **2001**, *52*, 561–591. [CrossRef] [PubMed]
3. Foreman, J.; Demidchik, V.; Bothwell, J.H.F.; Mylona, P.; Miedema, H.; Torres, M.A.; Linstead, P.; Costa, S.; Brownlee, C.; Jones, J.D.G.; et al. Reactive oxygen species produced by NADPH oxidase regulate plant cell growth. *Nature* **2003**, *422*, 442–446. [CrossRef] [PubMed]
4. Liu, X.; Williams, C.E.; Nemacheck, J.A.; Wang, H.; Subramanyam, S.; Zheng, C.; Chen, M.-S. Reactive oxygen species are involved in plant defense against a gall midge. *Plant Physiol.* **2010**, *152*, 985–999. [CrossRef]
5. Apel, K.; Hirt, H. Reactive oxygen species: Metabolism, oxidative stress, and signal transduction. *Annu. Rev. Plant Biol.* **2004**, *55*, 373–399. [CrossRef]
6. Dat, J.F.; Pellinen, R.; Beeckman, T.; Van De Cotte, B.; Langebartels, C.; Kangasjarvi, J.; Inze, D.; Breusegem, F. Changes in hydrogen peroxide homeostasis trigger an active cell death process in tobacco. *Plant J.* **2003**, *33*, 621–632. [CrossRef]
7. Prasad, T.K.; Anderson, M.D.; Martin, B.A.; Stewart, C.R. Evidence for chilling-induced oxidative stress in maize seedlings and a regulatory role for hydrogen peroxide. *Plant Cell* **1994**, *6*, 65–74. [CrossRef]
8. Bashandy, T.; Guilleminot, J.; Vernoux, T.; Caparros-Ruiz, D.; Ljung, K.; Meyer, Y.; Reichheld, J.P. Interplay between the NADP-linked thioredoxin and glutathione systems in *Arabidopsis* auxin signaling. *Plant Cell* **2010**, *22*, 376–391. [CrossRef]

9. Iglesias, M.J.; Terrile, M.C.; Bartoli, C.G.; D'Ippólito, S.; Casalongué, C.A. Auxin signaling participates in the adaptative response against oxidative stress and salinity by interacting with redox metabolism in *Arabidopsis*. *Plant Molec. Biol.* **2010**, *74*, 215–222. [CrossRef]
10. Jabs, T. Reactive oxygen intermediates as mediators of programmed cell death in plants and animals. *Biochem. Pharm.* **1999**, *57*, 231–245. [CrossRef]
11. Pennell, R.I.; Lamb, C. Programmed cell death in plants. *Plant Cell* **1997**, *9*, 1157–1168. [CrossRef] [PubMed]
12. Bowler, C.; van Montagu, M.; Inze, D. Superoxide dismutase and stress tolerance. *Annu. Rev. Plant Physiol. Plant Mol. Biol.* **1992**, *43*, 83–116. [CrossRef]
13. Vandenabeele, S.; Vanderauwera, S.; Vuylsteke, M.; Tombauts, S.; Langebartels, C.; Seidlitz, H.K.; Zabeau, M.; Van Montagu, M.; Inzé, D.; Van Breusegem, F. Catalase deficiency drastically affects high light-induced gene expression in *Arabidopsis thaliana*. *Plant J.* **2004**, *39*, 45–58. [CrossRef]
14. Foyer, C.H.; Souriau, N.; Perret, S.; Lelandais, M.; Kunert, K.J.; Pruvost, C.; Jouanin, L. Overexpression of glutathione reductase but not glutathione synthetase leads to increases in antioxidant capacity and resistance to photoinhibition in poplar trees. *Plant Physiol.* **1995**, *109*, 1047–1057. [CrossRef] [PubMed]
15. Noctor, G.; Foyer, C.H. Ascorbate and glutathione: Keeping active oxygen under control. *Annu. Rev. Plant Physiol. Plant Mol. Biol.* **1999**, *49*, 249–279. [CrossRef]
16. Narendra, D.; Venkataramani, S.; Shen, G.; Wang, J.; Pasapula, V.; Lin, Y.; Kornyeyev, D.; Holaday, A.S.; Zhang, H. The Arabidopsis ascorbate peroxidase 3 is a peroxisomal membrane-bound antioxidant enzyme and is dispensable for Arabidopsis growth and development. *J. Exp. Bot.* **2006**, *57*, 3033–3042. [CrossRef]
17. Hiraga, S.; Sasaki, K.; Ito, H.; Ohashi, Y.; Matsui, H. A family of class III plant peroxidases. *Plant Cell Physiol.* **2001**, *42*, 462–468. [CrossRef]
18. Tognolli, M.; Penel, C.; Greppin, H.; Simon, P. Analysis and expression of the class III peroxidase large gene family in *Arabidopsis thaliana*. *Gene* **2002**, *288*, 129–138. [CrossRef]
19. Cosio, C.; Dunand, C. Specific functions of individual class III peroxidase genes. *J. Exp. Bot.* **2009**, *60*, 391–408. [CrossRef]
20. Welinder, K.G.; Justesen, A.F.; Kjaersgard, I.V.H.; Jensen, R.B.; Rasmussen, S.K.; Jespersen, H.M.; Duroux, L. Structural diversity and transcription of class III peroxidases from Arabidopsis thaliana. *Eur. J. Biochem.* **2002**, *269*, 6063–6081. [CrossRef]
21. Veitch, N.C. Horseradish peroxidase: A modern view of a classic enzyme. *Phytochemistry* **2004**, *65*, 249–259. [CrossRef]
22. Lam, E. Controlled cell death, plant survival and development. *Nat. Rev. Mol. Cell Biol.* **2004**, *5*, 305–315. [CrossRef]
23. Gechev, T.S.; van Breusegem, F.; Stone, J.M.; Denev, I.; Laloi, C. Reactive oxygen species as signals that modulate plant stress responses and programmed cell death. *Bioessays* **2006**, *28*, 1091–1101. [CrossRef]
24. Bolwell, G.P.; Wojtaszek, P. Mechanisms for the generation of reactive oxygen species in plant defence—Broad perspective. *Physiol. Mol. Plant Pathol.* **2004**, *51*, 347–366. [CrossRef]
25. Hasegawa, P.M.; Bressan, R.A.; Zhu, J.-K.; Bohnert, H.J. Plant cellular and molecular responses to high salinity. *Annu. Rev. Plant Physiol. Plant Mol. Biol.* **2004**, *51*, 463–499. [CrossRef]
26. Sun, K.; Hunt, K.; Hauser, B.A. Ovule abortion in Arabidopsis triggered by stress. *Plant Physiol.* **2004**, *135*, 2358–2367. [CrossRef]
27. Zhang, F.; Wang, Y.; Yang, Y.L.; Wu, H.; Wang, D.; Liu, J.Q. Involvement of hydrogen peroxide and nitric oxide in salt resistance in the calluses from *Populus euphratica*. *Plant Cell Environ.* **2007**, *30*, 775–785. [CrossRef]
28. Sun, K.; Cui, Y.; Hauser, B.A. Environmental stress alters gene expression and induces ovule abortion: Reactive oxygen species appear as ovules commit to abort. *Planta* **2005**, *222*, 632–642. [CrossRef]
29. Schneitz, K.; Hulskamp, M.; Pruitt, R.E. Wild-type ovule development in *Arabidopsis thaliana—A* light microscope study of cleared whole-mount tissue. *Plant J.* **1995**, *7*, 731–749. [CrossRef]
30. Seltman, H.J. Two-way ANOVA. In *Experimental Design and Analysis*; Carnegie Mellon Press: Pittsburgh, PA, USA, 2012; pp. 267–292.
31. Neuberger, G.; Maurer-Stroh, S.; Eisenhaber, B.; Hartig, A.; Eisenhaber, F. Prediction of peroxisomal targeting signal 1 containing proteins from amino acid sequence. *J. Mol. Biol.* **2003**, *328*, 581–592. [CrossRef]
32. Winter, D.; Vinegar, B.; Nahal, H.; Ammar, R.; Wilson, G.V.; Provart, N.J. An "electronic fluorescent pictograph" browser for exploring and analyzing large-scale biological data sets. *PLoS ONE* **2007**, *2*, e718. [CrossRef]
33. Hauser, B.A.; Sun, K.; Oppenheimer, D.G.; Sage, T. Changes in mitochondrial membrane potential and accumulation of reactive oxygen species precede ultrastructural changes during ovule abortion. *Planta* **2006**, *223*, 292–299. [CrossRef]
34. Alonso, J.M.; Stepanova, A.N.; Leisse, T.J. Genome-wide insertional mutagenesis of *Arabidopsis thaliana*. *Science* **2003**, *301*, 653–657. [CrossRef]
35. Wagner, A. Distributed robustness versus redundancy as causes of mutational robustness. *BioEssays* **2005**, *27*, 176–188. [CrossRef]
36. Bailly, C. Active oxygen species and antioxidants in seed biology. *Seed Sci. Res.* **2004**, *14*, 93–107. [CrossRef]
37. Wang, Y.Y.; Hecker, A.G.; Hauser, B.A. The *APX4* locus regulates seed vigor and seedling growth in *Arabidopsis thaliana*. *Planta* **2014**, *239*, 909–919. [CrossRef]
38. Chaves, M.M.; Flexas, C.; Pinheiro, C. Photosynthesis under drought and salt stress: Regulation mechanisms from whole plant to cell. *Ann. Botany* **2009**, *103*, 551–560. [CrossRef]

39. Myouga, F.; Hosoda, C.; Umezawa, T.; Iizumi, H.; Kuromori, T.; Motohashi, R.; Shono, Y.; Nagata, N.; Ikeuchi, M.; Shinozaki, K. A heterocomplex of iron superoxide dismutases defends chloroplast nucleoids against oxidative stress and is essential for chloroplast development in *Arabidopsis*. *Plant Cell* **2008**, *20*, 20–3148. [CrossRef]
40. Tung, C.W.; Dwyer, K.G.; Nasrallah, M.E.; Nasrallah, J.B. Genome-wide identification of genes expressed in Arabidopsis pistils specifically along the path of pollen tube growth. *Plant Physiol.* **2005**, *138*, 977–989. [CrossRef]
41. Blee, K.A.; Jupe, S.C.; Richard, G.; Davies, D.R.; Bolwell, G.P. Molecular identification and expression of the peroxidase responsible for the oxidative burst in French bean (*Phaseolus vulgaris* L.) and related members of the gene family. *Plant Mol. Biol.* **2001**, *47*, 607–620. [CrossRef]
42. Liszkay, A.; Kenk, B.; Schopfer, P. Evidence for the involvement of cell wall peroxidase in the generation of hydroxyl radicals mediating extension growth. *Planta* **2003**, *217*, 658–667. [CrossRef]
43. Mei, W.; Qin, Y.; Song, W.; Li, J.; Zhu, Y. Cotton GhPOX1 encoding plant class III peroxidase may be responsible for the high level of reactive oxygen species production that is related to cotton fiber elongation. *J. Genet. Genomics* **2009**, *36*, 141–150. [CrossRef]
44. Potocky, M.; Jones, M.A.; Bezvoda, R.; Smirnoff, N.; Zarsky, V. Reactive oxygen species produced by NADPH oxidase are involved in pollen tube growth. *New Phytol.* **2007**, *174*, 742–751. [CrossRef]
45. Kandasamy, M.K.; Nasrallah, J.B.; Nasrallah, M.E. Pollen-pistil interactions and developmental regulation of pollen tube growth in Arabidopsis. *Development* **1994**, *120*, 3405–3418. [CrossRef]
46. Kámán-Tóth, E.; Dankó, T.; Gullner, G.; Bozsó, Z.; Palkovics, L.; Pogány, M. Contribution of cell wall peroxidase- and NADPH oxidase-derived reactive oxygen species to *Alternaria brassicicola*-induced oxidative burst in *Arabidopsis*. *Mol. Plant Pathol.* **2019**, *20*, 485–499. [CrossRef]
47. Escobar-Restrepo, J.M.; Huck, N.; Kessler, S.; Gagliardini, V.; Gheyselinck, J.; Yang, W.-C.; Grossniklaus, U. The FERONIA receptor-like kinase mediates male-female interactions during pollen tube reception. *Science* **2007**, *317*, 656–660. [CrossRef]
48. Völz, R.; Harris, W.; Hirt, H.; Lee, Y.W. ROS homeostasis mediated by MPK4 and SUMM2 determines synergid cell death. *Nat. Commun.* **2022**, *13*, 1746–1758. [CrossRef]
49. Baranova, E.N.; Chaban, I.A.; Kurenina, L.V.; Konovalova, L.N.; Varlamova, N.V.; Khaliluev, M.R.; Gulevich, A.A. Possible role of crystal-bearing cells in tomato fertility and formation of seedless fruits. *Int. J. Mol. Sci.* **2020**, *13*, 9480. [CrossRef]
50. Khaliluev, M.R.; Chaban, I.A.; Kononenko, N.V.; Baranova, E.N.; Dolgov, S.V.; Kharchenko, P.N.; Poliakov, V.I. Abnormal floral meristem development in transgenic tomato plants do not depend on the expression of genes encoding defense-related PR-proteins and antimicrobial peptides. *Ontogenez* **2014**, *45*, 28–41. [CrossRef]
51. Hu, C.H.; Zeng, Q.D.; Tai, L.; Li, B.B.; Zhang, P.P.; Nie, X.M.; Wang, P.Q.; Liu, W.T.; Li, W.Q.; Kang, Z.S.; et al. Interaction between TaNOX7 and TaCDPK13 contributes to plant fertility and drought tolerance by regulating ROS production. *J. Agric. Food Chem.* **2020**, *68*, 7333–7347. [CrossRef]
52. Gibeaut, D.M.; Hulett, J.; Cramer, G.R.; Seemann, J.R. Maximal biomass of *Arabidopsis thaliana* using a simple, low-maintenance hydroponic method and favorable environmental conditions. *Plant Physiol.* **1997**, *115*, 317–319. [CrossRef] [PubMed]
53. Clough, S.J.; Bent, A.F. Floral dip: A simplified method for Agrobacterium-mediated transformation of *Arabidopsis thaliana*. *Plant J.* **1998**, *16*, 735–743. [CrossRef]
54. Pratt, L.H.; McCurdy, D.W.; Shimazaki, Y.; Cordonnier, M.-M. Immunodetection of phytochrome: Immunocytochemistry, immunoblotting, and immunoquantitation. In *Immunology in Plant Science*; Linskens, H.F., Jackson, J.F., Eds.; Springer: New York, NY, USA, 1986; pp. 50–74.
55. Jefferson, R.A.; Kavanagh, T.A.; Bevan, M.W. GUS fusions: B-glucuronidase as a sensitive and versatile gene fusion marker in higher plants. *EMBO J.* **1987**, *6*, 3901–3907. [CrossRef] [PubMed]
56. Park, S.O.; Zheng, Z.; Oppenheimer, D.G.; Hauser, B.A. The *PRETTY FEW SEEDS2* gene encodes an Arabidopsis homeodomain protein that regulates ovule development. *Development* **2005**, *132*, 841–849. [CrossRef] [PubMed]

plants

MDPI

Article

Salt Stress Inhibits Photosynthesis and Destroys Chloroplast Structure by Downregulating Chloroplast Development–Related Genes in *Robinia pseudoacacia* Seedlings

Chaoxia Lu [1,†], Lingyu Li [2,3,†], Xiuling Liu [2], Min Chen [2], Shubo Wan [1,*] and Guowei Li [1,*]

1 Shandong Provincial Key Laboratory of Crop Genetic Improvement, Ecology and Physiology, Shandong Academy of Agricultural Sciences, Jinan 250100, China
2 Shandong Provincial Key Laboratory of Plant Stress Research, College of Life Science, Shandong Normal University, Jinan 250014, China
3 Dezhou Graduate School, North University of China, Kangbo Road, Dezhou 253034, China
* Correspondence: wanshubo2016@163.com (S.W.); liguowei@sdnu.edu.cn (G.L.); Tel.: +86-531-86180745 (S.W.); Fax: +86-531-86180107 (S.W.)
† These authors contributed equally to this work.

Abstract: Soil salinization is an important factor limiting food security and ecological stability. As a commonly used greening tree species, *Robinia pseudoacacia* often suffers from salt stress that can manifest as leaf yellowing, decreased photosynthesis, disintegrated chloroplasts, growth stagnation, and even death. To elucidate how salt stress decreases photosynthesis and damages photosynthetic structures, we treated *R. pseudoacacia* seedlings with different concentrations of NaCl (0, 50, 100, 150, and 200 mM) for 2 weeks and then measured their biomass, ion content, organic soluble substance content, reactive oxygen species (ROS) content, antioxidant enzyme activity, photosynthetic parameters, chloroplast ultrastructure, and chloroplast development-related gene expression. NaCl treatment significantly decreased biomass and photosynthetic parameters, but increased ion content, organic soluble substances, and ROS content. High NaCl concentrations (100–200 mM) also led to distorted chloroplasts, scattered and deformed grana lamellae, disintegrated thylakoid structures, irregularly swollen starch granules, and larger, more numerous lipid spheres. Compared to control (0 mM NaCl), the 50 mM NaCl treatment significantly increased antioxidant enzyme activity while upregulating the expression of the ion transport-related genes Na^+/H^+ exchanger 1(*NHX 1*) and salt overly sensitive 1 (*SOS 1*) and the chloroplast development-related genes *psaA*, *psbA*, *psaB*, *psbD*, *psaC*, *psbC*, *ndhH*, *ndhE*, *rps7*, and *ropA*. Additionally, high concentrations of NaCl (100–200 mM) decreased antioxidant enzyme activity and downregulated the expression of ion transport- and chloroplast development-related genes. These results showed that although *R. pseudoacacia* can tolerate low concentrations of NaCl, high concentrations (100–200 mM) can damage chloroplast structure and disturb metabolic processes by downregulating gene expression.

Keywords: *Robinia pseudoacacia* seedlings; photosynthesis; chloroplast; salt stress

Citation: Lu, C.; Li, L.; Liu, X.; Chen, M.; Wan, S.; Li, G. Salt Stress Inhibits Photosynthesis and Destroys Chloroplast Structure by Downregulating Chloroplast Development–Related Genes in *Robinia pseudoacacia* Seedlings. *Plants* **2023**, *12*, 1283. https://doi.org/10.3390/plants12061283

Academic Editors: Małgorzata Nykiel, Mateusz Labudda, Beata Prabucka, Marta Gietler and Justyna Fidler

Received: 19 February 2023
Revised: 7 March 2023
Accepted: 8 March 2023
Published: 11 March 2023

1. Introduction

Soil salinization is already a global problem that climate change, inappropriate irrigation methods, and natural disasters, such as hurricanes and tsunamis, threaten to exacerbate [1,2]. It is estimated that about 70% of the world's agricultural drylands (5.2 billion hectares) are affected by erosion, soil degradation, and salinization [3]. Saline environments account for nearly 10% of the world's total land area (950 Mha), and 50% of irrigated land (230 Mha), which seriously impedes agricultural development [4,5]. It is therefore urgent that we develop and utilize saline areas to solve the problem of land shortages and ensure food security [6,7].

High concentrations of salt ions in soil cause osmotic stress, ion toxicity, and nutrient imbalance in plants, which can seriously disrupt normal growth [8,9]. Salt ions in plants can also damage organelle structure; affect cellular processes including photosynthesis; alter mRNA and protein synthesis; and disrupt energy metabolism, amino acid biosynthesis, and lipid metabolism [10,11]. In addition, salt stress causes reactive oxygen species (ROS) accumulation, causing oxidative stress in plants. Some halophytes and salt-tolerant plants can eliminate ROS by increasing the content and activity of antioxidant enzymes under salt stress, thus reducing damage to the plant. However, most crop plants cannot remove ROS effectively, leading to metabolic disorders and inhibited plant growth [12–14].

Photosynthesis, as the source of energy for plant growth and development, is the most basic and critical physiological process in green plants [15–17]. Salt stress seriously affects photosynthesis in plants [18,19]. Large amounts of Na^+ and Cl^- in the leaves leads to water loss from guard cells and therefore to changes in guard cell and stomata morphology, activity, and/or density numbers, which can reduce or prevent CO_2 entry and limit normal photosynthesis [20–22]. Moreover, salt stress leads to ion toxicity and oxidative stress, particularly in chloroplasts, the main sites of ROS production. Salt stress affects the electron transport chain (ETC) from the photosynthetic complex photosystem II (PSII) to PSI, resulting in the copious production of ROS that causes oxidative damage to nucleic acids, proteins, lipids, membranes, and some photosynthetic enzymes, reducing CO_2 oxidation and crop yield [23,24]. Salt stress can also damage the structural integrity of chloroplasts, which is essential to the photosynthetic light reactions and carbon assimilation processes [25]. Under salt stress, salt-tolerant plants can maintain the normal morphology of photosynthetic structures to sustain their functions [26]. However, when salt stress intensity exceeds the tolerance capacity of the plants, various organelles are damaged, accompanied by changed ultrastructure. Studies showed that under salt stress, the chloroplast membrane system was damaged, with the shape of the chloroplasts changing from ovoid to spherical, the inner lamellae becoming slightly swollen, the number of basal lamellae decreasing, and the number of starch grains, lipid droplets, and osmiophilic globules in the chloroplasts increasing [27]. Many genes involved in the development and maintenance of the normal chloroplast structure that is integral to photosynthesis are known, and salt stress has been shown to affect the expression of such genes, leading to aberrant chloroplast structure. Huang et al. (2019) found that the expression of *Ndhf* genes related to light response, *Rbcl* genes related to dark response, and *Matk* genes related to chloroplastic intron splicing were downregulated under 150 mM NaCl in *Eucalyptus robusta* chloroplasts, which led to decreased photosynthesis and deformed chloroplasts [28–30].

Robinia pseudoacacia L. also known as black locust or false acacia, is native to the United States and belongs to the Faboideae subfamily of the Fabaceae. *R. pseudoacacia* is often used as an afforestation species in saline areas because it reproduces easily and grows rapidly, while also having a high tolerance for salinity, drought, soot, and dust. Therefore, it plays an important role in improving saline environments, preventing soil erosion, and regulating hydrology [31,32]. However, Lu et al. (2021) found the severe degradation of black locust forests in the Yellow River Delta, manifested by yellowing leaves and dead tree tips [33]. This suggests that salt stress can affect photosynthesis in black locusts, but there is no report on how salt stress affects chloroplast structure or how chloroplasts adapt to changes during salt stress. In this study, we treated *R. pseudoacacia* seedlings with different concentrations of NaCl to elucidate these mechanisms by measuring photosynthetic indicators, chloroplast structure, and changes in gene expression.

2. Materials and Methods

2.1. Plant Material and Seed Germination

The *R. pseudoacacia* seeds were Luci 103 (provided by Daqingshan forest farm in Feixian County, Shandong Province). Seeds were sterilized in 75% ethanol for 15 min, then in NaClO (3%) solution for 15 min, and washed with sterile water 3 times. Sterilized, washed seeds were placed in a culture bottle with sterile water, which was put into an 85 °C water bath for 15 min followed by a 45 °C water bath for 12 h.

Seeds were evenly sown in nutrient soil in a greenhouse (25 ± 3 °C/22 ± 3 °C, day/night) at a light intensity of 600 mol/m^2/s (with a 15 h photoperiod) and a relative humidity of 60/80% (day/night). The seedlings were irrigated with water, and germinated for 3–5 days, and NaCl treatment was started when the seedlings had grown to 13–15 cm in length. Five NaCl concentrations of 0, 50, 100, 150, and 200 mM were used, with five replicates per treatment. Plants were watered regularly with 200 mL per pot at 8:30 am daily. To avoid salt shock, salt treatments were increased gradually (increasing by 50 mM per day) until the final desired concentration was reached (the fourth day). Measurements were taken after 2 weeks at that concentration.

2.2. Measurement of Plant Height, Root Length, Dry Weight, Fresh Weight and Water Content

Plant height and root length of *R. pseudoacacia* seedlings were measured to the nearest millimeter. Fresh weights of the shoots and the roots (FW1 and FW2, respectively) were taken, and then the samples were heated to 105 °C for 20 min. Samples were then kept in an 80 °C oven until weights remained constant (about two days), and then their dry weights (DW1, DW2) were measured.

$$\text{Water content of shoot} = (\text{FW1} - \text{DW1})/\text{FW1} \times 100\%$$

$$\text{Water content of root} = (\text{FW2} - \text{DW2})/\text{FW2} \times 100\%$$

2.3. Extraction and Quantification of Ion Contents

Leaf samples (0.3 g) were weighed, boiled in water for 3.5 h, and then filtered to a constant volume of 25 mL. Na^+ and K^+ contents were measured using a flame spectrophotometer (Model 2655-00 Digital Flame Analyzer; Cole-Parmer Instrument Co. Vernon Hills, IL, USA). Cl^- content was measured using an ICS-1100 ion chromatography system (Dionex Corporation, Sunnyvale, CA, USA).

2.4. Determination of Soluble Organic Matter

A 0.3 g subsample of fresh leaves was ground together with 2.4 mL of prepared phosphate-buffered saline (PBS) solution (pH 7.8) and then centrifuged at 10,000× *g* for 10 min at 4 °C.

The concentration of total soluble sugars was measured using the anthracene-sulfur colorimetric method [26]. Soluble protein content was determined using Coomassie Brilliant Blue G250 and BSA [34]. Proline content was determined using ninhydrin [35]. Fresh leaves from the same plants were also ground with 5 mL dH_2O, heated in a water bath at 85 °C for 30 min, allowed to cool, and filtered, and the solution volumes were standardized to 0.1 L. Six drops of phenolphthalein solution (1%) were added to 0.05 L of the extract and titrated with NaOH solution (10 mM) until it turned pink. The volume of NaOH required was recorded and used to calculate the content of organic acids.

2.5. Determination of MDA, O_2^-, H_2O_2

The fully expanded leaves of the third branches of *R. pseudoacacia* seedlings subjected to different NaCl treatments were used to measure H_2O_2, O_2^-, and malondialdehyde (MDA). Detailed experimental procedures were described below.

Leaves of 0.3 g FW were cut into pieces and ground with 2 mL 1% trichloroacetic acid solution and placed in a test tube containing 0.5% thiobarbituric acid solution. The test tube was boiled for 10 min and then the solution was filtered. OD values of the filtrate were measured at 532 nm and 600 nm to determine malondialdehyde content [36].

Leaves of 0.5 g FW were cut into pieces and ground with 5 mL 50 mM PBS (pH 7.8) at 4 °C. The solution was filtered, and the filtrate was placed in a test tube containing 5 mL reaction solution (17 mM aminobenzenesulfonic acid: 7 mM α-naphthylamine, 1:2 (v/v)). This was allowed to stand for 30 min after mixing and then the OD value at 530 nm was measured to determine O_2^- content [37,38].

Leaves of 0.5 g FW were cut into pieces and ground with 5 mL acetone at 4 °C. The solution was centrifugated at 5000× *g* for 8 min. A 2 mL supernatant was placed in a test tube containing 0.1 mL 20% $TiCl_4$ and 0.4 mL strong ammonia solution and then mixed and centrifugated at 8000× *g* for 7 min. After being washed 3–5 times with acetone, the precipitate was dissolved with 5 mL 2 M H_2SO_4 and the OD value 415 nm was measured to determine the H_2O_2 content [39,40].

2.6. Determination of Electrical Conductance

The relative electric conductivity (REC) was operated according to Guo et al. (2015) [41]. Leaves were cleaned and punched into leaf discs. The leaf discs were soaked in dH_2O for 2 h, which was called solution 1. The initial conductance of solution 1 was measured with a conductivity measuring instrument (DDG-5205A, Leici, Shanghai, China) and the conductance value is represented by L1. Solution 1 then was placed in a boiling water bath for 15–20 min and was then called solution 2. Solution 2 was cooled to room temperature and the leaf discs were removed. The conductance of solution 2 was represented by L2. The REC was calculated using the following formula:

$$REC = (L1/L2) \times 100\%$$

2.7. Antioxidant Enzyme Activity Assays

Fresh leaves (0.3 g) were ground in PBS solution (2.4 mL) and then centrifuged at 10,000 rpm for 10 min at 4 °C; the supernatant was then used as the enzyme solution for testing.

The superoxide dismutase (SOD) activity of the enzyme solution was measured by the nitro blue tetrazolium (NBT) photochemical reduction method [42,43], and was detected using an Total Superoxide Dismutase Assay Kit (S0109, Biyuntian, Shanghai, China) according to the manufacturer's protocols. Peroxidase (POD) activity was measured using guaiacol [44]. With guaiacol as a substrate, the changes of absorbance at 470 nm were measured after an enzymatic reaction, the amount of enzyme required for each 1 min absorbance change of 0.01 is one unit of activity. Catalase (CAT) activity was measured in 0.2 mL enzyme solution by adding 6 mL of reaction solution, which consisted of 200 mL PBS with 0.3092 mL of 30% H_2O_2 [43]. Ascorbate peroxidase (APX) activity was determined using 0.1 mL enzyme solution by adding 1.7 mL of PBS containing EDTA-Na_2 (0.1 mM), followed by 5 mM ascorbate (ASA) (0.1 mL), adding 20 mM H_2O_2 (0.1 mL), and immediately (within 40 s) measuring the light absorption at 290 nm wavelengths [45].

2.8. Measurement of Gas Exchange Parameters

After exposure to different NaCl concentrations for 15 days, photosynthetic parameters were measured from 14:00 h to 17:00 h using a Li-6000 portable photosynthesis system (Li-Cor, Lincoln, NE, USA). Five replicates were used for each treatment.

2.9. Preparation of Temporary Mount on the Leaf Surface of *R. pseudoacacia*

The leaves of *R. pseudoacacia* were put into Carnot fixative fluid (3:1 (*v*/*v*) ethanol:glacial acetic acid) for 1 h, and the temporary mount was prepared by adding Hoyers solution (40 mL lactic acid with chloral hydrate until saturation) for more than 30 min, which was observed under a differential interference (DIC) microscope (ECLIPSE80i Differential Interference Fluorescence Microscope, Nikon, Japan).

2.10. Ultrastructure of Chloroplasts in Seedlings of *R. pseudoacacia*

A 1 cm^2 sample of leaf tissue was placed in 2.5% glutaraldehyde (pH = 6.8) fixative fluid. The samples were immersed in the fixative by evacuating the air and left for 2 h at 4 °C. The fixed samples were removed and rinsed with 0.1 mol/L phosphoric acid buffer (pH = 6.8) at 2 h intervals (4 °C). This was followed by sequential dehydration with 50%, 70%, 95%, and 100% ethanol. Dehydrated samples were transferred to 100% acetone for displacement, then epoxy-impregnated, encapsulated, and polymerized. Sections were made with an ultrathin sectioning machine (Leica EM UC7, Weztlar, Germany). The sections were then double stained with uranyl acetate dioxide; lemon leaf was used and examined under a transmission electron microscope (Hitachi HT7800, Tokyo, Japan).

2.11. qRT-PCR Analysis

After NaCl treatment, the leaves of *R. pseudoacacia* seedlings were collected at varying time points (0 h, 24 h, 48 h, 72 h) for measuring gene expression and then immediately frozen in liquid nitrogen and stored at −80 °C. Samples for each time point were taken from the same plant and five biological replicates at a time. The chloroplast development-related gene sequences (*RppsbA*, *RppsbD*, *RppsaA*, *RppsaB*, *RpndhE*, *RpndhH*, *RppsaC*, *RppsbC*, *Rprps7*, and *RpropA*) and ion transporter genes (*NHX1*, *SOS1*) of *R. pseudoacacia* were obtained from the NCBI website, and primers were designed using primer 5.0 (Supplementary Table S1). Total RNA was extracted using the Rapid Universal Plant RNA Extraction Kit 3.0 (Hua Yueyang, Beijing, China). cDNA synthesis was performed using the *Evo M-MLV* RT Mix Kit with gDNA Clean for qPCR (Code No. AG11728). Real-time quantitative reverse-transcription PCR (qRT-PCR) was performed using the SYBR Green Premix *Pro Taq* HS qPCR Kit (Accurate Biology, Changsha, China). The relative expression of each gene was calculated using the $2^{-\Delta\Delta Ct}$ method to calculate the relative expression of each gene [46]. The housekeeping gene *TUBULIN* was used as a control.

2.12. Statistical Analysis

Statistical analyses were made using SPSS software (version 19.0; IBM, Armonk, New York, NY, USA). Significance ($p < 0.05$) was determined using a Duncan's multiple range test [47,48].

3. Results

3.1. Effect of Nacl Stress on the Growth, Ions Content and Organic Soluble Substances Contents of R. pseudoacacia Seedlings

NaCl stress significantly inhibited the growth of *R. pseudoacacia* seedlings. Although we did not detect significant differences from the 0 mM NaCl control at 50 mM NaCl, at higher NaCl concentrations (100, 150, and 200 mM), plant height, root length, shoot fresh weight, shoot dry weight, root fresh weight, and root dry weight all significantly decreased with increasing NaCl (Figure 1).

Compared with those in the control, the Na^+ content, Cl^- content, and Na^+/K^+ ratio significantly rose with increasing NaCl concentrations, while the K^+ content fell significantly (Figure 2A–D).

Compared with the control values, soluble protein, soluble sugar, and proline contents significantly increased with increasing NaCl concentration (Figure 3). Organic acid content showed no significant difference at 50 mM NaCl compared to the control, but rose at higher salt concentrations (100 mM, 150 mM, 200 mM). The increase in proline content was significantly greater than the increases for other osmotic regulators. At 50 mM, 100 mM, 150 mM, and 200 mM treatments, the proline contents were 1.31 times, 2.17 times, 3.36 times, and 3.68 times those of the control, respectively.

3.2. Effect of NaCl Stress on MDA, H_2O_2, and $O^-{}_2$ Contents and Electrical Conductance and Antioxidant Enzyme Activities of R. pseudoacacia Seedlings

Compared with those in the control, the MDA, H_2O_2, and O_2^- contents and the electrical conductivity significantly increased along with increasing salt concentrations (Figure 4), indicating that a large amount of ROS was produced and suggesting the plasma membrane may have suffered damage.

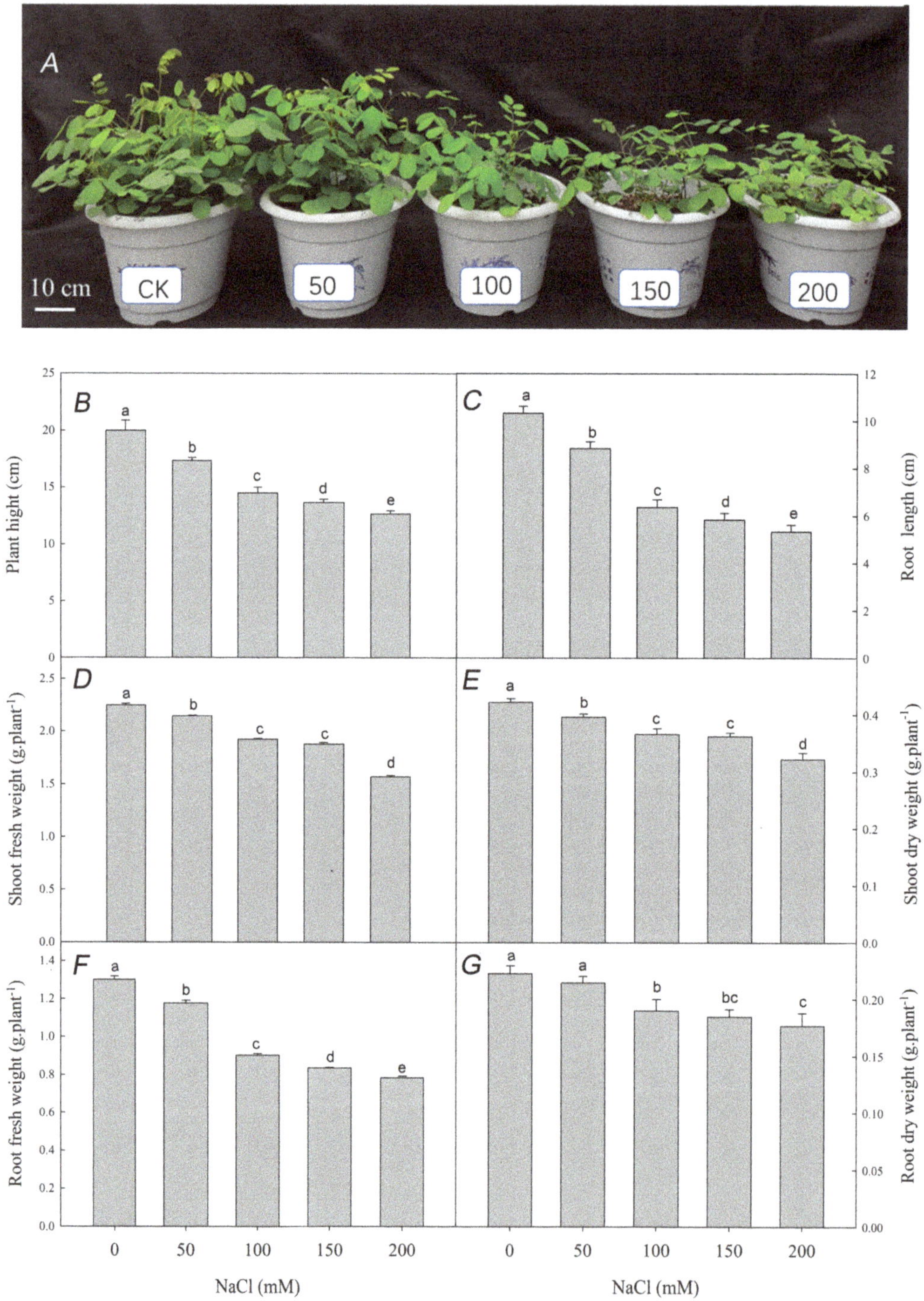

Figure 1. Effect of NaCl stress on the growth of *R. pseudoacacia* seedlings. (**A**) Seedlings under different NaCl concentrations (0 mM, 50 mM, 100 mM, 150 mM, 200 mM); (**B**) plant height; (**C**) root

length; (**D**) shoot fresh weight; (**E**) shoot dry weight; (**F**) root fresh weight; (**G**) root dry weight. Values are mean ± SD of five biological replicates. Bars with different letters are significantly different at $p < 0.05$ according to Duncan's multiple range tests.

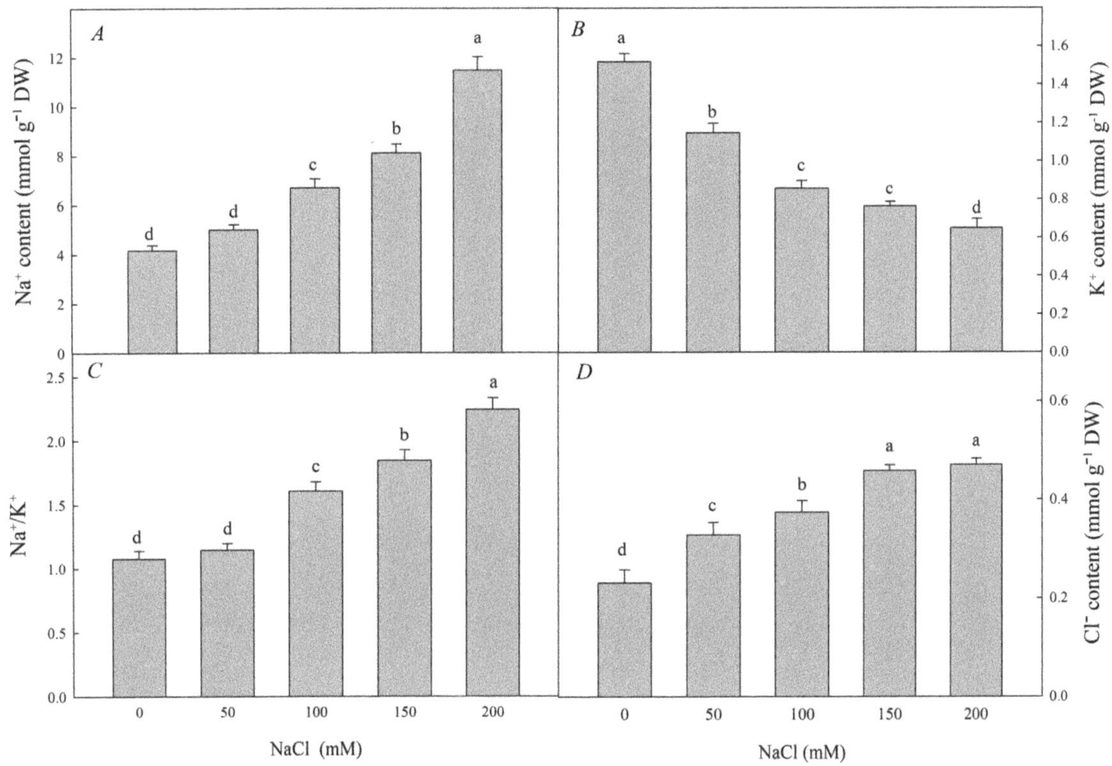

Figure 2. Effect of NaCl stress on Na^+, K^+, Cl^-, and Na^+/K^+ ratio and of *R. pseudoacacia* seedlings. (**A**) Na^+ content; (**B**) K^+ content; (**C**) Na^+/K^+; (**D**) Cl^- content. Values are mean ± SD of five biological replicates. Bars with different letters are significantly different at $p < 0.05$ according to Duncan's multiple range tests.

With increasing salt concentrations, the activities of the antioxidant enzymes CAT, APX, SOD, and POD first increased and then decreased compared with the control (Figure 5). All four enzymes showed their highest activities at 50 mM NaCl. The activities of CAT and APX were also elevated at 100 mM NaCl, but reduced under higher-salt conditions (150 mM NaCl and 200 mM NaCl), while the activity of SOD increased at 100 and 150 mM NaCl and then decreased at 200 mM NaCl. Notably, the activity of POD showed the sharpest drop-off under high-salt conditions, being decreased at 100, 150, and 200 mM NaCl.

3.3. Effect of Nacl Stress on Photosynthesis, Chlorophyll Content and Leaf Grease Content and Chloroplast Structure of R. pseudoacacia Seedlings

Compared with those in the control, the net photosynthetic rate (Pn), transpiration rate (Tr), and stomatal conductance (Gs) significantly decreased with increasing NaCl concentrations. When treated with 200 mM NaCl, Tr, Pn, and Gs were 43.2%, 40.1%, and 41.1% of those of the control group, respectively. It can be seen that the inhibition effect of salt stress on photosynthetic parameters is obvious. However, intercellular CO_2 concentration (Ci) increased with increasing NaCl concentration. When treated with 200 mM NaCl, Ci was 1.61 times higher than that of the control group. This indicates that the ability of gas exchange between plants and the outside world is inhibited under salt stress. Ci significantly increased, suggesting that NaCl stress inhibited the reduction of CO_2 (Figure 6).

With increasing salt concentrations, total chlorophyll, chlorophyll a, and chlorophyll b contents significantly decreased compared to the control (Figure 7). At 200 mM NaCl, the total chlorophyll, chlorophyll a, and chlorophyll b contents were 55.2%, 47.1%, and 48.7% of the control group abundances, respectively.

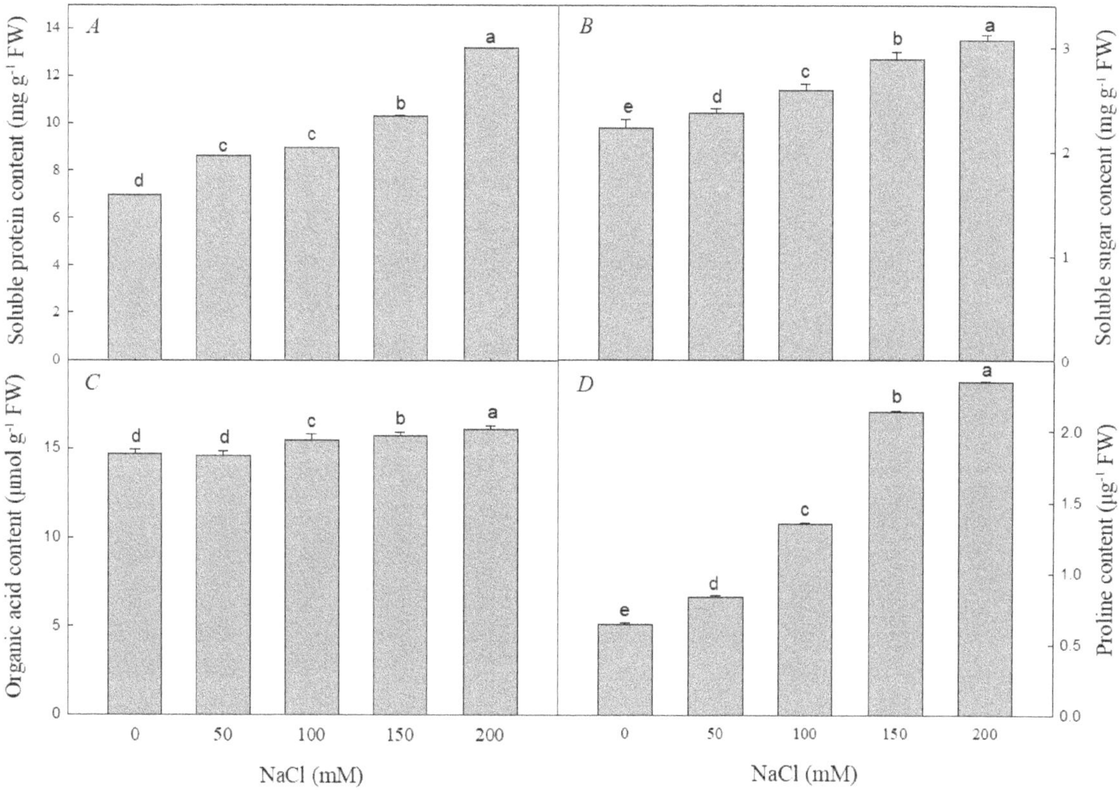

Figure 3. Effect of NaCl stress on the contents of soluble protein (**A**), soluble sugar (**B**), organic acids (**C**), and proline (**D**) of *R. pseudoacacia* seedlings. Values are mean ± SD of five biological replicates. Bars with different letters are significantly different at $p < 0.05$ according to Duncan's multiple range tests.

With increasing salt concentrations, both the grease content and the degree of chloroplast structural damage increased compared to the control (Figure 8A,B). After 0 mM, 100 mM, and 200 mM NaCl treatment, there was grease content in the leaves of *R. pseudoacacia* seedlings. Under the control condition, the leaves of black locust seedlings had a small amount of grease content. However, with the increase in NaCl concentration, the grease content increased significantly. Further observation showed that the chloroplast structure changed with the increase in salt concentration. At 100 mM NaCl, the chloroplasts began to distort, the grana lamellae scattered and deformed, the structure of the thylakoids began to disintegrate, the starch granules started to become larger, swollen, and irregular, and the number and volume of lipid spheres increased. At 200 mM NaCl, the chloroplast structure further swelled and deformed, the thylakoids continued to expand and became disorderly, distorted, and deformed, filamentous lamellae appeared, starch granules continued to become larger and more irregular, and the number and volume of lipid spheres increased further.

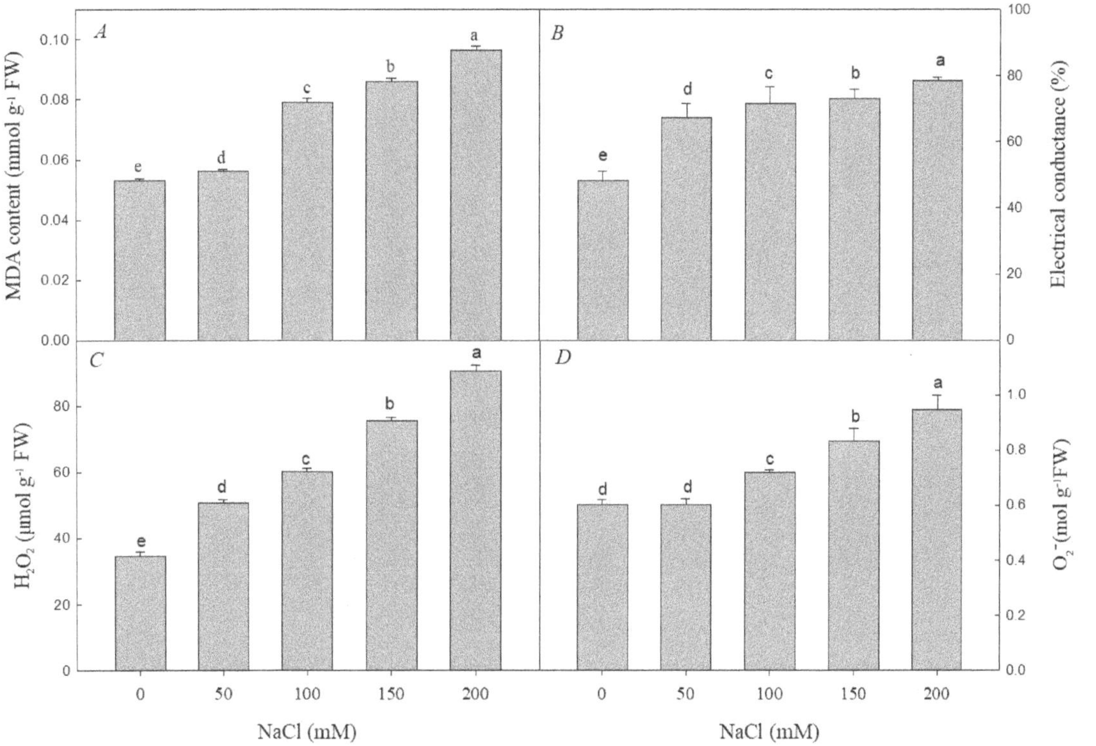

Figure 4. Effect of NaCl stress on the MDA content (**A**), electrical conductance (**B**), H_2O_2 content (**C**), and O_2^- content (**D**) of *R. pseudoacacia* seedlings. Values are mean ± SD of five biological replicates. Bars with different letters are significantly different at $p < 0.05$ according to Duncan's multiple range tests.

3.4. Effect of NaCl Stress on the Expression of Key Genes in Chloroplast Development and Ions Transporter of R. pseudoacacia Seedlings

The relative expression of key genes (*RppsbA*, *RppsbD*, *RppsaA*, *RppsaB*, *RpndhE*, *RpndhH*, *RppsaC*, *RppsbC*, *Rprps7*, and *RpropA*) during the chloroplast development of *R. pseudoacacia* seedlings were significantly upregulated under 50 mM NaCl treatment for 24, 48, and 72 h, but was significantly downregulated at higher NaCl levels (100 mM, 150 mM, 200 mM) (Figure 9). At 50 mM NaCl, the relative expression of *NHX1* and *SOS1* was significantly upregulated compared to the control, whereas it was significantly downregulated at higher NaCl concentrations (100 mM, 150 mM, and 200 mM) (Figure 9).

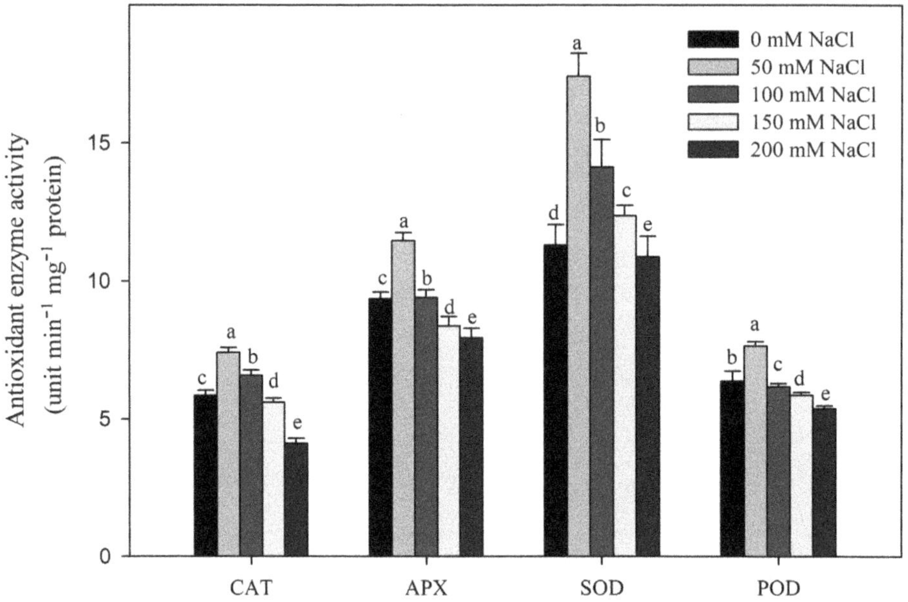

Figure 5. Effect of NaCl stress on activities of CAT, SOD, APX, and POD in *R. pseudoacacia* seedlings. Values are mean ± SD of five biological replicates. Bars with different letters are significantly different at $p < 0.05$ according to Duncan's multiple range tests.

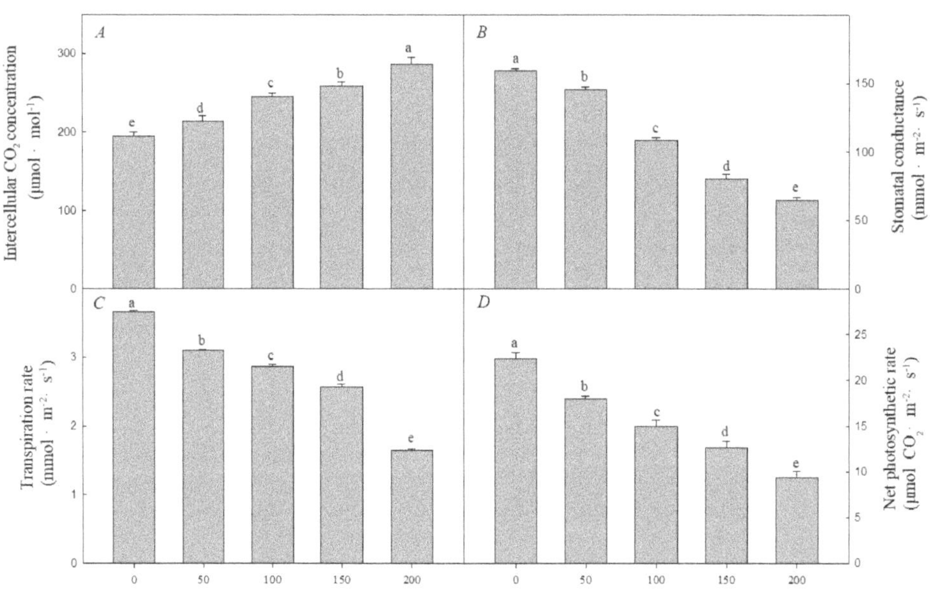

Figure 6. Effect of NaCl stress on photosynthesis of *R. pseudoacacia* seedlings. (**A**) Intercellular CO_2 concentration; (**B**) stomatal conductance; (**C**) transpiration rate; (**D**) net photosynthetic rate. Values are mean ± SD of five biological replicates. Bars with different letters are significantly different at $p < 0.05$ according to Duncan's multiple range tests.

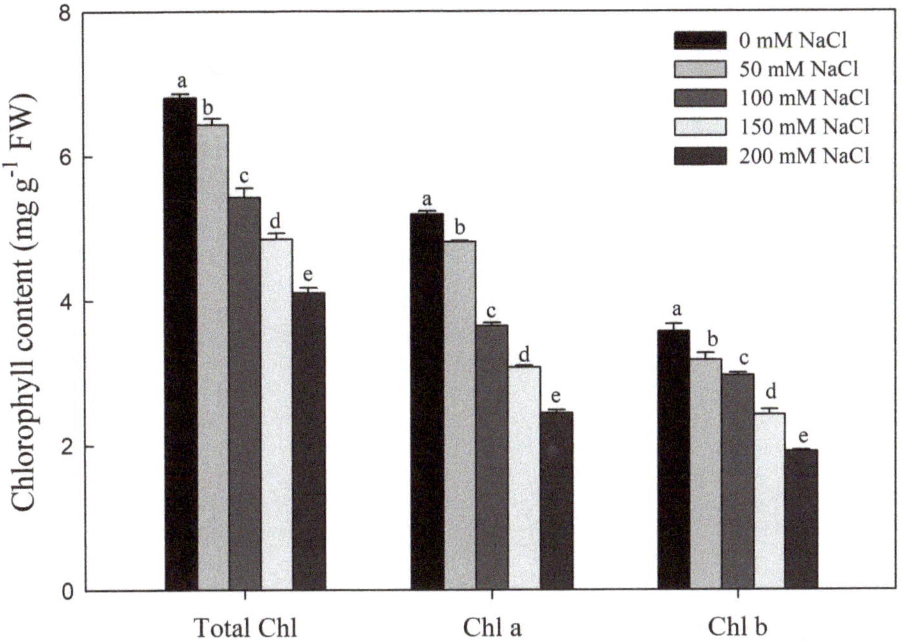

Figure 7. Effect of NaCl stress on the content of total chlorophyll, chlorophyll a, and chlorophyll b in *R. pseudoacacia* seedlings. Values are mean ± SD of five biological replicates. Bars with different letters are significantly different at $p < 0.05$ according to Duncan's multiple range tests.

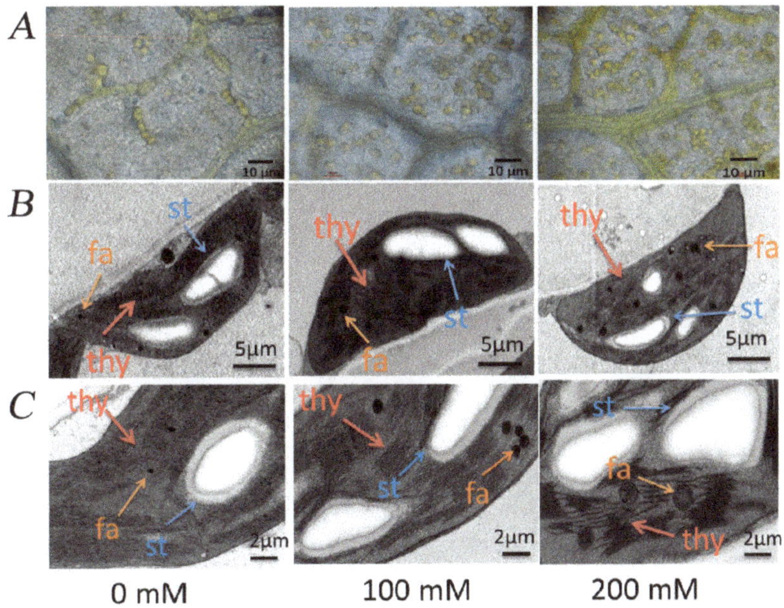

Figure 8. Effect of NaCl stress on the grease content (**A**) and chloroplast structure (**B**,**C**) of *R. pseudoacacia* seedlings. thy: thylakoid; st: starch granule; fa: liposphere. Bars in (**A**) are 10 μm; bars in (**B**) are 5 μm; bars in (**C**) are 2 μm.

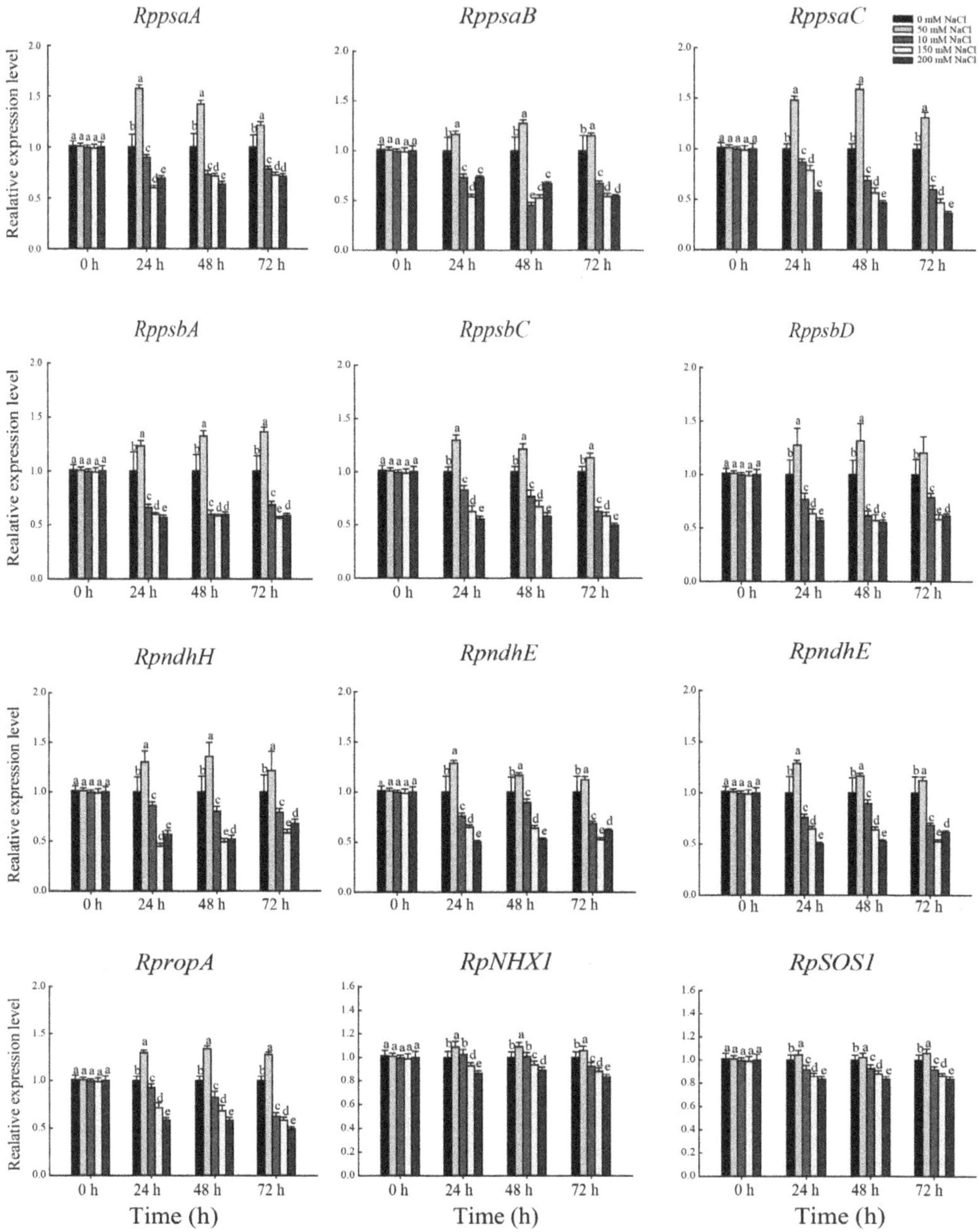

Figure 9. Effect of NaCl stress on the expression of key genes in chloroplast development (*RppsaA*, *RppsaB*, *RppsaC*, *RppsbA*, *RppsbC*, *RppsbD*, *RpndhE*, *RpndhH*, *Rprps7*, and *RpropA*) and ion transporters (*RpNHX1* and *RpSOS1*) of *R. pseudoacacia* seedlings. Values are mean ± SD of five biological replicates. Bars with different letters are significantly different at $p < 0.05$ according to Duncan's multiple range tests.

4. Discussion

Salt stress is an important environmental factor that restricts plant growth and development and reduces crop yield [49]. Under salt stress, plants exhibit slow growth and development, metabolic inhibition, and, in severe cases, wilting and even death. Here, we showed that NaCl treatment significantly inhibited the growth of *R. pseudoacacia* seedlings (Figure 1). Plant height and root length decreased, dry and fresh weight also decreased with the increase of salt concentration.

Salt stress has a broad range of effects on plant metabolism, including the contents of other ions and organic soluble substances, various aspects of cell metabolism, and the expression of corresponding genes [50]. For example, in this experiment, with increased NaCl concentrations, Na^+ and Cl^- accumulated in large quantities in plants, while K^+ decreased, resulting in a high Na^+/K^+ ratio (Figure 2A–D). K^+ is a key ion to ensure the normal metabolism of plants. Because of Na^+, K^+ competing K^+/Na^+ transporters are at the same binding site and the accumulation of too much Na^+ inhibits K^+/Na^+ exchange, thus significantly reducing the K^+ content in the seedlings [51]. Previous studies have shown that under NaCl stress, and more Na^+ was retained in the roots, the K^+/Na^+ ratio in the aboveground part of *R. pseudoacacia* was lower, while less Na^+ is distributed in the leaves, thus reducing the damage of Na^+ to the leaves [52]. Under salt stress, plants tend to upregulate Na^+/H^+ antiporter genes to maintain Na^+/K^+ homeostasis and avoid Na^+ accumulation in the cytoplasm. *SOS1*, one Na^+/H^+ antiporter located on the plasma membrane, can be activated by phosphorylation to transport Na^+ to the extracellular matrix. Similarly, *NHX1*, located on the tonoplast, can be activated by phosphorylation to pump excess intracellular Na^+ into the vacuole [53–55]. An increasing number of experiments have demonstrated the role of Na^+/H^+ antiporters in salt resistance: for example, heterologous expression of *AoNHX1* (from *Avicennia officinalis* L.) increases salt tolerance in rice and Arabidopsis [56], *RtNHX1* (from *Reaumuria trigyna*) enhances salt tolerance in transgenic Arabidopsis plants by sequestering Na^+ into the vacuole and decreasing the Na^+/K^+ ratio in the cytoplasm [57]. In this experiment, the relative expression of *RpNHXl* and *RpSOS1* was higher than in the control under 50 mM NaCl treatment, but lower at higher NaCl concentrations (100 mM, 150 mM, and 200 mM) (Figure 9). This indicates that *R. pseudoacacia* seedlings under 50 mM NaCl treatment can effectively maintain ion homeostasis in the cytoplasm by either compartmentalizing Na^+ into the vacuole or transporting Na^+ to the extracellular matrix, thereby reducing damage caused by salt stress, but this mechanism becomes insufficient at higher salt concentrations. This is consistent with past findings [56,57].

Salt ions in soil can cause osmotic stress in plants, which prompts them to accumulate various organic substances in their roots to reduce osmotic potential, improve cellular water retention capacity, and mitigate the damage caused by osmotic stress. These substances are either small molecules such as proline and betaine, or structural substances such as sucrose and starch [51]. Organic osmoregulatory substances play an important role in plant tolerance to salt stress. Under salt stress, the *P5CS1* (*Delta1-pyrroline-5-carboxylate synthase 1*) gene in the Arabidopsis *myc2* mutant is upregulated to synthesize proline, thereby enhancing salt tolerance [58]. The enrichment of soluble sugars in rice can also effectively relieve osmoregulatory effects and enhance salt tolerance [59,60]. In this experiment, concentrations of organic substances increased under salt stress in *R. pseudoacacia* seedlings (Figure 3). The most significant increase was in the proline content, suggesting that proline plays a major role in osmotic adjustment [58,59].

Salt stress is a complex process, and almost all physiological and biochemical pathways in plants will be affected [8,61]. It not only causes direct primary damage to plants, but also causes secondary damage, such as peroxide stress [62]. In *R. pseudoacacia* seedlings subjected to salt stress, the O_2^-, H_2O_2, and MDA contents increased, as did electrical conductivity (Figure 4), which are signs of oxidative stress [63,64]. However, salt-tolerant plants tend to adapt to salt stress by increasing ROS scavenging capacity using enzymatic antioxidants (SOD, POD, APX, CAT, etc.) and non-enzymatic antioxidants (AsA, GSH, etc.) [65]. Among them, SOD is the key enzyme for O_2^- scavenging, and CAT, APX, and POD are the key enzymes for H_2O_2 scavenging [10,66]. Under salt stress, cucumber responded by adjusting CAT and APX antioxidant enzyme activities [67], whereas the CAT and SOD activities in sweet sorghum increased and then decreased under NaCl stress [68]. In our experiments, the activities of SOD, POD, CAT, and APX increased at low concentrations of NaCl, but decreased at high concentrations of NaCl (Figure 5). This indicates that *R. pseudoacacia* seedlings can tolerate low concentrations of NaCl and effectively scavenge ROS by increasing antioxidant enzyme activity. However, at high concentrations of NaCl, the ROS scavenging ability of seedlings decreased, resulting in ROS accumulation and subsequent damage to the plants' growth and development [67–69].

Photosynthesis is sensitive to salt stress [16]. Pn, Ci, and Gs are important for understanding physiological processes of leaves in nature; changes in photosynthetic parameters are direct reflections of photosynthetic function [70]. In this experiment, Pn, Tr, and Gs of *R. pseudoacacia* seedlings decreased under NaCl treatment, indicating that photosynthesis was inhibited (Figure 6). When plants are exposed to salt stress, the first reaction is stomatal closure caused by osmotic stress, the decrease of stomatal conductance Gs causes the decrease of stomatal factor and Ci, thus limiting Pn [71]. Furthermore, chlorophyll content is an important indicator of leaf senescence. Our study showed that chlorophyll content was significantly reduced under NaCl treatment (Figure 7), which likely contributed to reduced photosynthesis. Under salt stress, there is a large amount of Na^+ uptake, chlorophyll loss, and stomatal closure in *R. pseudoacacia* seedlings. The resulting restriction of CO_2 entry and salt accumulation led to the inhibition of CO_2 assimilation. Because chloroplasts serve as the main site of photosynthesis, the structural and functional integrity of chloroplasts is another prerequisite for photosynthesis [72]. Chloroplasts are also the organelles most sensitive to salt stress, which often damages the chloroplast membrane system and deforms the chloroplast structure. In *Cornus hongkongensis* subsp. *elegans*, *Vitis amurensis* Rupr, and *Podocarpus macrophyllus*, salt stress led to disorganized cystoid, basidiome, and stroma lamellae structures, which seriously damaged the integrity of the chloroplast ultrastructure [17,27,73]. The results of our experiments showed that the grease content increased as the salt concentration increased. Additionally, the chloroplast ultrastructure in *R. pseudoacacia* seedlings was clearly damaged by salt stress, with the chloroplasts being distorted and deformed, and the starch granules swollen and irregular (Figure 8). Previous reports showed that *Atriplex halimus* may form lipid deposition to resist the harmful effects of salt-induced toxicity [74,75]. The chloroplast ultrastructure was consistent with previous reports of salt-treated diploid black locust [71].

The photosynthetic complex PSII is an important site for the absorption, transmission, and conversion of light energy in the light reactions, and the core proteins D1, V2, and D2 are encoded by the chloroplast genes *psbA*, *psbC*, and *psbD*. In this experiment, the expression of *psbA*, *psbC*, and *psbD* was significantly upregulated under 50 mM NaCl treatment (Figure 9), indicating that *R. pseudoacacia* seedlings tolerated low concentration NaCl stress. However, these genes were downregulated at elevated salt concentrations, suggesting that these core PSII proteins are important factors affecting photosynthetic efficiency during salt stress. *psaA* has a light-induced regulatory function during the transition from proplastid to chloroplast in C4 plants (such as sorghum) [76]. In the present study, the expression of *psaA*, *psaB*, and *psaC* was also significantly upregulated under low-concentration NaCl treatment but was significantly downregulated under high-concentration NaCl treatment (Figure 9). We speculate that the upregulation of these genes at lower NaCl concentrations can compensate for the damage to chloroplasts caused by salt stress, but this compensatory mechanism becomes insufficient at higher NaCl concentrations.

We also found that the expression of *rps7*, which encodes a chloroplast ribosomal protein that plays an important role in maintaining chloroplast function, was significantly upregulated under low-concentration NaCl treatment but significantly downregulated under high-concentration NaCl treatment (Figure 9), indicating that excess NaCl severely affected the synthesis of chloroplast ribosomal proteins. The NADH dehydrogenase-like (NDH) complex is involved in photosynthetic electron transport chains and catalyzes the transfer of electrons to drive the production of adenosine triphosphate [77]. It is encoded by the chloroplast genome and contributes to the adaptation of chloroplasts to environmental stresses and plays an important role in photosynthetic efficiency and response to stress [28]. The expression of *ndhE* and *ndhH*, encoding an NADH subunit and the PSI psaC subunit, respectively [78], was significantly downregulated under high concentrations of NaCl (Figure 9). We presume that these high NaCl concentrations severely disrupted signaling processes regulated by *ndhE* and *ndhH*, preventing the synthesis of the NDH complex. Several studies have demonstrated a strong relationship between *ropA* expression and abiotic stresses. In Arabidopsis, low levels of oxidative stress activate *ropA*, leading to the production of H_2O_2 and ethanol dehydrogenase to resist hypoxic stress [79,80]. In the present experiment, the expression of *ropA* was upregulated at low concentrations of NaCl, indicating that *R. pseudoacacia* seedlings are tolerant of low-concentration NaCl treatment (Figure 9), which is consistent with the antioxidant results. However, at high concentrations of NaCl (100 mM, 150 mM, and 200 mM), the above genes regulating chloroplast development were significantly downregulated, affecting the development and structure of chloroplasts and leading to structural abnormalities and functional damage.

5. Conclusions

In all, our results showed that *R. pseudoacacia* seedlings can tolerate low levels of NaCl. Although biomass decreased, photosynthetic structures were not damaged, and the plants continued to grow in the presence of low concentrations of NaCl. However, the high-concentration NaCl treatment downregulated ion transport- and chloroplast development-related gene expression, leading to ion accumulation and damage to photosynthetic structures, and thus resulting in growth arrest and sometimes death. These results will provide theoretical guidance for planting *R. pseudoacacia* on saline-alkali land.

Supplementary Materials: The following supporting information can be downloaded at: https://www.mdpi.com/article/10.3390/plants12061283/s1, Table S1: Primer pairs for real-time quantitative PCR.

Author Contributions: S.W. and G.L. designed the research and revised the paper; C.L. and L.L. performed the experiments; C.L. and L.L. wrote the paper with contributions from the other authors; X.L. and M.C. analyzed the data. All authors have read and agreed to the published version of the manuscript.

Funding: This work was supported by the Shandong Province key research and development plan (2021TZXD005), Special Project of Centra Government for Local Science and Technology Development of Shandong Province (YDZX20203700003989).

Data Availability Statement: The data are contained within the article.

Conflicts of Interest: The authors declare no conflict of interest.

References

1. Mukhopadhyay, R.; Sarkar, B.; Jat, H.S.; Sharma, P.C.; Bolan, N.S. Soil salinity under climate change: Challenges for sustainable agriculture and food security. *J. Environ. Manag.* **2021**, *280*, 111736. [CrossRef]
2. Hassani, A.; Azapagic, A.; Shokri, N. Global predictions of primary soil salinization under changing climate in the 21st century. *Nat. Commun.* **2021**, *12*, 6663. [CrossRef]
3. Riadh, K.; Wided, M.; Hans-Werner, K.; Chedly, A. Responses of Halophytes to Environmental Stresses with Special Emphasis to Salinity. In *Advances in Botanical Research*; Kader, J.-C., Delseny, M., Eds.; Academic Press: Cambridge, MA, USA, 2010; Volume 53, pp. 117–145.
4. Liu, L.; Wang, B. Protection of Halophytes and Their Uses for Cultivation of Saline-Alkali Soil in China. *Biology* **2021**, *10*, 353. [CrossRef] [PubMed]
5. Ruan, C.-J.; da Silva, J.A.T.; Mopper, S.; Qin, P.; Lutts, S. Halophyte Improvement for a Salinized World. *Crit. Rev. Plant Sci.* **2010**, *29*, 329–359. [CrossRef]
6. Egamberdieva, D.; Wirth, S.; Bellingrath-Kimura, S.D.; Mishra, J.; Arora, N.K. Salt-Tolerant Plant Growth Promoting Rhizobacteria for Enhancing Crop Productivity of Saline Soils. *Front. Microbiol.* **2019**, *10*, 2791. [CrossRef] [PubMed]

7. Shahid, S.A.; Zaman, M.; Heng, L. Soil Salinity: Historical Perspectives and a World Overview of the Problem. In *Guideline for Salinity Assessment, Mitigation and Adaptation Using Nuclear and Related Techniques*; Zaman, M., Shahid, S.A., Heng, L., Eds.; Springer International Publishing: Cham, Switzerland, 2018; pp. 43–53. [CrossRef]
8. Abbasi, H.; Jamil, M.; Haq, A.; Ali, S.; Ahmad, R.; Malik, Z.; Parveen. Salt stress manifestation on plants, mechanism of salt tolerance and potassium role in alleviating it: A review. *Zemdirbyste-Agriculture* **2016**, *103*, 229–238. [CrossRef]
9. Ghars, M.A.; Parre, E.; Debez, A.; Bordenave, M.; Richard, L.; Leport, L.; Bouchereau, A.; Savouré, A.; Abdelly, C. Comparative salt tolerance analysis between Arabidopsis thaliana and Thellungiella halophila, with special emphasis on K(+)/Na(+) selectivity and proline accumulation. *J. Plant Physiol.* **2008**, *165*, 588–599. [CrossRef] [PubMed]
10. Abdelhamid, M.T.; Sekara, A.; Pessarakli, M.; Alarcón, J.J.; Brestic, M.; El-Ramady, H.; Gad, N.; Mohamed, H.I.; Fares, W.M.; Heba, S.S.; et al. New Approaches for Improving Salt Stress Tolerance in Rice. In *Rice Research for Quality Improvement: Genomics and Genetic Engineering: Volume 1: Breeding Techniques and Abiotic Stress Tolerance*; Roychoudhury, A., Ed.; Springer: Singapore, 2020; pp. 247–268. [CrossRef]
11. Tseng, M.J.; Liu, C.W.; Yiu, J.C. Enhanced tolerance to sulfur dioxide and salt stress of transgenic Chinese cabbage plants expressing both superoxide dismutase and catalase in chloroplasts. *Plant Physiol. Biochem.* **2007**, *45*, 822–833. [CrossRef]
12. Liu, Y.; He, C. A review of redox signaling and the control of MAP kinase pathway in plants. *Redox Biol.* **2017**, *11*, 192–204. [CrossRef]
13. Lu, Y.; Lei, J.; Zeng, F. NaCl salinity-induced changes in growth, photosynthetic properties, water status and enzymatic antioxidant system of Nitraria Roborowskii kom. *Pak. J. Bot.* **2016**, *48*, 843–851.
14. Siddiqui, M.N.; Mostofa, M.G.; Akter, M.M.; Srivastava, A.K.; Sayed, M.A.; Hasan, M.S.; Tran, L.P. Impact of salt-induced toxicity on growth and yield-potential of local wheat cultivars: Oxidative stress and ion toxicity are among the major determinants of salt-tolerant capacity. *Chemosphere* **2017**, *187*, 385–394. [CrossRef] [PubMed]
15. Xu, Z.; Jiang, Y.; Zhou, G. Response and adaptation of photosynthesis, respiration, and antioxidant systems to elevated CO_2 with environmental stress in plants. *Front. Plant Sci.* **2015**, *6*, 701. [CrossRef] [PubMed]
16. Yang, Z.; Li, J.L.; Liu, L.N.; Xie, Q.; Sui, N. Photosynthetic Regulation Under Salt Stress and Salt-Tolerance Mechanism of Sweet Sorghum. *Front. Plant Sci.* **2019**, *10*, 1722. [CrossRef]
17. Zahra, N.; Al Hinai, M.S.; Hafeez, M.B.; Rehman, A.; Wahid, A.; Siddique, K.H.M.; Farooq, M. Regulation of photosynthesis under salt stress and associated tolerance mechanisms. *Plant Physiol. Biochem.* **2022**, *178*, 55–69. [CrossRef]
18. Chaves, M.M.; Flexas, J.; Pinheiro, C. Photosynthesis under drought and salt stress: Regulation mechanisms from whole plant to cell. *Ann. Bot.* **2009**, *103*, 551–560. [CrossRef] [PubMed]
19. Swoczyna, T.; Kalaji, H.M.; Bussotti, F.; Mojski, J.; Pollastrini, M. Environmental stress—What can we learn from chlorophyll a fluorescence analysis in woody plants? A review. *Front. Plant Sci.* **2022**, *13*, 1048582. [CrossRef]
20. Ben Amor, N.; Jiménez, A.; Boudabbous, M.; Sevilla, F.; Abdelly, C. Chloroplast Implication in the Tolerance to Salinity of the Halophyte Cakile maritima. *Russ. J. Plant Physiol.* **2020**, *67*, 507–514. [CrossRef]
21. Parida, A.K.; Das, A.B.; Mittra, B. Effects of NaCl Stress on the Structure, Pigment Complex Composition, and Photosynthetic Activity of Mangrove Bruguiera parviflora Chloroplasts. *Photosynthetica* **2003**, *41*, 191–200. [CrossRef]
22. Sui, N.; Yang, Z.; Liu, M.; Wang, B.J.B.G. Identification and transcriptomic profiling of genes involved in increasing sugar content during salt stress in sweet sorghum leaves. *BMC Genom.* **2015**, *16*, 534. [CrossRef]
23. Gulzar, S.; Hussain, T.; Gul, B.; Hameed, A. Photosynthetic Adaptations and Oxidative Stress Tolerance in Halophytes from Warm Subtropical Region. In *Handbook of Halophytes: From Molecules to Ecosystems towards Biosaline Agriculture*; Grigore, M.-N., Ed.; Springer International Publishing: Cham, Switzerland, 2021; pp. 1515–1545. [CrossRef]
24. Hameed, A.; Ahmed, M.Z.; Hussain, T.; Aziz, I.; Ahmad, N.; Gul, B.; Nielsen, B.L. Effects of Salinity Stress on Chloroplast Structure and Function. *Cells* **2021**, *10*, 2023. [CrossRef]
25. Bose, J.; Munns, R.; Shabala, S.; Gilliham, M.; Pogson, B.; Tyerman, S.D. Chloroplast function and ion regulation in plants growing on saline soils: Lessons from halophytes. *J. Exp. Bot.* **2017**, *68*, 3129–3143. [CrossRef] [PubMed]
26. Lin, J.; Li, J.P.; Yuan, F.; Yang, Z.; Wang, B.S.; Chen, M. Transcriptome profiling of genes involved in photosynthesis in Elaeagnus angustifolia L. under salt stress. *Photosynthetica* **2018**, *56*, 998–1009. [CrossRef]
27. Goussi, R.; Manaa, A.; Derbali, W.; Cantamessa, S.; Abdelly, C.; Barbato, R. Comparative analysis of salt stress, duration and intensity, on the chloroplast ultrastructure and photosynthetic apparatus in Thellungiella salsuginea. *J. Photochem. Photobiol. B Biol.* **2018**, *183*, 275–287. [CrossRef] [PubMed]
28. Zhang, Y.; Zhang, A.; Li, X.; Lu, C. The Role of Chloroplast Gene Expression in Plant Responses to Environmental Stress. *Int. J. Mol. Sci.* **2020**, *21*, 6082. [CrossRef] [PubMed]
29. Martín, M.; Funk, H.T.; Serrot, P.H.; Poltnigg, P.; Sabater, B. Functional characterization of the thylakoid Ndh complex phosphorylation by site-directed mutations in the ndhF gene. *Biochim. Biophys. Acta* **2009**, *1787*, 920–928. [CrossRef]
30. Huang, Y.; Cai, G.; Pan, D.; Huang, Z. Effects of salt stress on expression of four chloroplast genes in eucalyptus. *J. North. For. Uni.* **2019**, *47*, 4. [CrossRef]
31. Kleinbauer, I.; Dullinger, S.; Peterseil, J.; Essl, F. Climate change might drive the invasive tree Robinia pseudacacia into nature reserves and endangered habitats. *Biol. Conserv.* **2010**, *143*, 382–390. [CrossRef]
32. Buzhdygan, O.Y.; Rudenko, S.S.; Kazanci, C.; Patten, B.C. Effect of invasive black locust (Robinia pseudoacacia L.) on nitrogen cycle in floodplain ecosystem. *Ecol. Model.* **2016**, *319*, 170–177. [CrossRef]

33. Lu, C.; Zhao, C.; Liu, J.; Li, K.; Wang, B.; Chen, M. Increased salinity and groundwater levels lead to degradation of the Robinia pseudoacacia forest in the Yellow River Delta. *J. For. Res.* **2021**, *33*, 1233–1245. [CrossRef]
34. Bradford, M.M. A rapid and sensitive method for the quantitation of microgram quantities of protein utilizing the principle of protein-dye binding. *Anal. Biochem.* **1976**, *72*, 248–254. [CrossRef] [PubMed]
35. Tamás, L.; Dudíková, J.; Durceková, K.; Halusková, L.; Huttová, J.; Mistrík, I.; Ollé, M. Alterations of the gene expression, lipid peroxidation, proline and thiol content along the barley root exposed to cadmium. *J. Plant Physiol.* **2008**, *165*, 1193–1203. [CrossRef] [PubMed]
36. Draper, H.H.; Hadley, M. Malondialdehyde determination as index of lipid peroxidation. *Methods Enzymol.* **1990**, *186*, 421–431. [CrossRef] [PubMed]
37. Wang, A.G.; Luo, G.H. Quantitative Relation between the Reaction of Hydroxylamine and Superoxide Anion Radicals in Plants. *Plant Physiol. Commun.* **1990**, *84*, 2895–2898. [CrossRef]
38. Sui, N.; Wang, Y.; Liu, S.; Yang, Z.; Wang, F.; Wan, S. Transcriptomic and Physiological Evidence for the Relationship between Unsaturated Fatty Acid and Salt Stress in Peanut. *Front. Plant Sci.* **2018**, *9*, 7. [CrossRef] [PubMed]
39. Sairam, R.; Srivastava, G.C. Changes in antioxidant activity in sub-cellular fractions of tolerant and susceptible wheat genotypes in response to long term salt stress. *Plant Sci.* **2002**, *162*, 897–904. [CrossRef]
40. Li, J.; Liu, Y.; Zhang, M.; Xu, H.; Ning, K.; Wang, B.; Chen, M. Melatonin increases growth and salt tolerance of Limonium bicolor by improving photosynthetic and antioxidant capacity. *BMC Plant Biol.* **2022**, *22*, 16. [CrossRef]
41. Guo, J.; Shi, G.; Guo, X.; Zhang, L.; Xu, W.; Wang, Y.; Su, Z.; Hua, J. Transcriptome analysis reveals that distinct metabolic pathways operate in salt-tolerant and salt-sensitive upland cotton varieties subjected to salinity stress. *Plant Sci.* **2015**, *238*, 33–45. [CrossRef]
42. Giannopolitis, C.N.; Ries, S.K. Superoxide dismutases: I. Occurrence in higher plants. *Plant Physiol.* **1977**, *59*, 309–314. [CrossRef] [PubMed]
43. Tao, L.I.; Liu, R.J.; Xin-Hua, H.E.; Wang, B.S. Enhancement of Superoxide Dismutase and Catalase Activities and Salt Tolerance of Euhalophyte Suaeda salsa L. by Mycorrhizal Fungus Glomus mosseae. *Pedosphere* **2012**, *22*, 8. [CrossRef]
44. Bouteraa, M.T.; Mishra, A.; Romdhane, W.B.; Hsouna, A.B.; Siddique, K.H.M.; Saad, R.B. Bio-Stimulating Effect of Natural Polysaccharides from Lobularia maritima on Durum Wheat Seedlings: Improved Plant Growth, Salt Stress Tolerance by Modulating Biochemical Responses and Ion Homeostasis. *Plants* **2022**, *11*, 1991. [CrossRef]
45. Jimenez, A.; Hernandez, J.A.; Del Rio, L.A.; Sevilla, F. Evidence for the Presence of the Ascorbate-Glutathione Cycle in Mitochondria and Peroxisomes of Pea Leaves. *Plant Physiol.* **1997**, *114*, 275–284. [CrossRef] [PubMed]
46. Gao, W.; Zhang, Y.; Feng, Z.; Bai, Q.; He, J.; Wang, Y. Effects of Melatonin on Antioxidant Capacity in Naked Oat Seedlings under Drought Stress. *Molecules* **2018**, *23*, 1580. [CrossRef] [PubMed]
47. Mukri, G.; Patil, M.S.; Motagi, B.N.; Bhat, J.S.; Singh, C.; Jeevan Kumar, S.P.; Gadag, R.N.; Gupta, N.C.; Simal-Gandara, J. Genetic variability, combining ability and molecular diversity-based parental line selection for heterosis breeding in field corn (Zea mays L.). *Mol. Biol. Rep.* **2022**, *49*, 4517–4524. [CrossRef]
48. Lawrence, R.A. A pocket calculator program for Duncan's New Multiple Range Test and analysis of variance. *Comput. Biol. Med.* **1984**, *14*, 357–362. [CrossRef]
49. Wang, F.; Xu, Y.G.; Wang, S.; Shi, W.; Liu, R.; Feng, G.; Song, J. Salinity affects production and salt tolerance of dimorphic seeds of Suaeda salsa. *Plant Physiol. Biochem.* **2015**, *95*, 41–48. [CrossRef] [PubMed]
50. Liu, Q.; Liu, R.; Ma, Y.; Song, J. Physiological and molecular evidence for Na + and Cl − exclusion in the roots of two Suaeda salsa populations. *Aquat. Bot.* **2018**, *146*, 1–7. [CrossRef]
51. Gong, Z.; Xiong, L.; Shi, H.; Yang, S.; Herrera-Estrella, L.R.; Xu, G.; Chao, D.-Y.; Li, J.; Wang, P.-Y.; Qin, F.; et al. Plant abiotic stress response and nutrient use efficiency. *Sci. China Life Sci.* **2020**, *63*, 635–674. [CrossRef]
52. Mo, H.; Yin, Y.; Lu, Z.; Wei, X.; Xu, J. Effects of NaCl stress on the seedling growth and K+ and Na+ allocation of four leguminous tree species. *J. Appl. Ecol.* **2011**, *22*, 1155–1161. [CrossRef]
53. Miranda, R.S.; Mesquita, R.O.; Costa, J.H.; Alvarez-Pizarro, J.C.; Prisco, J.T.; Gomes-Filho, E. Integrative Control Between Proton Pumps and SOS1 Antiporters in Roots is Crucial for Maintaining Low Na+ Accumulation and Salt Tolerance in Ammonium-Supplied Sorghum bicolor. *Plant Cell Physiol.* **2017**, *58*, 522–536. [CrossRef]
54. Qiu, Q.S.; Guo, Y.; Dietrich, M.A.; Schumaker, K.S.; Zhu, J.K. Regulation of SOS1, a plasma membrane Na+/H+ exchanger in Arabidopsis thaliana, by SOS2 and SOS3. *Proc. Natl. Acad. Sci. USA* **2002**, *99*, 8436–8441. [CrossRef]
55. Zhang, W.-D.; Wang, P.; Bao, Z.; Ma, Q.; Duan, L.-J.; Bao, A.-K.; Zhang, J.-L.; Wang, S.-M. SOS1, HKT1;5, and NHX1 Synergistically Modulate Na(+) Homeostasis in the Halophytic Grass Puccinellia tenuiflora. *Front. Plant Sci.* **2017**, *8*, 576. [CrossRef]
56. Krishnamurthy, P.; Vishal, B.; Khoo, K.; Rajappa, S.; Loh, C.S.; Kumar, P.P. Expression of AoNHX1 increases salt tolerance of rice and Arabidopsis, and bHLH transcription factors regulate AtNHX1 and AtNHX6 in Arabidopsis. *Plant Cell Rep.* **2019**, *38*, 1299–1315. [CrossRef]
57. Li, N.; Wang, X.; Ma, B.; Du, C.; Zheng, L.; Wang, Y. Expression of a Na(+)/H(+) antiporter RtNHX1 from a recretohalophyte Reaumuria trigyna improved salt tolerance of transgenic Arabidopsis thaliana. *J. Plant Physiol.* **2017**, *218*, 109–120. [CrossRef]
58. Verma, D.; Jalmi, S.K.; Bhagat, P.K.; Verma, N.; Sinha, A.K. A bHLH transcription factor, MYC2, imparts salt intolerance by regulating proline biosynthesis in Arabidopsis. *FEBS J.* **2020**, *287*, 2560–2576. [CrossRef] [PubMed]

59. Boriboonkaset, T.; Theerawitaya, C.; Yamada, N.; Pichakum, A.; Supaibulwatana, K.; Cha-Um, S.; Takabe, T.; Kirdmanee, C. Regulation of some carbohydrate metabolism-related genes, starch and soluble sugar contents, photosynthetic activities and yield attributes of two contrasting rice genotypes subjected to salt stress. *Protoplasma* **2013**, *250*, 1157–1167. [CrossRef] [PubMed]
60. Cheng, B.; Hassan, M.J.; Feng, G.; Zhao, J.; Liu, W.; Peng, Y.; Li, Z. Metabolites Reprogramming and Na(+)/K(+) Transportation Associated With Putrescine-Regulated White Clover Seed Germination and Seedling Tolerance to Salt Toxicity. *Front. Plant Sci.* **2022**, *13*, 856007. [CrossRef] [PubMed]
61. Nabati, J.; Kafi, M.; Nezami, A.; Moghaddam, P.R.; Masomi, A.; Mehrjerdi, M.Z. Effect of salinity on biomass production and activities of some key enzymatic antioxidants in kochia (Kochia scoparia). *Pak. J. Bot.* **2011**, *43*, 539–548. [CrossRef]
62. Yang, Y.; Guo, Y. Unraveling salt stress signaling in plants. *J. Integr. Plant Biol.* **2018**, *60*, 796–804. [CrossRef]
63. Gao, Y.; Long, R.; Kang, J.; Wang, Z.; Zhang, T.; Sun, H.; Li, X.; Yang, Q. Comparative Proteomic Analysis Reveals That Antioxidant System and Soluble Sugar Metabolism Contribute to Salt Tolerance in Alfalfa (*Medicago sativa* L.) Leaves. *J. Proteome Res.* **2019**, *18*, 191–203. [CrossRef]
64. Siddiqui, M.H.; Alamri, S.; Al-Khaishany, M.Y.; Khan, M.N.; Al-Amri, A.; Ali, H.M.; Alaraidh, I.A.; Alsahli, A.A. Exogenous Melatonin Counteracts NaCl-Induced Damage by Regulating the Antioxidant System, Proline and Carbohydrates Metabolism in Tomato Seedlings. *Int. J. Mol. Sci.* **2019**, *20*, 353. [CrossRef]
65. Castro, B.; Citterico, M.; Kimura, S.; Stevens, D.; Wrzaczek, M.; Coaker, G. Stress-induced reactive oxygen species compartmentalization, perception and signalling. *Nat. Plants* **2021**, *7*, 403–412. [CrossRef] [PubMed]
66. Liang, W.; Ma, X.; Wan, P.; Liu, L. Plant salt-tolerance mechanism: A review. *Biochem. Biophys. Res. Commun.* **2018**, *495*, 286–291. [CrossRef] [PubMed]
67. Naliwajski, M.; Skłodowska, M. The Relationship between the Antioxidant System and Proline Metabolism in the Leaves of Cucumber Plants Acclimated to Salt Stress. *Cells* **2021**, *10*, 609. [CrossRef] [PubMed]
68. Wang, X.; He, X.; Fan, Y.; Guo, D. Effects of salt stress on seed germination and seeding antioxidant enzyme activities of Sweet Sorgjum. *Mol. Plant Breed.* **2022**, *20*, 4462–4467. [CrossRef]
69. Noctor, G.; Foyer, C.H. Intracellular Redox Compartmentation and ROS-Related Communication in Regulation and Signaling. *Plant Physiol.* **2016**, *171*, 1581–1592. [CrossRef]
70. Poór, P.; Borbély, P.; Czékus, Z.; Takács, Z.; Ördög, A.; Popović, B.; Tari, I. Comparison of changes in water status and photosynthetic parameters in wild type and abscisic acid-deficient sitiens mutant of tomato (Solanum lycopersicum cv. Rheinlands Ruhm) exposed to sublethal and lethal salt stress. *J. Plant Physiol.* **2019**, *232*, 130–140. [CrossRef]
71. Wang, Z.; Wang, M.; Liu, L.; Meng, F. Physiological and Proteomic Responses of Diploid and Tetraploid Black Locust (*Robinia pseudoacacia* L.) Subjected to Salt Stress. *Int. J. Mol. Sci.* **2013**, *14*, 20299–20325. [CrossRef]
72. Kirchhoff, H. Chloroplast ultrastructure in plants. *New Phytol.* **2019**, *223*, 565–574. [CrossRef] [PubMed]
73. Zou, L.; Li, T.; Li, B.; He, J.; Liao, C.; Wang, L.; Xue, S.; Sun, T.; Ma, X.; Wu, Q. De novo transcriptome analysis provides insights into the salt tolerance of Podocarpus macrophyllus under salinity stress. *BMC Plant Biol.* **2021**, *21*, 489. [CrossRef] [PubMed]
74. Kelley, D. *Salinity Effects on Growth and Fine Structure of Atriplex Halimus 1*; Texas Tech University: Lubbock, TX, USA, 1974.
75. Peharec Štefanić, P.; Koffler, T.; Adler, G.; Bar-Zvi, D. Chloroplasts of salt-grown Arabidopsis seedlings are impaired in structure, genome copy number and transcript levels. *PLoS ONE* **2013**, *8*, e82548. [CrossRef]
76. She, D.; Zhang, H.; Wang, Y.; Hou, H. Sequence analyses of chloroplastic psaA gene fragment from Porphyra haitanensis. *Fish. Sci. Res.* **2007**, *26*, 289–291. [CrossRef]
77. Ma, M.; Zhong, M.; Zhang, Q.; Zhao, W.; Wang, M.; Luo, C.; Xu, B. A genome-wide analysis of the chloroplast NADH dehydrogenase-like genes in Zostera marina. *J. Oceanol. Limnol.* **2022**, *40*, 656–677. [CrossRef]
78. Yukawa, M.; Tsudzuki, T.; Sugiura, M. The 2005 version of the chloroplast DNA sequence from tobacco (Nicotiana tabacum). *Plant Mol. Biol. Rep.* **2005**, *23*, 359–365. [CrossRef]
79. Baxter-Burrell, A.; Yang, Z.; Springer, P.S.; Bailey-Serres, J. RopGAP4-dependent Rop GTPase rheostat control of Arabidopsis oxygen deprivation tolerance. *Science* **2002**, *296*, 2026–2028. [CrossRef]
80. Chen, Y.; Wang, S.; Liu, X.; Wang, D.; Liu, Y.; Hu, L.; Meng, S. Analysis of Rac/Rop Small GTPase Family Expression in Santalum album L. and Their Potential Roles in Drought Stress and Hormone Treatments. *Life* **2022**, *12*, 1980. [CrossRef] [PubMed]

plants

MDPI

Article

The ABCISIC ACID INSENSITIVE (ABI) 4 Transcription Factor Is Stabilized by Stress, ABA and Phosphorylation

Tzofia Maymon [1,2], Nadav Eisner [1,2] and Dudy Bar-Zvi [1,2,*]

1 Department of Life Sciences, Ben-Gurion University of the Negev, 1 Ben-Gurion Blvd, Beer-Sheva 8410501, Israel
2 The Doris and Bertie I. Black Center for Bioenergetics in Life Sciences, Ben-Gurion University of the Negev, 1 Ben-Gurion Blvd, Beer-Sheva 8410501, Israel
* Correspondence: barzvi@bgu.ac.il

Abstract: The Arabidopsis transcription factor ABSCISIC ACID INSENSITIVE 4 (ABI4) is a key player in the plant hormone abscisic acid (ABA) signaling pathway and is involved in plant response to abiotic stress and development. Expression of the *ABI4* gene is tightly regulated, with low basal expression. Maximal transcript levels occur during the seed maturation and early seed germination stages. Moreover, ABI4 is an unstable, lowly expressed protein. Here, we studied factors affecting the stability of the ABI4 protein using transgenic Arabidopsis plants expressing *35S::HA-FLAG-ABI4-eGFP*. Despite the expression of eGFP-tagged ABI4 being driven by the highly active 35S CaMV promoter, low steady-state levels of ABI4 were detected in the roots of seedlings grown under optimal conditions. These levels were markedly enhanced upon exposure of the seedlings to abiotic stress and ABA. ABI4 is degraded rapidly by the 26S proteasome, and we report on the role of phosphorylation of ABI4-serine 114 in regulating ABI4 stability. Our results indicate that ABI4 is tightly regulated both post-transcriptionally and post-translationally. Moreover, abiotic factors and plant hormones have similar effects on ABI4 transcripts and ABI4 protein levels. This double-check mechanism for controlling ABI4 reflects its central role in plant development and cellular metabolism.

Keywords: *Arabidopsis thaliana*; ABI4; MAPK; salinity; ABA; proteasome; transcription factor

Citation: Maymon, T.; Eisner, N.; Bar-Zvi, D. The ABCISIC ACID INSENSITIVE (ABI) 4 Transcription Factor Is Stabilized by Stress, ABA and Phosphorylation. *Plants* **2022**, *11*, 2179. https://doi.org/10.3390/plants11162179

Academic Editors: Małgorzata Nykiel, Mateusz Labudda, Beata Prabucka, Marta Gietler and Justyna Fidler

Received: 31 July 2022
Accepted: 18 August 2022
Published: 22 August 2022

1. Introduction

Plant development and response to environmental cues involve signaling pathways in which the last components are often transcription factors (reviewed by [1]). As a result, these signaling pathways affect the transcription of a large number of genes, the expression of which is affected by the respective transcription factors.

The Arabidopsis *ABSCISIC ACID INSENSITIVE 4 (ABI4)* gene encodes an APETALA 2 (AP2) family transcription factor [2]. APETALA 2 is a plant-specific DNA-binding domain with a length of ~60 amino acids first characterized in the Arabidopsis *APETALA2* homeotic gene [3,4]. The *ABI4* gene was identified by screening gamma-irradiated Arabidopsis seeds for mutants capable of germination in the presence of inhibitory concentrations of the plant hormone abscisic acid (ABA) [5]. ABI4 alleles were isolated by screening for germination in the presence of high concentrations of salt and sugar [6–10]. ABI4 also plays a central role in other plant signaling pathways, including lipid mobilization, lateral root development, regulation of light-modulated genes, redox signaling, pathogen response, and mitochondrial retrograde signaling (reviewed in [11]). Its role in chloroplast retrograde signaling is disputed [12,13].

ABI4 expression is tightly developmentally regulated; the highest steady-state levels of the ABI4 transcript are found in embryos, maturing pollen, and early germination stages [14–16]. The transcript levels of *ABI4* are significantly reduced in other developmental stages; its expression is restricted to root phloem companion cells and parenchyma and, to some extent, to the vascular system of the shoot [17–19]. In addition, steady-state levels of the *ABI4* transcript are enhanced by ABA, NaCl, and glucose and repressed by auxin [17,18,20].

ABI4 is a highly unstable protein [21,22]. Several protein motifs, such as PEST and AP2-associated [21,22], destabilize ABI4 via degradation by the proteasome. Other regions of the protein destabilize it in a proteasome-independent manner [21]. ABI4 is stabilized by high concentrations of salt and sugar [21,22] and by preventing light exposure [23]. COP1 is involved in the light-mediated degradation of ABI4 [23]; levels of ABI4 were enhanced in light-exposed *cop1* mutant seedlings and further increased by treating *cop1* mutants with the MG132 proteasome inhibitor, suggesting that COP1, as well as additional E3s, modulate ABI4 stability [23].

Being downstream of the signaling pathway cascades, transcription factors are frequently modulated by phosphorylation, resulting in their activation or inhibition. ABI4 was phosphorylated in vitro by MPK3, MPK4, and MPK6 [24–27]. Phosphorylation of ABI4 by MAPKs repressed the expression of the *LHCB* gene [25] and inhibited the emergence of adventitious roots [26]. In addition, the phosphorylation of S114 is essential for the biological activity of ABI4, as shown in studies of the complementation of the *abi4* mutant phenotype [27].

Here, we studied factors affecting the stability of the ABI4 protein in Arabidopsis plants by expressing *HA-FLAG-ABI4-eGFP* driven by the constitutive highly active 35S promoter. The tagged ABI4 was detected in embryos rescued from imbibed seeds but not in seedlings. Treatment of the seedlings with NaCl resulted in a transient stabilization of ABI4, peaking at 2–4 h. ABA and high glucose also stabilized ABI4-eGFP but with slower kinetics, reaching lower levels than with NaCl treatment. The phosphomimetic ABI4 (S114E) protein was more stable than the wild-type ABI4 and the phosphorylation-null ABI4 (S114A) mutant in salt-treated plants, suggesting that phosphorylation of ABI4 by MAPKs results in stabilization of ABI4. Interestingly, NaCl, ABA, and glucose are known to similarly affect the steady-state levels of ABI4 transcripts [17,18,20]. We thus propose that the MAPK signaling cascade also activates ABI4 transcription via the phosphorylation of MYB, WRKY, and ABI4 transcription factors known to transactivate the transcription of the *ABI4* gene. As a result, similar cues regulate *ABI4* in terms of transcriptional and post-transcriptional levels, resulting in a very tight regulation of this key factor.

2. Results

2.1. The 35S::HA-FLAG-ABI4-eGFP Construct Encodes a Biologically Active Protein

To study ABI4 *in planta*, we used the enhanced green fluorescent protein (eGFP) [28] fused to the carboxy terminus of ABI4 and transcription driven by the highly active cauliflower mosaic virus constitutive 35S promoter (35S) [29]. We previously found that overexpressing *35S::ABI4* in Arabidopsis resulted in seedling death within three days of germination, whereas fusing the HA_3-$FLAG_3$ tag to the N terminus of ABI4 resulted in viable plants [18]. Therefore, constructed *35S::HA-FLAG-ABI4-eGFP* and used it for the transformation of Arabidopsis. To determine whether HA-FLAG-ABI4-eGFP protein is biologically active, we tested whether tagged ABI4 can complement the *abi4-1* mutant. *abi4-1* is a frameshift mutant resulting from a single-bp deletion at codon 157 [2]; the expressed protein has the AP2 DNA binding domain but lacks the transactivation domain. The resulting transgenic plants did not have any visible phenotype when grown on agar plates with 0.5 × MS, 0.5% sucrose medium or in pots containing potting mix. To determine whether the expressed HA-FLAG-ABI4-eGFP (ABI4-eGFP) protein retains the biological activity of ABI4, we examined its ability to complement the phenotype of *abi4* mutants by assaying seed germination in the presence of ABA, the most extensively studied phenotype of these mutants [30]. Figure 1 shows that expressing HA-FLAG-ABI4-eGFP in *abi4-1* plants restored the ABA sensitivity, indicating that tagging ABI4 at neither its amino- nor carboxy-termini impairs its biological activity.

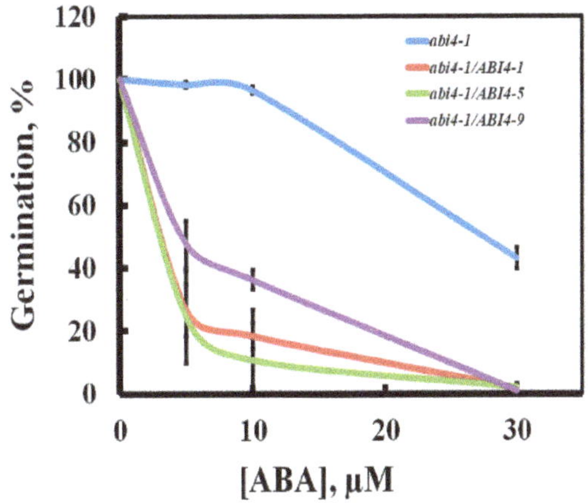

Figure 1. Complementation of the *abi4-1* mutant by *35S::HA*$_3$*-FLAG*$_3$*-ABI4-eGFP*. Seeds of the homozygous plants of the indicated genotypes were plated on agar-solidified 0.5 × MS, 0.5% sucrose medium supplemented with the indicated concentrations of ABA. Germination was scored 7 days later. *abi4*, *abi4-1* mutant; *abi4/ABI4-1*, *-5*, *-9*, transgenic lines 1, 5, and 9 of *abi4-1* plants transformed with the *35S::HA*$_3$*-FLAG*$_3$*-ABI4-eGFP* construct. Data represent means ± SE; $n = 3$ biological replicates.

The 35S promoter is a commonly used strong constitutive promoter that is active in most plant tissues [31]. We therefore expected to detect high eGFP fluorescence signals in seedlings of WT plants transformed with the *35S::HA-FLAG-ABI4-eGFP* construct (WT/ABI4-eGFP). Surprisingly, we did not detect significant fluorescent signals in these plants (Figure 2A,B). To confirm the construct, we examined the fluorescence in embryos prepared from imbibed seeds and detected a highly fluorescent signal in the entire embryo (Figure 2C,D). These results suggest that ABI4 levels may be subject to post-transcriptional regulation.

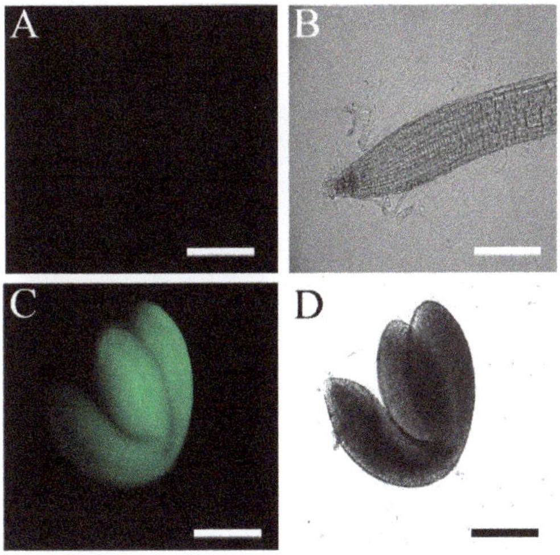

Figure 2. Fluorescence levels of *35S::HA*$_3$*-FLAG*$_3$*-ABI4-GFP* seedling roots and embryosroots. The fluorescence of plants transformed with the *35S::HA*$_3$*-FLAG*$_3$*-ABI4-GFP* construct was examined by microscopy. (**A**,**B**) Ten-day-old root; (**C**,**D**) embryo extracted from a seed imbibed for 24 h; (**A**,**C**) fluorescence images; (**B**,**D**) bright-field images. Scale bar = 100 μm.

2.2. Accumulation of ABI4-eGFP Is NaCl-Dependent

Previous studies showed that environmental signals post-transcriptionally regulate ABI4. To examine whether varying external and internal cues affect the steady-state levels of ABI4, we tested cues known to affect the activity of the *ABI4* promoter. The steady-state levels of *ABI4* mRNA driven by its endogenous promoter are enhanced by NaCl [17]. We therefore examined whether NaCl also affects protein levels of ABI4 when transcription is driven by the 35S promoter. Exposing WT plants expressing *ABI4-eGFP* to 0.3 M NaCl resulted in a transient increase in the eGFP fluorescence signal, with the maximal signal observed 2–3 h following seedling exposure to salt (Figure 3A). The signal was NaCl-dose-dependent, with the maximum at 0.3 M NaCl (Figure 3B). No fluorescence was observed in seedlings transferred to fresh 0.5 × MS, 0.5% sucrose medium, suggesting that the transient increase in fluorescence observed in the NaCl-treated seedlings did not result from transferring the seedlings from the agar plates to the buffer-soaked filter paper. To confirm the observed fluorescence signals, protein extracts of the roots of salt-treated WT/*ABI4-eGFP* seedlings were subjected to western blot analysis using an anti-GFP antibody. The results confirmed that ABI4-eGFP was essentially undetectable in control untreated roots, whereas a transient increase in ABI4-eGFP was observed following exposure to NaCl, peaking 2 h after the application of NaCl (Figure 3C). The protein levels of ABI4-eGFP were low, even at maximal values, and were detected with a high-sensitivity detection assay.

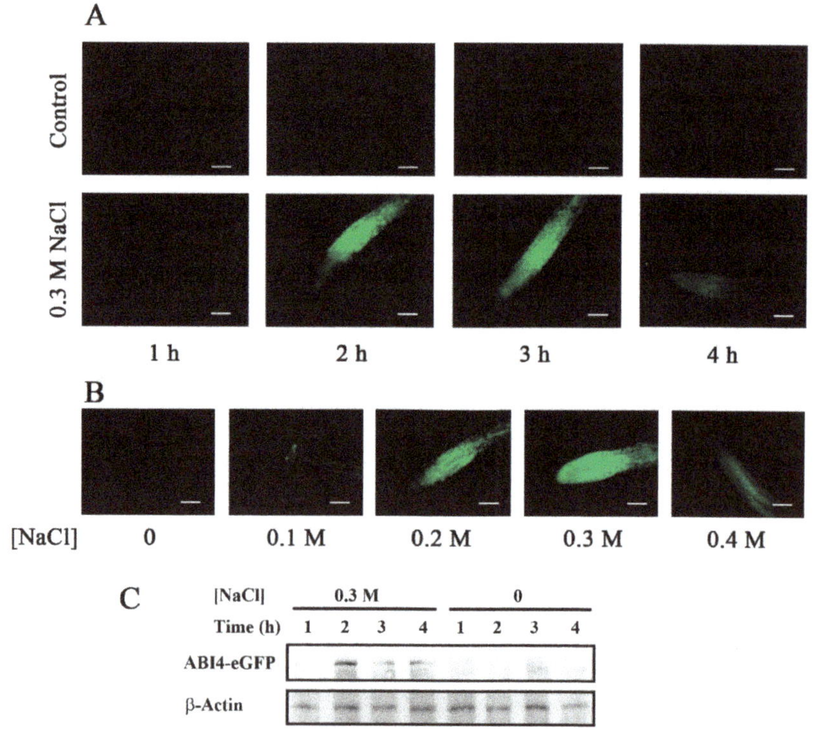

Figure 3. NaCl treatment transiently enhances ABI4-eGFP protein levels. Ten-day-old Arabidopsis plants expressing the *35S::HA$_3$-FLAG$_3$-ABI4-eGFP* construct incubated for the indicated times with 0.5 × MS salts, 0.5% sucrose without (**A**, control, upper row) or with 0.3 M NaCl (**A**, lower row) or for 2.5 h with growth medium containing the indicated incubated concentrations of NaCl (**B**). Roots were examined by fluorescence microscopy. Scale bar = 100 µm. (**C**) Western blot analysis showing the expression levels of ABI4-eGFP following NaCl treatment. β-actin was used as the loading control.

The NaCl-dependent increase in ABI4 protein levels may result from either a change in the transcript levels of the encoding mRNA or regulation of the protein levels. To assess this point, we quantified the *ABI4-eGFP* transcript and protein levels in roots of untreated and NaCl-treated seedlings. *ABI4-eGFP* transcript levels were determined by RT-qPCR using amplification primers from the sequence encoding the HA_3-$FLAG_3$ tag to avoid assaying the expression of the endogenous *ABI4* gene. The steady-state mRNA levels of *ABI4-eGFP* were increased by 2.0-and 1.6-fold at 2.5 and 4 h, respectively, after seedling exposure to high salt concentration. The ABI4-eGFP protein levels quantified using the fluorescence intensity of the roots were 32.7 and 7.3 times higher for roots of plants exposed to 0.3 M NaCl for 2.5 and 4 h, respectively, compared to control untreated roots (Figure 4). Control fluorescence signals of plants expressing *35S::GFP* were not affected by salt treatment (Figure S1). These results indicate that the NaCl-dependent increase in ABI4-GFP protein levels resulted from post-transcriptional regulation of *ABI4* rather than from changes in the transcript levels or the effect of salt on the GFP tag.

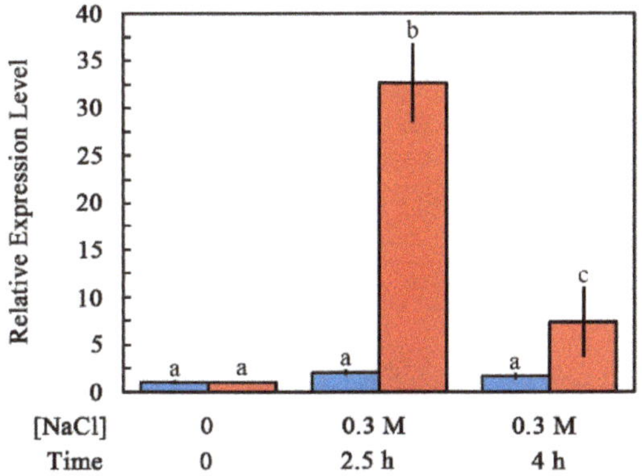

Figure 4. Effect of NaCl treatment on steady-state levels of *ABI4-eGFP* transcript and protein in the roots. Ten-day-old seedlings transformed with the *35S::HA3-FLAG3-ABI4-eGFP* construct were transferred onto a filter paper soaked with 0.5 × MS salts, 0.5% sucrose with or without 0.3 M NaCl. Roots were harvested at the indicated times, and the levels of HA_3-$FLAG_3$-ABI4-eGFP transcript (blue) or protein (red) were determined by RT-qPCR and fluorescence microscopy, respectively. Data represent mean ± SE. Bars with different letters indicate significant differences according to one-way ANOVA and Tukey's HSD post hoc test ($p \leq 0.01$).

2.3. Subcellular Localization of ABI4-GFP following NaCl Treatment Is Cell-Type-Specific

Although NaCl treatment of plants expressing the *35S::HA-FLAG-ABI4-eGFP* construct resulted in enhanced protein levels in most root cells, the observed fluorescence pattern of the ABI4-eGFP was diffusive in most cell types. In contrast, it was found in spherical structures mostly in the root stele, suggesting nuclear localization (Figure 5A). Staining nuclei of the roots of NaCl-treated *ABI4-eGFP* plants with the DNA fluorescence stain 4′,6-diamidino-2-phenylindole (DAPI) (Figure 5B) shows that the DAPI and eGFP fluorescence signals overlap, confirming that ABI4-eGFP is localized in the nuclei of root stele cells (Figure 5C). This pattern is specific to ABI4, as it was not observed in the roots expressing the eGFP tag alone (Figure S1).

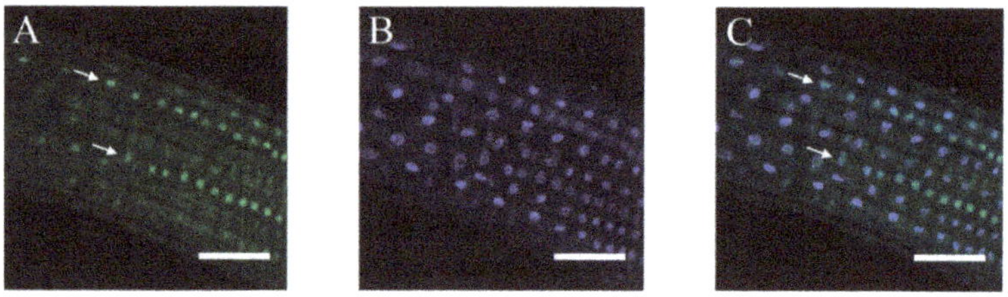

Figure 5. The subcellular localization of ABI4-GFP in the roots is cell-type-dependent. Ten-day old seedlings expressing the *35S::HA3-FLAG3-ABI4-GFP* construct were treated with 0.3 M NaCl for 2.5 h. Roots were stained with DAPI and examined by confocal microscopy. (**A**) GFP fluorescence; (**B**) DAPI fluorescence; (**C**) merged images of (**A**,**B**). Arrows mark columns of cells expressing ABI4-eGFP in the nuclei. Scale bar = 10 μm.

2.4. ABA and Glucose Treatment Enhance ABI4-eGFP Protein Levels

Transcript levels of endogenous *ABI4* are also induced by treatment with ABA or high concentrations of glucose [20]. To determine whether these treatments also affect the levels of the ABI4-eGFP protein, ten-day-old *35S::HA-FLAG-ABI4-eGFP* transgenic plants were transferred to media containing ABA or glucose, and ABI4-eGFP accumulation was followed by fluorescence microscopy. Enhanced ABI4-eGFP levels were detected in the root stele of seedlings approximately 24 h after treatment with ABA or glucose (Figure 6).

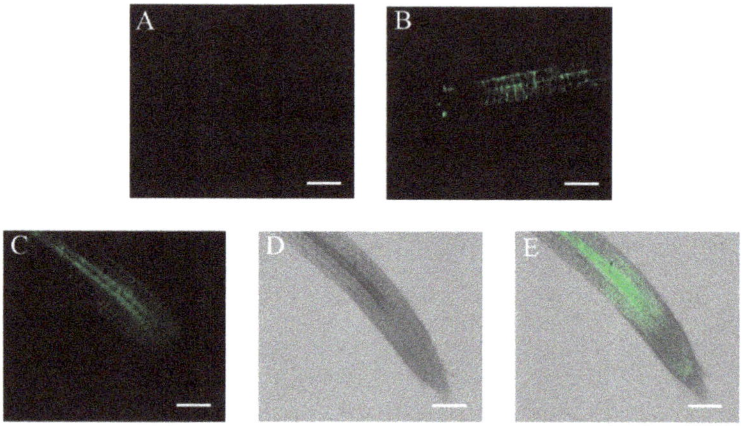

Figure 6. ABI4 expression is increased following glucose and ABA treatment. Ten-day-old *35S::HA3-FLAG3-ABI4-eGFP*-expressing plants were incubated for 24 h in 0.5 × MS, 0.5% sucrose growth medium (**A**) or in the same medium supplemented with 7% glucose (**B**) or 30 μM ABA (**C**). (**D**,**E**) Bright-field and merged image of the ABA treated root shown in (**C**). Scale bar = 100 μm.

2.5. Auxin Counteracts the NaCl-Induced Increase in ABI4-eGFP Levels

ABI4 mediates cytokinin inhibition of lateral root formation by reducing the polar transport of auxin, a plant hormone known to induce the formation of lateral roots [18]. Exogenous auxin also reduced the steady-state levels of *ABI4* transcripts in the roots [18]. To test whether auxin also post-transcriptionally regulates ABI4, we tested whether auxin counteracts the NaCl-induced enhancement of ABI4-eGFP. Figure 7 shows that when added together with NaCl, 3-indole acetic acid (IAA) prevented the NaCl-induced accumulation of ABI4-eGFP, indicating that auxin negatively regulates the steady-state levels of the ABI4 protein.

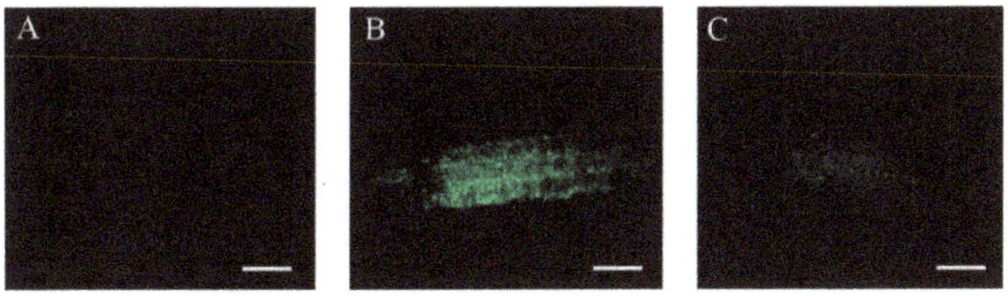

Figure 7. Auxin prevents the NaCl-induced increase in ABI4. Ten-day-old seedlings were treated with 0.5 × MS 0.5% sucrose (**A**) supplemented with 0.3 M NaCl (**B**) or 0.3 M NaCl and 20 μM IAA (**C**). Roots were examined 2.5 h later. Scale bar = 100 μm.

2.6. Steady-State Levels of ABI4-eGFP Are Controlled by De Novo Translation and Degradation by the 26S Proteasome

We used the protein synthesis inhibitor cycloheximide (CHX) and the proteasome inhibitor MG132 to further characterize the transient accumulation of ABI4-eGFP following exposure to NaCl. As expected, CHX prevented the NaCl-dependent accumulation of ABI4-eGFP protein (Figure 8A), suggesting that exposure to NaCl enhances de novo translation of ABI4-eGFP. Treatment with a mix of NaCl and MG132 resulted in increased stabilization of ABI4-eGFP, and a high signal was detected, even 6 h after the co-application of NaCl and MG132 (Figure 8B) but not in the roots of plants treated with NaCl alone.

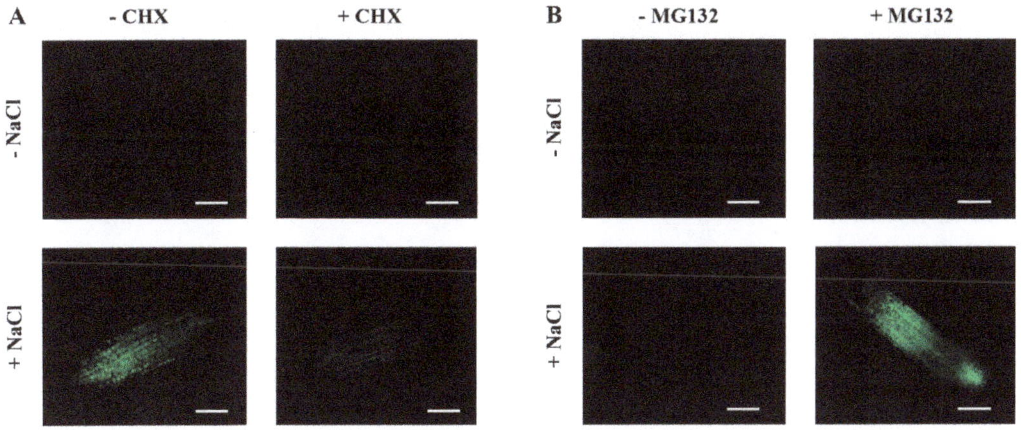

Figure 8. The NaCl-dependent transient increase in the ABI4-eGFP levels is a result of de novo protein synthesis and its degradation by the 26S proteasome. Ten-day-old ABI4-eGFP-expressing seedlings were incubated in light for 2.5 h (**A**) or 6 h (**B**) on filter paper soaked with 0.5 × MS, 0.5% sucrose solution supplemented, as indicated, with 0.3 M NaCl, 20 μg/mL cycloheximide (CHX) or 20 μg/mL MG132. Roots were then examined by fluorescence microscopy. Scale bar = 100 μm.

2.7. The Phosphorylation State of Serine 114 Affects the Stability of the ABI4 Protein

We recently showed that phosphorylation of serine 114 of ABI4 by MPK3 or MPK6 is essential for its biological activity [27]. Here, we tested whether the phosphorylation state of S114 of ABI4 also affects its stability; WT Arabidopsis plants were transformed with *35S::HA-FLAG-ABI4-eGFP* constructs encoding the ABI4 (S114A), phosphorylation-null mutant or ABI4 (S114E), phosphomimetic mutated proteins. Ten-day-old NaCl-treated seedlings were examined by fluorescent microscopy. Roots expressing WT (114S) ABI4-eGFP showed very low levels of fluorescence (Figure 9A), as shown in Figure 3. Fluorescence levels in roots of plants transformed with the ABI4-eGFP (S114A) phosphorylation-null mutant (Figure 9B) were similar to those of the WT ABI4-eGFP protein. In contrast, the

S114E phosphomimetic mutation stabilized the ABI4-eGFP protein, and high levels were observed, even 6 h following NaCl exposure (Figure 9C). The fluorescence signal was quantified (Figure 9D), and the ABI4-eGFP protein levels were also confirmed by western blot analysis using an anti-GFP antibody (Figure 9E). Our data indicate that the phosphorylation of serine 114 by MAPKs stabilizes ABI4, the active form of this transcription factor.

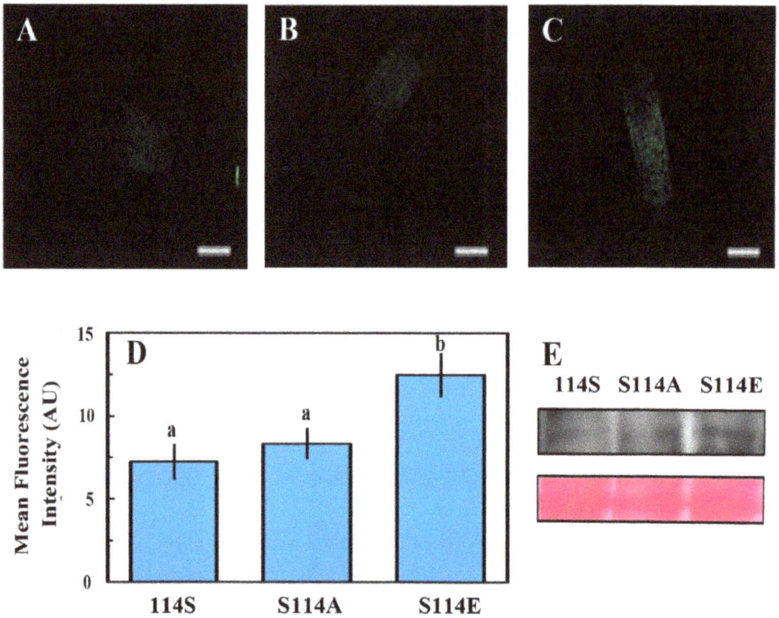

Figure 9. Phosphorylation of S114 stabilizes the ABI4 protein. Ten-day-old ABI4-eGFP expressing seedlings (**A**) WT ABI4-eGFP, (**B**) the phosphorylation-null (S114A) mutant, or (**C**) the phosphomimetic (S114E) mutant, were incubated for 6 h with 0.5 × MS, 0.5% sucrose and 0.3 M NaCl. The roots were examined by fluorescence microscopy. Scale bar = 100 μm. (**D**) The fluorescent signals of 70 plants were quantified. Data are expressed as average ± SE. Bars with different letters represent statistically significant differences according to Tukey's HSD post hoc test ($p < 0.05$). (**E**) Western blot analysis of seedling proteins using anti-GFP antibody showing the expression levels of the S114 ABI4-eGFP phosphorylation state mutants following 6 h of NaCl treatment (upper panel). Ponceau-stained RuBisCo large subunit was used as a loading control (lower panel).

3. Discussion

In this study, we showed that ABI4 protein is highly unstable and that it is degraded by the 26S proteasome. In the roots, ABI4 is transiently stabilized by salt, ABA, and high glucose. Phosphorylation of S114 of ABI4, a residue previously shown to be phosphorylated by MAP kinase, increases its stability. ABI4 is a master transcription regulator, acting as both activator and repressor in the regulation of developmental processes, such as seed development, germination, root development, response to stress and hormones, disease resistance, and lipid metabolism [11,32]. *ABI4* is evolutionarily conserved and is a single gene in Arabidopsis and in most plant genomes that encode ABI4 [11,33], suggesting that its biological role is non-redundant. As a result, *abi4* mutants display pronounced phenotypes, such as insensitivity to ABA inhibition of seed germination and reduced sensitivity to high glucose and salt [6,7,18,30].

3.1. *ABI4 Is a Lowly Expressed and Highly Regulated Gene*

As expected, as a key regulator, ABI4 levels and activity are tightly regulated. Maximal levels of *ABI4* transcripts are detected in developed seeds and in early germination stages, with very low levels present during other developmental stages [14–16,19] in which it is expressed in the phloem and parenchyma of the roots [18,19]. ABI4 expression is regulated by plant hormones: enhanced by ABA [14] and cytokinin [34] and reduced by auxin [18]. It is also enhanced in response to high glucose [20], as well as osmotic [20] and salt [17] stresses. Arabidopsis *ABI4* is encoded by an intronless gene. Intronless genes are characteristic of highly regulated TFs in both plants and animals [35,36]. Moreover, intronless genes are differentially expressed in response to drought and salt treatment [36].

3.2. *ABI4 Is a Post-Transcriptionally Regulated Low-Level Protein*

Because the transcription of *ABI4* is highly regulated, in order to study the post-transcriptional regulation of ABI4, we expressed *HA-FLAG-ABI4-eGFP* driven by the constitutive highly expressed CaMV 35S promoter (35S). eGFP is a GFP variant that is 35 times brighter than the original GFP [37], allowing for the detection of lower concentrations of tagged proteins than with the previously used ABI4-GFP [21]. This construct complemented the phenotype of the *abi4*-1 mutant, indicating that tagging ABI4 at its N and C termini does not eliminate its biological activity (Figure 1). We did not detect recombinant ABI4 protein in ten-day-old seedlings grown on plates under control conditions (Figure 2). In contrast, a high fluorescence signal was observed in imbibed embryos, confirming the performance of the construct (Figure 2). Although we used the highly active constitutive viral 35S CaMV promoter to express ABI4-eGFP, the resulting transgenic plants did not show any significant fluorescence of the eGFP tag (Figure 2). GFP-ABI4 was not detected in Arabidopsis plants transformed with 35S::GFP-ABI4 [21]. GUS activity staining identified the expression of ABI4-GUS recombinant protein driven by the same promoter. In contrast, ABI4-GFP fluorescence was detected in Arabidopsis protoplasts transfected with a *35S::ABI4-GFP* construct [22]. This discrepancy may be explained by protoplasts being under stress caused by enzymatic digestion of the cell wall [38].

3.3. *ABI4 Is Stabilized by External Signals*

Although the 35S promoter is active in most plant tissues, ABI4-GFP is expressed primarily in the roots following stress (Figure 3), confirming observations by Finkelstein et al., who expressed ABI4-GUS fusion protein [21]. ABI4-eGFP was observed mainly in the vascular system of the roots following ABA and glucose treatments (Figure 6). ABI4-eGFP accumulated in the cells in which the endogenous ABI4 promoter was active [18]. Although NaCl treatment resulted in the accumulation of ABI4-eGFP throughout the roots, it was targeted to the nuclei only in the vascular cells (Figure 5), suggesting that both the accumulation and subcellular localization of the ABI4 protein are regulated in a cell-specific manner. This is similar to the transcription factor ABI5, which also accumulated following NaCl and ABA treatments, although increased levels were observed only four days after exposure to 200 mM NaCl [39]. Furthermore, ABA stabilization of ABI5 was restricted to a narrow developmental window 2 days after germination [39]. The difference in kinetics and responsive window suggests that although ABI5 and ABI4 proteins are stabilized by similar agents (ABA and NaCl), each protein has different domains [2,40], and as such, they are likely to be stabilized through other mechanisms. Transient expression of stress-induced genes has been reported for many genes, whereby the steady-state levels of mRNA peak at a given time after application of the stress agent, followed by a decrease. For example, mRNA levels of the stress-induced Arabidopsis transcription factors DREB1A DREB2A and rd29A are transiently induced following exposure to cold, drought, and salt stresses [41].

3.4. *Phosphorylation of S114 Stabilizes ABI4*

The phosphomimetic S114E form of ABI4 was more stable than the WT or the non-phosphorylated S114A mutant (Figure 9), suggesting that phosphorylation of S114 may decrease its ubiquitylation by a yet unidentified ubiquitin ligase. The S114 residue is included in the serine/threonine (S/T) region motif of ABI4 [2]. Several domains are proposed to contribute to the instability of ABI4: the PEST domain located at the N terminus of ABI4 (amino-acids 22–40) enhances the degradation of ABI4 [21,22]. Furthermore, the N-terminal half of the ABI4 protein, including the PEST, APETALA2 (AP2), serine/threonine rich domain (S/T); the glutamine-rich domain (Q); and the C-terminal half containing the Q and proline-rich (P) domains, were shown to be highly unstable [21]. Degradation of the N-terminal half but not the C-terminal half of ABI4 was suppressed by the MG132 proteasome inhibitor, suggesting that although highly unstable, the C-terminal half of ABI4 may not be degraded by the proteasome [21]. The AP2-associated motif was also shown to destabilize ABI4 [22]. Although the S/T rich region was included in the labile N-terminal half of ABI4 [21], the instability of this region was mainly attributed to the PEST motif. Proteasomal degradation of ABI4 through the PEST motif is modulated by sugar levels [22].

Using the proteomic approach in human cell lines, Wu et al. [42] recently showed that phosphorylation delays the turnover of many proteins in growing cells. Moreover, the phosphomimetic mutated proteins catenin beta-1 (CTNNB1) S191D and the transcriptional receptor protein YY1 S118D were more stable than the WT proteins, and the phosphorylation-null in which the respective serine residues were mutated to alanine were destabilized [42]. In addition, phosphoserine residues had a larger stabilization effect than phosphothreonine, and phosphotyrosine had only a marginal stabilization effect.

Phosphorylation of type-A response regulator 5 (ARR5) by SnRK2s enhanced its stability [43]. Furthermore, overexpressing WT ARR5 but not the non-phosphorylatable mutated protein enhanced ABA hypersensitivity, suggesting that the phosphorylated form of ARR5 is biologically active. ABA suppressed the degradation of ARR5 [43]. Phosphorylation of the rate-limiting enzymes of ethylene biosynthesis, 1-aminocyclopropane-1-carboxylic acid synthase2 and 6 (ACS2 and ACS6), by MPK6 stabilizes the respective ACS proteins. Furthermore, the phosphomimetic ACS6 mutant was constitutively active, suggesting that phosphorylation of ACS6 by MPK6 is essential for its activity [44]. The RNA-binding protein tandem zinc finger 9 (TZF9) is destabilized by MAPK-mediated phosphorylation [45].

3.5. MAPK Regulates ABI4 Both Transcriptionally and Post-Transcriptionally

We showed that phosphorylation of S114 stabilizes ABI4 (Figure 9). We recently demonstrated that MPK3, MPK4, and MPK6 phosphorylate S114 of ABI4 and that this phosphorylation is essential for the biological activity of ABI4 and the complementation of *abi4* mutant plants [27]. MPK3, MPK4, and MP6 are involved in the abiotic and biotic stress response (reviewed in [46]). Treatments with NaCl, ABA, and high glucose, which result in stabilization of ABI4 (Figures 3 and 6), also enhance the steady-state levels of the *ABI4* transcripts [17,18,20]. Our results indicate that MAPK signaling affects both ABI4 transcription and protein stability.

The kinetics we observed for the transient stabilization of ABI4 following salt treatment (Figure 3) resemble the described transient activation of MKK5 following exposure of Arabidopsis plants to high salt, whereby increased activity of MKK5 was detected within 30 min of the treatment, reaching maximal activity at 2–4 h and declining at 6 h after exposure to NaCl to nearly basal activity levels [47]. MKK5 phosphorylates and activates several MPKs, including MPK3, MPK4, and MPK6. Therefore, the activity of these MPKs is also expected to be transient following salt treatment, resulting in a transient wave of phosphorylation of ABI4. MPK4 and MPK6 are rapidly activated by treatments such as high salt and osmotic stress but not by ABA treatment [48]. ABA activates the transcription of many genes encoding components of the MAPK cascade [49], suggesting that the slow kinetics leading to accumulation of ABI4-eGFP following ABA treatment may result from slow de novo synthesis of the MAPKs rather than fast activation of pre-existing latent enzymes.

MPK3, MPK4, and MPK6 also phosphorylate the transcription factors WRKY and MYB [24,50]. Several WRKY and MYB transcription factors may regulate ABI4 expression [51–60]. In addition, as ABI4 also activates the transcription of its own gene [61], its phosphorylation by these MAPKs also enhances its transcript levels.

In summary, our results show that phosphorylation of ABI4 by MAPK results in the stabilization of ABI4. Phosphorylation of S114 by MPKs may interfere with its binding to a yet unidentified E3 for proteasomal degradation. Alternatively, the catalytic efficiency of the E3 may be reduced toward phosphorylated ABI4. MAPK signaling also regulates ABI4 transcription. Thus, we suggest that regulation of both the *ABI4* transcript and ABI4 protein levels results in the tight regulation of the activity of this key transcription factor in the ABA signaling pathway.

4. Materials and Methods

4.1. Plant Material and Growth Conditions

Arabidopsis thaliana (Col) seeds of the indicated genotypes were surface-sterilized, cold-treated for 3 days, and plated in Petri dishes containing 0.5 × Murashige and Skoog medium (MS), 0.55% plant agar, and 0.5% (w/v) sucrose, as previously described [18]. Plates were incubated at 22–25 °C and 50% humidity under a circadian regime of 12 h light/12 h dark.

4.2. Constructs and Plant Transformation

The pGA-eGFP2 vector was constructed by replacing the sequences of the MCS and 35S::mGFP5 (9640–1038) in the pCAMBIA1302 vector (www.cambia.org accessed on 1 April 2010) with the 2 × 35S-MCS-eGFP DNA sequence (405–2332) from the pSAT4-eGFP-N1 plasmid using Gibson assembly cloning [62]. The DNA sequence encoding HA_3-$FLAG_3$-ABI4 was isolated by digesting the pJIM19-ABI4 plasmid [17] with restriction enzymes *Nco*I and *Pst*I and subcloning into the respective sites in pGA-eGFP2 to yield the pGA-HA3-FLAG3-ABI4 plasmid. To construct plasmids encoding ABI4 (S114A) and ABI4 (S114E) mutant proteins, the respective DNA sequences were amplified from the respective pRSET-ABI4 plasmid [27] using gene-specific primers flanked by the *Sal*I restriction sites and digesting the amplified sequences with *Sal*I. The DNA sequence encoding WT-ABI4 was removed from the pGA-HA3-FLAG3-ABI4 plasmid by digestion with *Sal*I, followed by subcloning of the DNA sequences encoding mutated ABI4 protein. Primers used for the construction of plasmids are shown in Table S1. The resulting plasmids were verified by PCR and DNA sequencing and were introduced into *Agrobacterium tumefaciens* strain GV3101. The transformed bacteria were used to transform WT Col or *abi4-1* Arabidopsis plants by the floral dip method [63]. Transgenic plants were selected on plates containing hygromycin and transferred to pots. Plant were grown at 22–25 °C and 50% humidity with 16 h light/8 h dark. Homozygous T2 and T3 generation plants were used in this study.

4.3. Germination Assay

Sterilized cold-treated seeds were plated on agar-solidified 0.5 × MS, 0.5% sucrose medium supplemented with the indicated concentrations of the phytohormone ABA. Germination was scored 7 days later.

4.4. Plant Treatment

For the various treatments, 10-day-old seedlings were transferred to Petri dishes containing Whatman No.1 filter papers soaked with 0.5 × MS medium and 0.5% (w/v) sucrose supplemented with the indicated stress agent, plant hormone, or inhibitors. Plants were incubated at room temperature in under light for the indicated times.

4.5. Microscopy

The indicated tissues were examined using a fluorescent microscope (ECLIPSE Ci-L; Nikon) with filters set for GFP. The images reflect GFP signals in all the cells of the examined tissue, thus representing the expression levels in all cell types. All images in each experimental repeat were taken using the same microscope, camera setup, and exposure times. Each experiment was repeated at least three times using at least four independent lines of the transgenic plants. Fluorescent signals were quantified using ImageJ software [64], with the black background set as zero for the measurement of the fluorescent intensity of the image. Subcellular localization images were taken with a Zeiss LSM-880 confocal microscope

4.6. Embryo Excision

Arabidopsis seeds imbibed for 24 h in water at room temperature were pressed gently between two microscope slides. Embryos released from seed coats were collected and rinsed briefly in water.

4.7. Protein Extraction, SDS-PAGE, and Western Blot Analysis

Ten-day-old seedlings were harvested into a 1.5 mL microcentrifuge tube, and their fresh weight was determined. Next, 2:1 (v/w) 4 × SDS-PAGE sample buffer [65] was added, and the seedlings were homogenized with a microcentrifuge pestle. To ensure efficient solubilization of plant proteins, homogenates were passed through 2 cycles of freezing in liquid nitrogen andboiling for 5 min. Tubes were centrifuged for 10 min at 12,000× g at room temperature, and supernatant samples were resolved by SDS PAGE. Proteins were electroblotted onto nitrocellulose membranes. ABI4-eGFP and β-actin were detected using the primary antibodies anti-GFP (Abcam, ab1218, Cambridge, UK) and anti-β-actin (Sigma, A4700, Saint Louis, MO, USA), respectively, and secondary peroxidase-coupled anti-mouse IgG antibody (Sera Care 5450–0011). Membranes were incubated in reaction mixes prepared from with a highly sensitive SuperSignal West Dura extended substrate kit (Thermo scientific, Waltham, MA, USA), and chemiluminescent signals were recorded using ImageQuant RT ECL Imager (GE Healthcare, Chicago, IL, USA).

4.8. Quantitative RT-PCR Analysis

Total RNA was isolated from roots using a ZR Plant RNA MiniPrep kit (Zymo research). The RNA concentration was estimated spectrally (Nano Drop ND-1000; Nano Drop Technologies). cDNA was synthesized using a qScript cDNA synthesis kit (Quanta). The reaction mixture contained 700 ng total RNA and random primers. Primer design and RT-qPCR assays for determining relative steady-state transcript levels were as previously described [17]. Primers are described in Table S1.

Supplementary Materials: The following supporting information can be downloaded at: https://www.mdpi.com/article/10.3390/plants11162179/s1, Figure S1: eGFP expression in roots of salt-treated 35s::eGFP plants; Table S1: Primers used in this study.

Author Contributions: Conceptualization, T.M. and D.B.-Z.; methodology, T.M. and N.E.; writing, T.M. and D.B.-Z.; supervision, D.B.-Z.; funding acquisition, D.B.-Z. All authors have read and agreed to the published version of the manuscript.

Funding: This research was funded by the US Israel Binational Science Foundation (BSF), grant number 2011097.

Acknowledgments: We thank Guy Adler for constructing the pGA-eGFP vector. Dudy Bar-Zvi is the incumbent of The Israel and Bernard Nichunsky Chair in Desert Agriculture, Ben-Gurion University of the Negev.

Conflicts of Interest: The authors declare no conflict of interest.

References

1. Chen, X.; Ding, Y.; Yang, Y.; Song, C.; Wang, B.; Yang, S.; Guo, Y.; Gong, Z. Protein kinases in plant responses to drought, salt, and cold stress. *J. Integr. Plant Biol.* **2021**, *63*, 53–78. [CrossRef] [PubMed]
2. Finkelstein, R.R.; Wang, M.L.; Lynch, T.J.; Rao, S.; Goodman, H.M. The Arabidopsis abscisic acid response locus ABI4 encodes an APETALA 2 domain protein. *Plant Cell* **1998**, *10*, 1043–1054. [CrossRef] [PubMed]

3. Jofuku, K.D.; Den Boer, B.; Van Montagu, M.; Okamuro, J.K. Control of Arabidopsis flower and seed development by the homeotic gene APETALA2. *Plant Cell* **1994**, *6*, 1211–1225. [PubMed]
4. Okamuro, J.K.; Caster, B.; Villarroel, R.; Van Montagu, M.; Jofuku, K.D. The AP2 domain of *APETALA2* defines a large new family of DNA binding proteins in Arabidopsis. *Proc. Natl. Acad. Sci. USA* **1997**, *94*, 7076–7081. [CrossRef]
5. Finkelstein, R.R. Mutations at two new Arabidopsis ABA response loci are similar to the *abi3* mutations. *Plant J.* **1994**, *6*, 765–771. [CrossRef]
6. Quesada, V.; Ponce, M.R.; Micol, J.L. Genetic analysis of salt-tolerant mutants in *Arabidopsis thaliana*. *Genetics* **2000**, *154*, 421–436. [CrossRef]
7. Arenas-Huertero, F.; Arroyo, A.; Zhou, L.; Sheen, J.; León, P. Analysis of Arabidopsis glucose insensitive mutants, *gin5* and *gin6*, reveals a central role of the plant hormone ABA in the regulation of plant vegetative development by sugar. *Genes Dev.* **2000**, *14*, 2085–2096. [CrossRef]
8. Laby, R.J.; Kincaid, M.S.; Kim, D.; Gibson, S.I. The Arabidopsis sugar-insensitive mutants *sis4* and *sis5* are defective in abscisic acid synthesis and response. *Plant J.* **2000**, *23*, 587–596. [CrossRef]
9. Huijser, C.; Kortstee, A.; Pego, J.; Weisbeek, P.; Wisman, E.; Smeekens, S. The Arabidopsis *SUCROSE UNCOUPLED-6* gene is identical to *ABSCISIC ACID INSENSITIVE-4*: Involvement of abscisic acid in sugar responses. *Plant J.* **2000**, *23*, 577–585. [CrossRef]
10. Rook, F.; Corke, F.; Card, R.; Munz, G.; Smith, C.; Bevan, M.W. Impaired sucrose-induction mutants reveal the modulation of sugar-induced starch biosynthetic gene expression by abscisic acid signalling. *Plant J.* **2001**, *26*, 421–433. [CrossRef]
11. Wind, J.J.; Peviani, A.; Snel, B.; Hanson, J.; Smeekens, S.C. ABI4: Versatile activator and repressor. *Trends Plant Sci.* **2013**, *18*, 125–132. [CrossRef] [PubMed]
12. Koussevitzky, S.; Nott, A.; Mockler, T.C.; Hong, F.; Sachetto-Martins, G.; Surpin, M.; Lim, J.; Mittler, R.; Chory, J. Signals from chloroplasts converge to regulate nuclear gene expression. *Science* **2007**, *316*, 715–719. [CrossRef] [PubMed]
13. Kacprzak, S.M.; Mochizuki, N.; Naranjo, B.; Xu, D.; Leister, D.; Kleine, T.; Okamoto, H.; Terry, M.J. Plastid-to-nucleus retrograde signalling during chloroplast biogenesis does not require ABI4. *Plant Physiol.* **2019**, *179*, 18–23. [CrossRef] [PubMed]
14. Söderman, E.M.; Brocard, I.M.; Lynch, T.J.; Finkelstein, R.R. Regulation and function of the Arabidopsis *ABA-insensitive4* gene in seed and abscisic acid response signaling networks. *Plant Physiol.* **2000**, *124*, 1752–1765. [CrossRef] [PubMed]
15. Nakabayashi, K.; Okamoto, M.; Koshiba, T.; Kamiya, Y.; Nambara, E. Genome-wide profiling of stored mRNA in *Arabidopsis thaliana* seed germination: Epigenetic and genetic regulation of transcription in seed. *Plant J.* **2005**, *41*, 697–709. [CrossRef]
16. Penfield, S.; Li, Y.; Gilday, A.D.; Graham, S.; Graham, I.A. Arabidopsis *ABA INSENSITIVE4* regulates lipid mobilization in the embryo and reveals repression of seed germination by the endosperm. *Plant Cell* **2006**, *18*, 1887–1899. [CrossRef]
17. Shkolnik-Inbar, D.; Adler, G.; Bar-Zvi, D. ABI4 downregulates expression of the sodium transporter *HKT1;1* in Arabidopsis roots and affects salt tolerance. *Plant J.* **2013**, *73*, 993–1005. [CrossRef]
18. Shkolnik-Inbar, D.; Bar-Zvi, D. ABI4 mediates abscisic acid and cytokinin inhibition of lateral root formation by reducing polar auxin transport in Arabidopsis. *Plant Cell* **2010**, *22*, 3560–3573. [CrossRef]
19. Shkolnik-Inbar, D.; Bar-Zvi, D. Expression of *ABSCISIC ACID INSENSITIVE 4 (ABI4)* in developing Arabidopsis seedlings. *Plant Signal. Behav.* **2011**, *6*, 694–696. [CrossRef]
20. Arroyo, A.; Bossi, F.; Finkelstein, R.R.; León, P. Three genes that affect sugar sensing (*Abscisic Acid Insensitive 4, Abscisic acid Insensitive 5,* and *Constitutive Triple Response 1*) are differentially regulated by glucose in Arabidopsis. *Plant Physiol.* **2003**, *133*, 231–242. [CrossRef]
21. Finkelstein, R.; Lynch, T.; Reeves, W.; Petitfils, M.; Mostachetti, M. Accumulation of the transcription factor ABA-insensitive (ABI)4 is tightly regulated post-transcriptionally. *J. Exp. Bot.* **2011**, *62*, 3971–3979. [CrossRef] [PubMed]
22. Gregorio, J.; Hernández-Bernal, A.F.; Cordoba, E.; León, P. Characterization of evolutionarily conserved motifs involved in activity and regulation of the ABA-INSENSITIVE (ABI) 4 transcription factor. *Mol. Plant* **2014**, *7*, 422–436. [CrossRef]
23. Xu, X.; Chi, W.; Sun, X.; Feng, P.; Guo, H.; Li, J.; Lin, R.; Lu, C.; Wang, H.; Leister, D.; et al. Convergence of light and chloroplast signals for de-etiolation through ABI4-HY5 and COP1. *Nat. Plants* **2016**, *2*, 16066. [CrossRef] [PubMed]
24. Popescu, S.C.; Popescu, G.V.; Bachan, S.; Zhang, Z.; Gerstein, M.; Snyder, M.; Dinesh-Kumar, S.P. MAPK target networks in *Arabidopsis thaliana* revealed using functional protein microarrays. *Genes Dev.* **2009**, *23*, 80–92. [CrossRef] [PubMed]
25. Guo, H.; Feng, P.; Chi, W.; Sun, X.; Xu, X.; Li, Y.; Ren, D.; Lu, C.; Rochaix, J.D.; Leister, D. Plastid-nucleus communication involves calcium-modulated MAPK signalling. *Nat. Commun.* **2016**, *7*, 12173. [CrossRef]
26. Bai, Z.; Zhang, J.; Ning, X.; Guo, H.; Xu, X.; Huang, X.; Wang, Y.; Hu, Z.; Lu, C.; Zhang, L. A kinase–phosphatase–transcription factor module regulates adventitious root emergence in Arabidopsis root–hypocotyl junctions. *Mol. Plant.* **2020**, *13*, 1162–1177. [CrossRef]
27. Eisner, N.; Maymon, T.; Sanchez, E.C.; Bar-Zvi, D.; Brodsky, S.; Finkelstein, R.; Bar-Zvi, D. Phosphorylation of serine 114 of the transcription factor ABSCISIC ACID INSENSITIVE 4 is essential for activity. *Plant Sci.* **2021**, *305*, 110847. [CrossRef]
28. Cinelli, R.A.; Ferrari, A.; Pellegrini, V.; Tyagi, M.; Giacca, M.; Beltram, F. The enhanced green fluorescent protein as a tool for the analysis of protein dynamics and localization: Local fluorescence study at the single-molecule level. *Photochem. Photobiol.* **2000**, *71*, 771–776. [CrossRef]
29. Odell, J.T.; Nagy, F.; Chua, N.H. Identification of DNA sequences required for activity of the cauliflower mosaic virus 35S promoter. *Nature* **1985**, *313*, 810–812. [CrossRef]

30. Finkelstein, R.R. Abscisic acid-insensitive mutations provide evidence for stage-specific signal pathways regulating expression of an Arabidopsis late embryogenesis-abundant (*lea*) gene. *Mol. Gen. Genet.* **1993**, *238*, 401–408. [CrossRef]
31. Amack, S.C.; Antunes, M.S. CaMV35S promoter—A plant biology and biotechnology workhorse in the era of synthetic biology. *Curr. Plant Biol.* **2020**, *24*, 100179. [CrossRef]
32. Chandrasekaran, U.; Luo, X.F.; Zhou, W.G.; Shu, K. Multifaceted signaling networks mediated by Abscisic Acid Insensitive 4. *Plant Commun.* **2020**, *1*, 100040. [CrossRef] [PubMed]
33. Xie, Z.; Nolan, T.M.; Jiang, H.; Yin, Y. AP2/ERF transcription factor regulatory networks in hormone and abiotic stress responses in Arabidopsis. *Front. Plant Sci.* **2019**, *10*, 228. [CrossRef] [PubMed]
34. Huang, X.; Zhang, X.; Gong, Z.; Yang, S.; Shi, Y. ABI4 represses the expression of type-A *ARR*s to inhibit seed germination in Arabidopsis. *Plant J.* **2017**, *89*, 354–365. [CrossRef]
35. Aviña-Padilla, K.; Ramírez-Rafael, J.A.; Herrera-Oropeza, G.E.; Muley, V.Y.; Valdivia, D.I.; Díaz-Valenzuela, E.; García-García, A.; Varela-Echavarría, A.; Hernández-Rosales, M. Evolutionary perspective and expression analysis of intronless genes highlight the conservation of their regulatory role. *Front. Genet.* **2021**, *12*, 654256. [CrossRef]
36. Liu, H.; Lyu, H.-M.; Zhu, K.; Van de Peer, Y.; Cheng, Z.-M. The emergence and evolution of intron-poor and intronless genes in intron-rich plant gene families. *Plant J.* **2021**, *105*, 1072–1082. [CrossRef]
37. Zhang, G.; Gurtu, V.; Kain, S.R. An enhanced green fluorescent protein allows sensitive detection of gene transfer in mammalian cells. *Biochem. Biophys. Res. Commun.* **1996**, *227*, 707–711. [CrossRef]
38. Birnbaum, K.; Shasha, D.E.; Wang, J.Y.; Jung, J.W.; Lambert, G.M.; Galbraith, D.W.; Benfey, P.N. A gene expression map of the Arabidopsis root. *Science* **2003**, *302*, 1956–1960. [CrossRef]
39. Lopez-Molina, L.; Mongrand, S.; Chua, N.H. A postgermination developmental arrest checkpoint is mediated by abscisic acid and requires the ABI5 transcription factor in Arabidopsis. *Proc. Natl. Acad. Sci. USA* **2001**, *98*, 4782–4787. [CrossRef]
40. Finkelstein, R.R.; Lynch, T.J. The Arabidopsis abscisic acid response gene ABI5 encodes a basic leucine zipper transcription factor. *Plant Cell* **2000**, *12*, 599–609. [CrossRef]
41. Liu, Q.; Kasuga, M.; Sakuma, Y.; Abe, H.; Miura, S.; Yamaguchi-Shinozaki, K.; Shinozaki, K. Two transcription factors, DREB1 and DREB2, with an EREBP/AP2 DNA binding domain separate two cellular signal transduction pathways in drought- and low-temperature-responsive gene expression, respectively, in Arabidopsis. *Plant Cell* **1998**, *10*, 1391–1406. [CrossRef] [PubMed]
42. Wu, C.; Ba, Q.; Lu, D.; Li, W.; Salovska, B.; Hou, P.; Mueller, T.; Rosenberger, G.; Gao, E.; Di, Y.; et al. Global and site-specific effect of phosphorylation on protein turnover. *Dev. Cell* **2021**, *56*, 111–124. [CrossRef] [PubMed]
43. Huang, X.; Hou, L.; Meng, J.; You, H.; Li, Z.; Gong, Z.; Yang, S.; Shi, Y. The Antagonistic action of abscisic acid and cytokinin signaling mediates drought stress response in Arabidopsis. *Mol. Plant* **2018**, *11*, 970–982. [CrossRef] [PubMed]
44. Liu, Y.; Zhang, S. Phosphorylation of 1-aminocyclopropane-1-carboxylic acid synthase by MPK6, a stress-responsive mitogen-activated protein kinase, induces ethylene biosynthesis in Arabidopsis. *Plant Cell* **2004**, *16*, 3386–3399. [CrossRef] [PubMed]
45. Maldonado-Bonilla, L.D.; Eschen-Lippold, L.; Gago-Zachert, S.; Tabassum, N.; Bauer, N.; Scheel, D.; Lee, J. The Arabidopsis tandem zinc finger 9 protein binds RNA and mediates pathogen-associated molecular pattern-triggered immune responses. *Plant Cell Physiol.* **2014**, *55*, 412–425. [CrossRef]
46. Bigeard, J.; Hirt, H. Nuclear Signaling of Plant MAPKs. *Front. Plant Sci.* **2018**, *9*, 469. [CrossRef]
47. Xing, Y.; Chen, W.H.; Jia, W.; Zhang, J. Mitogen-activated protein kinase kinase 5 (MKK5)-mediated signalling cascade regulates expression of iron superoxide dismutase gene in Arabidopsis under salinity stress. *J. Exp. Bot.* **2015**, *66*, 5971–5981. [CrossRef]
48. Ichimura, K.; Mizoguchi, T.; Yoshida, R.; Yuasa, T.; Shinozaki, K. Various abiotic stresses rapidly activate Arabidopsis MAP kinases ATMPK4 and ATMPK6. *Plant J.* **2000**, *24*, 655–665. [CrossRef]
49. Menges, M.; Dóczi, R.; Ökrész, L.; Morandini, P.; Mizzi, L.; Soloviev, M.; Murray, J.A.; Bögre, L. Comprehensive gene expression atlas for the Arabidopsis MAP kinase signalling pathways. *New Phytol.* **2008**, *179*, 643–662. [CrossRef]
50. Sheikh, A.H.; Eschen-Lippold, L.; Pecher, P.; Hoehenwarter, W.; Sinha, A.K.; Scheel, D.; Lee, J. Regulation of WRKY46 transcription factor function by mitogen-activated protein kinases in *Arabidopsis thaliana*. *Front. Plant Sci.* **2016**, *7*, 61. [CrossRef]
51. Ma, Q.; Xia, Z.; Cai, Z.; Li, L.; Cheng, Y.; Liu, J.; Nian, H. GmWRKY16 enhances drought and salt tolerance through an ABA-mediated pathway in *Arabidopsis thaliana*. *Front. Plant Sci.* **2019**, *9*, 1979. [CrossRef] [PubMed]
52. Chen, Y.-S.; Chao, Y.-C.; Tseng, T.-W.; Huang, C.-K.; Lo, P.-C.; Lu, C.-A. Two MYB-related transcription factors play opposite roles in sugar signaling in Arabidopsis. *Plant Mol. Biol.* **2017**, *93*, 299–311. [CrossRef] [PubMed]
53. Huang, Y.; Feng, C.Z.; Ye, Q.; Wu, W.H.; Chen, Y.F. Arabidopsis WRKY6 transcription factor acts as a positive regulator of abscisic acid signaling during seed germination and early seedling development. *PLoS Genet.* **2016**, *12*, e1005833. [CrossRef] [PubMed]
54. Ding, Z.J.; Yan, J.Y.; Li, C.X.; Li, G.X.; Wu, Y.R.; Zheng, S.J. Transcription factor WRKY46 modulates the development of Arabidopsis lateral roots in osmotic/salt stress conditions via regulation of ABA signaling and auxin homeostasis. *Plant J.* **2015**, *84*, 56–69. [CrossRef] [PubMed]
55. Li, X.; Zhong, M.; Qu, L.; Yang, J.; Liu, X.; Zhao, Q.; Liu, X.; Zhao, X. AtMYB32 regulates the ABA response by targeting *ABI3*, *ABI4* and *ABI5* and the drought response by targeting CBF4 in Arabidopsis. *Plant Sci.* **2021**, *310*, 110983. [CrossRef]
56. Guo, Z.; Xu, H.; Lei, Q.; Du, J.; Li, C.; Wang, C.; Yang, Y.; Yang, Y.; Sun, X. The Arabidopsis transcription factor LBD15 mediates ABA signaling and tolerance of water-deficit stress by regulating *ABI4* expression. *Plant J.* **2020**, *104*, 510–521. [CrossRef]

57. Feng, C.Z.; Chen, Y.; Wang, C.; Kong, Y.H.; Wu, W.H.; Chen, Y.F. Arabidopsis RAV1 transcription factor, phosphorylated by SnRK2 kinases, regulates the expression of *ABI3*, *ABI4*, and *ABI5* during seed germination and early seedling development. *Plant J.* **2014**, *80*, 654–668. [CrossRef]
58. Shang, Y.; Yan, L.; Liu, Z.-Q.; Cao, Z.; Mei, C.; Xin, Q.; Wu, F.-Q.; Wang, X.-F.; Du, S.-Y.; Jiang, T. The Mg-chelatase H subunit of Arabidopsis antagonizes a group of WRKY transcription repressors to relieve ABA-responsive genes of inhibition. *Plant Cell* **2010**, *22*, 1909–1935. [CrossRef]
59. Lee, K.; Seo, P.J. Coordination of seed dormancy and germination processes by MYB96. *Plant Signal. Behav.* **2015**, *10*, e1056423. [CrossRef]
60. Reeves, W.M.; Lynch, T.J.; Mobin, R.; Finkelstein, R.R. Direct targets of the transcription factors ABA-Insensitive(ABI)4 and ABI5 reveal synergistic action by ABI4 and several bZIP ABA response factors. *Plant Mol. Biol.* **2011**, *75*, 347–363. [CrossRef]
61. Bossi, F.; Cordoba, E.; Dupré, P.; Mendoza, M.S.; Román, C.S.; León, P. The Arabidopsis ABA-INSENSITIVE (ABI) 4 factor acts as a central transcription activator of the expression of its own gene, and for the induction of *ABI5* and *SBE2.2* genes during sugar signaling. *Plant J.* **2009**, *59*, 359–374. [CrossRef] [PubMed]
62. Gibson, D.G.; Young, L.; Chuang, R.Y.; Venter, J.C.; Hutchison, C.A., 3rd; Smith, H.O. Enzymatic assembly of DNA molecules up to several hundred kilobases. *Nat. Methods* **2009**, *6*, 343–345. [CrossRef] [PubMed]
63. Clough, S.J.; Bent, A.F. Floral dip: A simplified method for *Agrobacterium*-mediated transformation of *Arabidopsis thaliana*. *Plant J.* **1998**, *16*, 735–743. [CrossRef]
64. Schneider, C.A.; Rasband, W.S.; Eliceiri, K.W. NIH Image to ImageJ: 25 years of image analysis. *Nat. Methods* **2012**, *9*, 671–675. [CrossRef] [PubMed]
65. Laemmli, U.K. Cleavage of structural proteins during the assembly of the head of bacteriophage T4. *Nature* **1970**, *227*, 680–685. [CrossRef]

plants

MDPI

Article

Transcriptional Expression of Nitrogen Metabolism Genes and Primary Metabolic Variations in Rice Affected by Different Water Status

Gahyun Kim and Jwakyung Sung *

Department of Crop Science, Chungbuk National University, Cheongju 28644, Republic of Korea; 1996rlark@gmail.com
* Correspondence: jksung73@chungbuk.ac.kr; Tel.: +82-43-261-2512

Abstract: The era of climate change strongly requires higher efficiency of energies, such as light, water, nutrients, etc., during crop production. Rice is the world's greatest water-consuming plant, and, thus, water-saving practices such as alternative wetting and drying (AWD) are widely recommended worldwide. However the AWD still has concerns such as lower tillering, shallow rooting, and an unexpected water deficit. The AWD is a possibility to not only save water consumption but also utilize various nitrogen forms from the soil. The current study tried to investigate the transcriptional expression of genes in relation to the acquisition-transportation-assimilation process of nitrogen using qRT-PCR at the tillering and heading stages and to profile tissue-specific primary metabolites. We employed two water supply systems, continuous flooding (CF) and alternative wetting and drying (AWD), during rice growth (seeding to heading). The AWD system is effective at acquiring soil nitrate; however, nitrogen assimilation was predominant in the root during the shift from the vegetative to the reproductive stage. In addition, as a result of the greater amino acids in the shoot, the AWD was likely to rearrange amino acid pools to produce proteins in accordance with phase transition. Accordingly, it is suggested that the AWD (1) actively acquired nitrate from soil and (2) resulted in an abundance of amino acid pools, which are considered a rearrangement under limited N availability. Based on the current study, further steps are necessary to evaluate form-dependent N metabolism and root development under the AWD condition and a possible practice in the rice production system.

Keywords: rice; water management; nitrogen metabolism genes; primary metabolites

Citation: Kim, G.; Sung, J. Transcriptional Expression of Nitrogen Metabolism Genes and Primary Metabolic Variations in Rice Affected by Different Water Status. *Plants* **2023**, *12*, 1649. https://doi.org/10.3390/plants12081649

Academic Editors: Małgorzata Nykiel, Mateusz Labudda, Beata Prabucka, Marta Gietler and Justyna Fidler

Received: 18 March 2023
Revised: 9 April 2023
Accepted: 12 April 2023
Published: 14 April 2023

1. Introduction

The climate change is strongly raising the stakes for a paradigm shift in all industrial structures, including agriculture, and, especially, the elevating temperature and greenhouse gas emissions (GHG) are positioned as negative factors threatening sustainable food security [1]. It is forecast that a yield loss of major crops, including rice, could be an inevitable consequence, and to ensure sustainable crop production and an agricultural ecosystem, experimental approaches are constantly being attempted via expanding agricultural knowledge and practices.

Rice is one of the most important food crops, providing 20% of daily calories to more than 3.5 billion people worldwide [1]. Rice is almost exclusively produced in Asia, accounting for 90.7% of global production, and is from East Asia, where it occupies approximately 33% [2]. Unlike other crops, rice can critically grow under the flooded field environment due to the development of aerenchyma cells, which transport oxygen from the leaf to the root [3]. However, limited water irrigation owing to climate change is forcing efforts to enhance water use efficiency (WUE), and water-saving practices also lead to significant reductions in greenhouse gas emissions (CO_2) from the rhizosphere.

One practice that is considered to achieve effective water use in rice systems is an irrigation management technique referred to as alternate wetting (saturated) and drying (unsaturated) (AWD) [4], and it was reported that an application of this practice could save water input by 23% [5] compared to continuous flooding (CF). In addition to water savings, the beneficial effect of AWD is to reduce the generation of greenhouse gases, especially methane (a 45~90% reduction) [6]. Since ammonium is a preferential form of nitrogen, adjusting water management in the rice production system may also affect nitrogen availability.

Nitrogen (N) is an essential element requiring substantially for plant growth and development, and all plant species preferentially absorb nitrogen in both inorganic and organic forms, such as nitrate, ammonium, or amino acids, from the rhizosphere. Due to the importance of nitrogen, nitrogen fertilizer has been widely used to attain better crop yields. Despite the fact that ammonium is the prevailing form in rice paddy fields, oxygen moved from the shoot via aerenchyma cells generates an aerobic environment in the rhizosphere, and, in turn, ammonium is nitrified to nitrate by nitrifying bacteria, which accounts for 40% of the gross N taken up by rice [7]. Therefore, there is evidence that rice root takes up both ammonium and nitrate, with findings as transporter genes: 84 NRTs and 12 AMTs [8]. Considering both N availability and water management in the rice production system, it could be hypothesized how the rice plant acquires and assimilates N under different water managements and, as a result, alters the primary metabolism. In addition, it remains unclear how the AWD affects N metabolism at different rice growth stages (i.e., vegetative and reproductive stages). The limited water supply perturbs the cell turgor, and, therefore, plants preferentially promote the synthesis and transport of compatible solutes, which play an essential role as osmotic protectants, such as soluble sugars and amino acids (i.e., proline and glycine-betaine) [9,10]. Thus, primary metabolites, including sugars and amino acids, are closely involved in coping with various environmental impacts.

To do this, we focused on understanding the transcriptional differences in key genes directly involved in nitrogen metabolism like uptake, transportation, assimilation, and remobilization in the CF and AWD at tillering and heading stages, and profiling targeted primary metabolites from both water management systems at the heading stage.

2. Results

The growth of rice plants was significantly different between continuous flooding (CF) and alternative wetting and drying (AWD), and the difference was obvious in root development (Figure 1). The AWD resulted in limited root growth (less volume), which triggered fewer tillers (17.7/plant, n = 3), compared to the CF (19.7/plant, n = 3), and slightly promoted leaf senescence. By contrast, there was no difference in the heading time. Therefore, we investigated nitrogen metabolism and metabolic alteration to see how different water management practices affected physiological responses, which caused distinct phenotypic differences.

2.1. Nitrogen Metabolism

The selected thirteen genes directly involved in the uptake and assimilation of nitrogen were compared between CF and AWD from the leaf blade, leaf sheath, and root of rice plants at both tillering and heading stages (Figure 2). The AWD greatly promoted the uptake of nitrate (NO_3-N) via activating the nitrate transporter (*OsNRT2.1*) at the tillering stage, although other groups (*OsNPF2.4*, *OsNPF8.20*, and *OsAMT1;1*) were slightly expressed. Whilst the expression of OsNPF genes was obvious at the heading stage. In the AWD, an absorbed nitrate was not only quickly assimilated into ammonium in the root but also transported toward the shoot. A glutamic acid in the root was finally assimilated into glutamine by the gene expression *OsGS1;2*, encoding glutamine synthetase (GS), while the reverse reaction by *OsNADH-GOGAT1*, encoding glutamate synthase (GOGAT), was limited in the AWD compared to the CF. It was observed that nitrate-N from the root to the shoot was better transported by NRT (*OsNRT2.1* and *OsNRT2.3a*) and NPF (*OsNPF6.5*) proteins in the AWD at tillering stage. In the leaf sheath, the assimilation of nitrogen into amino acids was not significantly different between both treatments at the tillering stage; however, the expression of genes (*OsGS1;1* and *OsFd-GOGAT*) encoding GS-GOGAT metabolism was extremely reduced by the AWD treatment at the heading stage. Moreover, the expression of genes *OsNPF2.4* and *OsNPF8.2*, encoding nitrate transporter enzymes regulating nitrate influx toward leaf blades, was greatly decreased in AWD treatment at the heading stage. The gene expression encoding nitrate reductase (NR) in leaf blades was significantly enhanced by AWD treatment, whereas nitrite reductase (NIR) remained unchanged. In addition, based on the expression levels of *OsGS1;1* and *OsFd-GOGAT*, the reversibility of the GS-GOGAT pathway was higher in the AWD treatment at the tillering stage. At the heading stage, however, their expression was similar at both water managements.

2.2. Principle Component Analysis (PCA) of Polar Metabolites

To characterize primary metabolism in different tissues (i.e., leaf blades, leaf sheaths, and roots) by different types of water management, the targeted primary metabolites were measured using GC−TOFMS at the heading stage. The data was employed in PCA and identified major differences between water management and plant parts (Figure 3). The first two principal components explained 60.2 and 22.3% of the variability (Figure 3–top). The PC1 and PC2 clearly revealed a strong difference between leaf blade and root, and differential water management showed no effect on the PC1, indicating that the composition of targeted primary metabolites was definitely affected by plant part rather than water management. We also analyzed positive or negative effects between metabolites using loading plots (Figure 3–bottom). A majority showed a positive correlation on an *X*–axis, whereas a *Y*–axis generally separated soluble carbohydrates and organic acids from amino acids. The results implied that carbon and nitrogen metabolisms in the tissues of rice plants were greatly affected by water supply conditions.

2.3. Primary Metabolites

Metabolite profiles provide a much broader view of systematic adjustment in metabolic processes compared to the conventional biochemical approaches and also abundant opportunities to reveal new insights on metabolism. The targeted polar metabolites (50 biomolecules) were compared to investigate the difference between two water treatments, CF and AWD (Figure 3). The level of soluble sugars was greatly affected by tissues and water treatments. The AWD resulted in a lower level of soluble sugar pools except for xylose in leaf blades, whereas levels were significantly higher in the leaf sheath and root. In contrast, glucose-6-P and fructose-6-P in the root were significantly reduced by the AWD. A mannitol, of sugar alcohol, was greatly accumulated in the AWD−grown root. Most amino acids showed a tendency for accumulation by the AWD, and the significantly higher level of glutamine and asparagine was notable, indicating 1.0- and 1.4-fold ($\log_2$ scale) greater in leaf blades compared to CF and 4.2- and 6.8-fold higher in the leaf sheath, respectively. Especially, the obviously accumulated amino acids in the leaf sheath were valine (2.3-fold), serine (2.3), leucine (2.3), isoleucine (3.0), proline (2.5), glycine (2.5), threonine (2.3), β-alanine (3.4), methionine (3.3), lysine (2.6), tryptophan (3.9), and putrescine (2.2), including glutamine and asparagine in the AWD treatment. In AWD-treated roots, the level of putrescine was markedly increased. A majority of organic acids, including the TCA intermediates, were not significantly different from the water treatments, whereas pyruvic and fumaric acids were greatly accumulated in leaf blades.

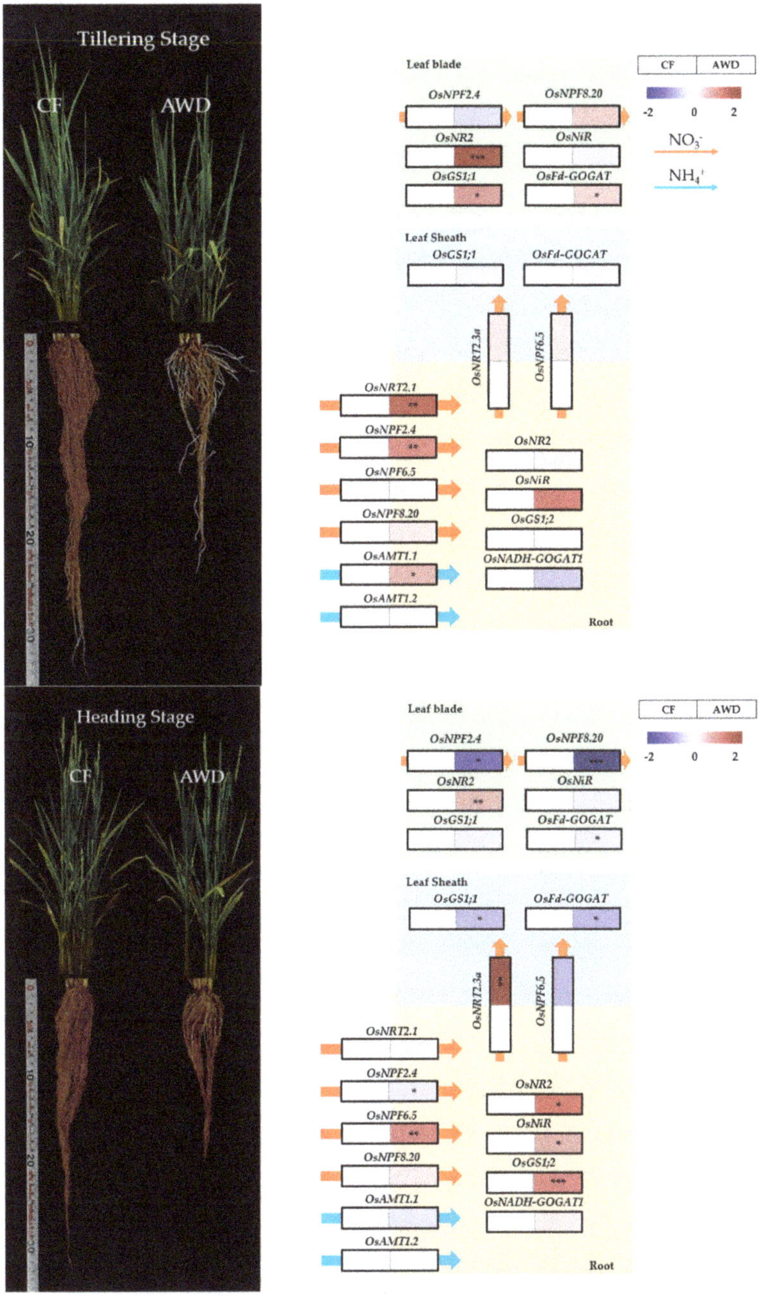

Figure 1. Rice (*O. sativa* cv. Jinbongbyeo) growth at both tillering and heading stages. The expression of genes (relative level of AWD to CF) involved in nitrogen metabolism in leaf blade, leaf sheath, and root at both tillering and heading stages subjected to different water managements; CF, continuous flooding; AWD, alternative wetting and drying (n = 3). The colors red and blue represent positive and negative expressions of genes, respectively. The qRT−PCR data was subjected to a *t*-test. *, **, and *** indicate the significance at $p < 0.05$, 0.01, and 0.001, respectively.

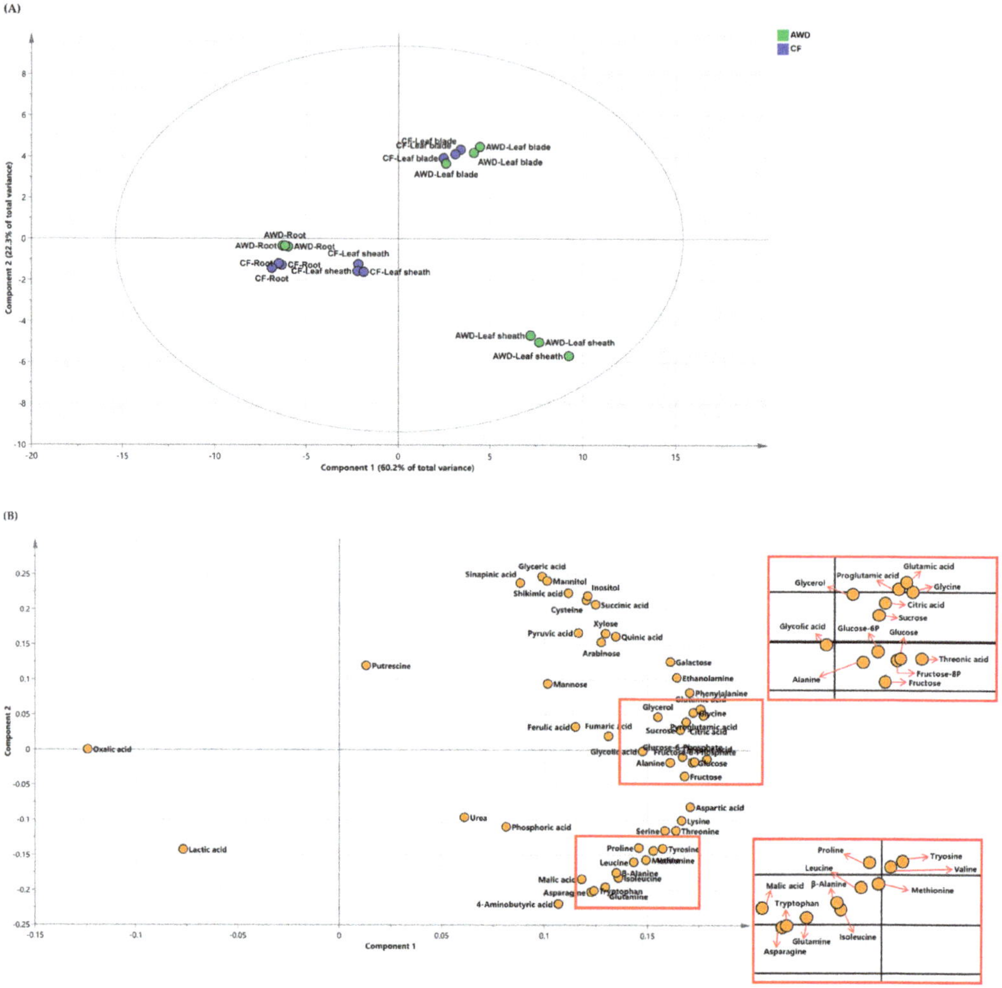

Figure 2. Principal component analysis (PCA) score (**A**) and loading (**B**) plots of metabolites identified from leaf blades, leaf sheaths, and roots of rice plants at heading stage.

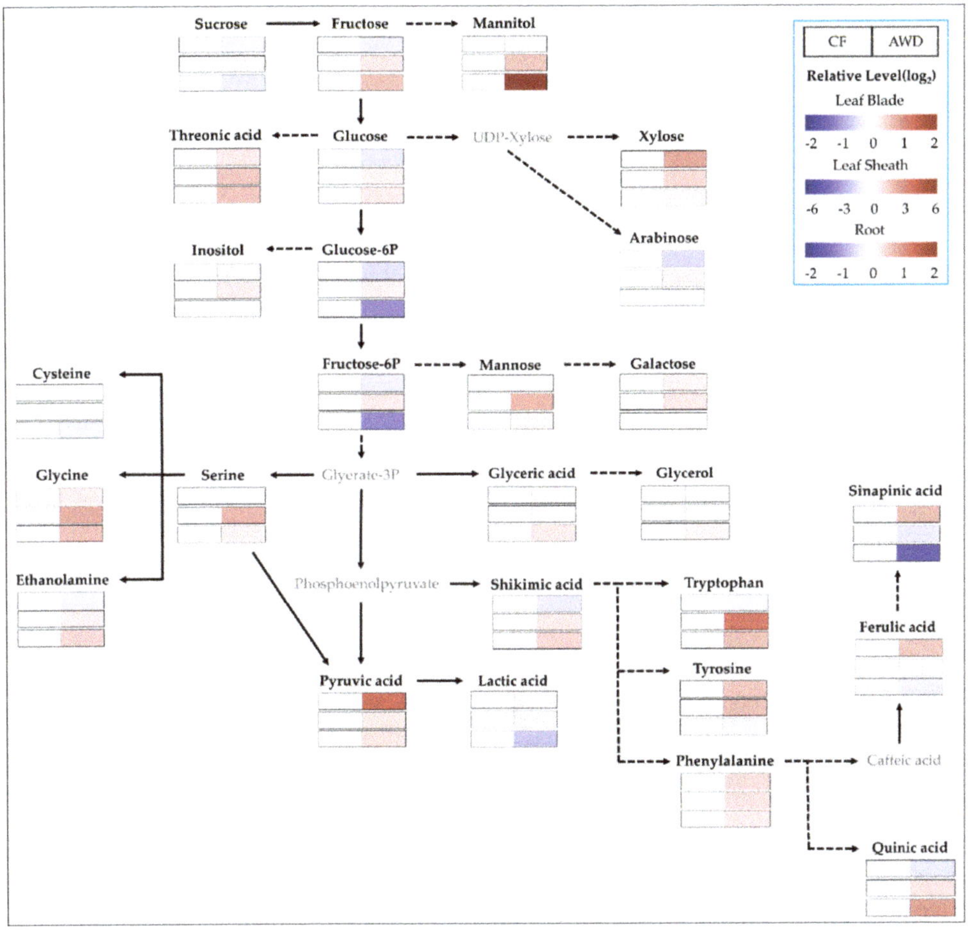

Figure 3. *Cont.*

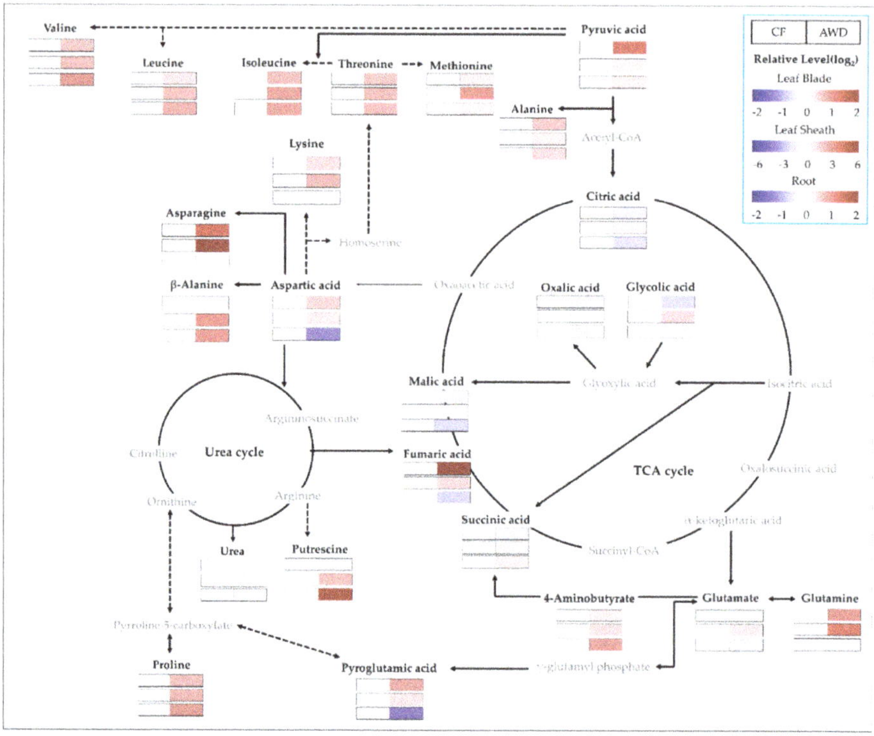

Figure 3. Primary metabolic changes (relative level of AWD to CF) in leaf blade, leaf sheath, and root at heading stage subjected to different water managements; CF, continuous flooding; AWD, alternative wetting and drying (n = 3).

3. Discussion

The current work has been studied to better understand a variation in nitrogen availability and primary metabolism under water−saving management (alternate wetting and drying, AWD), and is discussed with an emphasis on the significant findings. It has been widely verified that limited water supply contributed to the lower biomass production as a result of the trade−off between growth and stress defense [11]. The observations of root development under limited water supply differed from experimental conditions such as pot and/or field. Root development tended to increase as a result of the water stress response [12,13]. On the other hand, it was slightly affected by limited water [14] or shallower [15]. Our results showed that the AWD led to a reduction in biomass, tillering, and root development, which was partly in line with previous observations.

The transcriptional expression of selected genes functioning in the acquisition, transportation, assimilation, and remobilization of nitrogen in the leaf blade, leaf sheath, and root of the rice plant was compared between CF and AWD at both tillering and heading stages. We found that the uptake of nitrate was obviously promoted by the AWD, and, especially, it was dominant in high (*OsNRT2.1*) and low (*OsNPF2.4*, nitrate peptide family) affinity transporters at the tillering stage. In contrast, the dual affinity transporter of nitrate, *OsNPF6.5*, was significantly upregulated at the heading stage. The upregulation of *OsNRT2.1* enhanced nitrate uptake to increase assimilation efficiency in an AWD−employed rice system [16]. By contrast, a drought significantly decreased the expression of *OsNRT2.1* [16]. From our study, both NRT and NPF genes were positively promoted under moderate water limitation (AWD). Thus, it is implied that AWD practice might contribute to enhanced availability of nitrate and, finally, improved N use efficiency (NUE). Another very interesting finding was that the difference in N assimilation strongly depended on the growth stage, tillering vs. heading, in the AWD. At the tillering stage, soil N was preferentially transported to the shoot to be assimilated into amino acids, whereas at the heading stage, root N utilization was dominant.

Photosynthetic rate was not significantly affected under mild and/or moderate soil drying due to enhanced N assimilation in the rice leaves [17], indicating a collaboration of photosynthesis and N metabolism under limited watering. Matsunami et al. (2018) [18] reported that the expression of N metabolism genes (*OsAMT1;3*, *OsAMT2;1*; *OsGS1;2*, *OsGOGAT2*) was greater at heading than at tillering, probably reflecting long−term N demand during the growth and development of the rice plant. Reduced osmotic potential at reproductive stage resulted in a decrease in the expression of NR, NIR, and GS-GOGAT genes in flag leaves of the rice plant [19] due to inhibition of water and nitrate movement at root level [20,21]. Our results implied that the uptake and assimilation of nitrogen under moderately limited water conditions (e.g., AWD) greatly differed depending on the growth stage. At the vegetative stage, acquired nitrogen is primarily utilized to increase photosynthesis and leaf biomass, whereas at the reproductive stage, it is likely to be consumed to promote root development as a countermeasure to extend the soil N−acquiring zone. Further investigation is also required to clarify that an expansion of the rhizosphere via up-regulated expression of nitrogen metabolism genes is closely involved in the development of panicles under moderate water limitation (AWD).

Non-structural soluble carbohydrates (NSCs) are especially important within a group of osmoprotectants. Soluble carbohydrates, including sucrose, glucose, and fructose, are greatly accumulated in tissues during limited water conditions like drought [22–26]. An abundance of soluble sugars measured at the heading stage greatly differed from tissues, with a tendency to decrease in source (leaf blade) and increase in sink (leaf sheath and root). In particular, the level of glucose and fructose was significantly accumulated in the root in AWD. Our results suggested that the transportation from source to sink was considered (1) a typical response to water stress, (2) a temporary retention in the leaf sheath for transporting to grain (reproductive tissue), and/or (3) a promotion of root development to extend the water-/nutrient-acquiring zone. The perturbation of soluble sugars by limited water supply somewhat differed from plant species and a period of water limitation. Water stress, like drought, resulted in an accumulation of soluble sugars, including sucrose, in plant roots [27–29]. However, a short−term water limitation led to a decrease in glucose and fructose in a leaf of a cereal crop, whereas an increase in xylose [25]. In contrast, an accumulation of soluble sugars, including sucrose, remained unchanged under limited water conditions [23,30]. Therefore, we carefully suggest further work to understand the variation of carbohydrate metabolism during the vegetative to reproductive transition under the AWD.

A higher level of amino acids reflects sensitivity due to the increased degradation and decreased synthesis of proteins in a water−limited environment [31,32]. The significant accumulation of amino acids from our study, especially in leaf blade and sheath, may be closely associated with nitrogen storage for further metabolic remobilization [33] and was clearly explained with down−regulation of transportation/assimilation−involved genes (Figure 1). Indeed, the AWD resulted in higher ratios of glutamine (gln)/glutamate (glu) and asparagine (asn)/aspartate (asp) in both the leaf blade and sheath. This agrees with the findings that water limitation accumulated not only higher levels of glutamine and asparagine but also other amino acids, including glycine, valine, and alanine [23,25,34–36]. Accordingly, the ratio of gln/glu and asn/asp could be a reliable biochemical marker to evaluate N metabolism like the NUE in the AWD. Previous studies have highlighted that plants experiencing water stress accumulate proline [37–39], an osmotic adjuster, and γ−aminobutyric acid (GABA) [40], a stress marker. We also observed that proline and GABA showed significantly higher increases in the AWD compared to CF. A higher level of those amino acids in the roots indicates that the AWD−grown rice plants might experience moderate water stress. Polyamines are closely linked with proline in terms of their biosynthetic and catabolic pathways [41] and play a role in mitigating abiotic stresses such as drought [42]. The AWD showed a significant increase in the level of putrescine in roots. Many reports showed the opposite behavior of putrescine. Accumulating putrescine in rice [43] and maize [44] is a tolerant response to drought. On the other hand, its level decreased in rice [45,46] and tomato [47] during the drought, helping those sensitive organisms accumulate putrescine [48]. Therefore, it is carefully proposed that an abundance of putrescine in our study is a collaborative tolerance mechanism with proline and GABA under an intermittently water−limited environment such as the AWD.

4. Materials and Methods

4.1. Plant Material, Growth Condition, and Treatment

This study was conducted at an experimental greenhouse of Chungbuk National University, Cheongju, South Korea (36°37′48.6″ N, 127°27′05.3″ E) from April to August 2021. The soil was sandy loam with 6.2 of pH, 0.28 dS/m of EC, 0.1 g kg^{-1} of total nitrogen, 106 mg kg^{-1} available P_2O_5, 0.61 $cmol^+$ kg^{-1} of potassium (K), 2.7 $cmol^+$ kg^{-1} of calcium (Ca), 1.38 $cmol^+$ kg^{-1} of magnesium (Mg), and 5.6 $cmol^+$ kg^{-1} of CEC.

The seeds of rice (Oryza sativa cv. Jinbongbyeo) were sterilized for 48 h at 28 °C in 5 L of water, including 2.5 mL of seed sterilizer (Kimaen, Farm Hannong), transferred to an incubator (28 °C, darkness) for 5 days, then moved to a growth chamber (VS-91G09M-2600, Vision Scientific, Korea) with a 12/12 h photoperiod, 60% (*w*/*v*) relative humidity (RH), and 24/20 °C (day/night) after de-etiolation, in order to induce uniform germination. Uniformly growing rice seedlings (3rd to 4th-leaf stage) were transplanted into plastic containers (20 × 20 × 20 cm), including the soil.

N (urea), P (superphosphate), and K (KCl) were applied at a rate of 90-45-56 kg/ha (a standard fertilization recommendation, Rural Development Administration, South Korea). N and K were split into three stages (50-30-20% at basal−tillering−panicle differentiation) and two stages (60-40% at basal−panicle differentiation), respectively. Two different types of water management were employed for this study: (1) constant flooding (CF, 5 cm height from the topsoil surface) as the control group, and (2) alternative wetting and drying (AWD, 0~5 cm of water level by evapotranspiration). Water supply for all treatments was temporarily ceased during the non−productive tillering stage (no panicle tiller). The rice plants (n = 3, each treatment) were taken at tillering (approximately 30 days after transplanting) and heading (panicle emergence) stages, immediately washed out with ddH_2O, divided into leaf blades, leaf sheaths, and roots, and stored at −80 °C for further molecular and biological analysis.

4.2. RNA Extraction, cDNA Synthesis, and Relatively Quantitative Real−Time Polymerase Chain Reaction (qRT−PCR)

Total RNA was extracted from leaf blades, leaf sheaths, and roots of rice at the tillering and heading stages, using TRIzol reagent (Invitrogen, Carlsbad, CA, USA), according to the manufacturer's instructions. The extracted RNA purity and concentration were measured by NanoDrop (Thermo Fisher Scientific, Madison, WI, USA) and double−checked by electrophoresis on a 1% agarose gel. Complementary DNA (cDNA) was synthesized by using a Maxime RT PreMix Kit with oligo (dT) primers and 1 μg of total RNA. Quantitative Real−Time PCR was performed using SYBR Green Q Master mix ROX (SYBR) (LaboPass, Seoul, Republic of Korea) with a CFX Connect Optics Real−Time System (Bio-Rad, Hercules, CA, USA) according to the manufacturer's instructions. To analyze gene expression, put 1 μL of cDNA, 2 μL of forward primer (FW), 2 μL of reverse primer (RV), and 5 μL of SYBR in a PCR tube. Quantification method ($2^{-\Delta\Delta Ct}$) was used, and rice actin primer (*OsACT-1*, *Os03g0718100*, FW: 5′-TGTATGCCAGTGGTCGTACC-3′, RV: 5′-CCAGCAAGGTCGAGACGAA-3′) was used for the control gene to normalize the data. The expression of selected genes (Table S1) was validated using four biological replicates and two technical replicates. The PCR program consisted of initial denaturation at 95 °C for 10 min, followed by 40 cycles of denaturation at 95 °C for 20 s, annealing at 50–59 °C for

30 s, and extension at 72 °C for 30 s. After the last extension step, perform the melt curve step at 65 °C for 5 s and 95 °C for 0.5 s.

4.3. Extraction and Analysis of Polar Compounds

Polar compounds (e.g., sugars, sugar alcohols, amino acids, organic acids, and secondary metabolism intermediaries) were analyzed in accordance with a previously reported method [16]. The powdered samples (100 mg) were placed in microtubes. 1 mL of a methanol:water:chloroform (2.5:1:1, *v*/*v*/*v*) [17] solution was added to the tubes. Used ribitol (60 µL, 0.2 mg mL^{-1}) solution as an internal standard (IS). After vortexing it, the mixture was incubated at 37 °C and subjected to 1200 rpm for 30 min in a thermomixer (Eppendorf AG, Hamburg, Germany). was After centrifuging the mixture for 5 min at 4 °C and 16,000× *g*, 0.8 mL of the supernatant was transferred to new tubes, and 0.41 mL of deionized water was added. Centrifuged the mixture at the same conditions (4 °C, 16,000× *g* for 5 min). After that, 0.9 mL of the supernatant was transferred to new tubes and dried in a vacuum concentrator (VS-802F; Visionbionex, Gyeonggi, Republic of Korea) for at least 3 h, and then freeze−dried for 16 h. For derivatization, add 80 µL of methoxyamine hydrochloride (20 mg mL^{-1}) to pyridine and shake at 30 °C for 90 m. To execute trimethylsilylated etherification, 80 µL of N-methyl-N-(trimethylsilyl) trifluoroacetamide (MSTFA) was added to the tubes and incubated at 37 °C for 30 min. The samples were analyzed on a GC (Agilent 7890A, Agilent Technologies, Santa Ckara, CA, USA) equipped with a Pegasus TOF−MS (LECO, St. Joseph, MI, USA). A Rtx-5MS column (30 cm × 0.25 mm, 0.25-µm i.d. film thickness; Restek, Bellefonte, PA, USA) was used to separate the polar compounds. The oven temperature was programmed as follows: 80 °C for 2 m, followed by ramping to 320 °C at 15 °C/min and holding at this temperature for 50 min. Helium gas (He) was passed at a rate of 1 mL min^{-1}, and 1 µL of sample was injected at a 1:25 ratio in split mode. The inlet temperature was set at 230 °C, respectively, and spectral data were scanned over an m/z mass range of 85 to 600. Data were analyzed using ChromaTOF software (v5.5; LECO), and peaks were identified based on mass−spectral data compared to standards (i.e., NIST 11, Wiley 9, and in−house libraries). For quantification, the relative ratio of the peak area of the compounds to the peak area of the IS was acquired based on the selected ions.

4.4. Statistical Analysis

Statistical analysis was performed by the RStudio (version R−4.1.3, RStudio Team, 2022). The qRT−PCR data was subjected to a *t*-test. Polar compounds were analyzed by principal component analysis (PCA), subjected to a *t*-test, and visualized by a heatmap.

5. Conclusions

The current study aimed to extend our knowledge of carbon (primary metabolites) and nitrogen (acquisition and assimilation) metabolism under limited water supply (alternative wetting and drying, AWD) during rice growth and development. The AWD system promoted not only the acquisition of nitrate but also its transportation and assimilation, and, in particular, nitrogen assimilation was preferential in the root during the shift from the vegetative to reproductive stage. Based on the significant increase in amino acids in the shoot (leaf blade and sheath), the AWD was considered to rearrange amino acid pools in order to synthesize proteins in accordance with phase transition. AWD acquired more active nitrate from soil and resulted in an abundance of amino acid pools, which are considered a rearrangement for reproduction−related proteins under limited N availability. We have taken some fruitful information to provide directions for further study to clarify form−dependent N metabolism and root development under the AWD condition and a practical approach in the rice production system.

Supplementary Materials: The following supporting information can be downloaded at: https://www.mdpi.com/article/10.3390/plants12081649/s1, Table S1: Selected genes for this study and primer sequences used for qRT−PCR.

Author Contributions: Conceptualization, J.S.; validation, J.S.; investigation, G.K.; resources, J.S.; writing—original draft preparation, G.K.; writing—review and editing, J.S.; supervision, J.S.; project administration, J.S.; funding acquisition, J.S. All authors have read and agreed to the published version of the manuscript.

Funding: This research was funded by the Cooperative Research Program for Agriculture Science and Technology Development (Project No. PJ017002), RDA, Republic of Korea.

Data Availability Statement: Not applicable.

Acknowledgments: We thank the National Institute of Crop Science, RDA, for providing rice seed (cv. Jinbongbyeo).

Conflicts of Interest: The authors declare no conflict of interest.

References

1. Food and Agriculture Organization. *Climate Change and Food Security: Risks and Responses*; FAO: Rome, Italy, 2016.
2. Yearbook, F.F.S. *World Food and Agriculture 2020*; FAO: Rome, Italy, 2020.
3. Kirk, G.J. Rice root properties for internal aeration and efficient nutrient acquisition in submerged soil. *New Phytol.* **2003**, *159*, 185–194. [CrossRef]
4. Lampayan, R.M.; Rejesus, R.M.; Singleton, G.R.; Bouman, B.A. Adoption and economics of alternate wetting and drying water management for irrigated lowland rice. *Field Crops Res.* **2015**, *170*, 95–108. [CrossRef]
5. Bouman, B.; Tuong, T.P. Field water management to save water and increase its productivity in irrigated lowland rice. *Agric. Water Manag.* **2001**, *49*, 11–30. [CrossRef]
6. Montzka, S.A.; Dlugokencky, E.J.; Butler, J.H. Non-CO_2 greenhouse gases and climate change. *Nature* **2011**, *476*, 43–50. [CrossRef] [PubMed]
7. Kirk, G.; Kronzucker, H. The potential for nitrification and nitrate uptake in the rhizosphere of wetland plants: A modelling study. *Ann. Bot.* **2005**, *96*, 639. [CrossRef]
8. Nazish, T.; Arshad, M.; Jan, S.U.; Javaid, A.; Khan, M.H.; Naeem, M.A.; Baber, M.; Ali, M. Transporters and transcription factors gene families involved in improving nitrogen use efficiency (NUE) and assimilation in rice (*Oryza sativa* L.). *Transgenic Res.* **2021**, *31*, 23–42. [CrossRef]
9. Foyer, C.H.; Valadier, M.-H.; Migge, A.; Becker, T.W. Drought-induced effects on nitrate reductase activity and mRNA and on the coordination of nitrogen and carbon metabolism in maize leaves. *Plant Physiol.* **1998**, *117*, 283–292. [CrossRef]
10. Morison, J.; Baker, N.; Mullineaux, P.; Davies, W. Improving water use in crop production. *Philos. Trans. R. Soc. B Biol. Sci.* **2008**, *363*, 639–658. [CrossRef]
11. Karasov, T.L.; Chae, E.; Herman, J.J.; Bergelson, J. Mechanisms to mitigate the trade-off between growth and defense. *Plant Cell* **2017**, *29*, 666–680. [CrossRef]
12. Asch, F.; Dingkuhn, M.; Sow, A.; Audebert, A. Drought-induced changes in rooting patterns and assimilate partitioning between root and shoot in upland rice. *Field Crops Res.* **2005**, *93*, 223–236. [CrossRef]
13. Price, A.H.; Steele, K.; Moore, B.; Jones, R. Upland rice grown in soil-filled chambers and exposed to contrasting water-deficit regimes: II. Mapping quantitative trait loci for root morphology and distribution. *Field Crops Res.* **2002**, *76*, 25–43. [CrossRef]
14. Bouman, B.; Yang, X.; Wang, H.; Wang, Z.; Zhao, J.; Chen, B. Performance of aerobic rice varieties under irrigated conditions in North China. *Field Crops Res.* **2006**, *97*, 53–65. [CrossRef]
15. Kondo, M.; Pablico, P.; Aragones, D.; Agbisit, R.; Abe, J.; Morita, S.; Courtois, B. Genotypic and environmental variations in root morphology in rice genotypes under upland field conditions. In Proceedings of the Roots: The Dynamic Interface between Plants and the Earth: The 6th Symposium of the International Society of Root Research, Nagoya, Japan, 11–15 November 2001; pp. 189–200.
16. Luo, B.; Chen, J.; Zhu, L.; Liu, S.; Li, B.; Lu, H.; Ye, G.; Xu, G.; Fan, X. Overexpression of a high-affinity nitrate transporter OsNRT2. 1 increases yield and manganese accumulation in rice under alternating wet and dry condition. *Front. Plant Sci.* **2018**, *9*, 1192. [CrossRef]
17. Zhong, C.; Cao, X.; Bai, Z.; Zhang, J.; Zhu, L.; Huang, J.; Jin, Q. Nitrogen metabolism correlates with the acclimation of photosynthesis to short-term water stress in rice (*Oryza sativa* L.). *Plant Physiol. Biochem.* **2018**, *125*, 52–62. [CrossRef]
18. Alhaj Hamoud, Y.; Guo, X.; Wang, Z.; Shaghaleh, H.; Chen, S.; Hassan, A.; Bakour, A. Effects of irrigation regime and soil clay content and their interaction on the biological yield, nitrogen uptake and nitrogen-use efficiency of rice grown in southern China. *Agric. Water Manag.* **2019**, *213*, 934–946. [CrossRef]
19. Sathee, L.; Jha, S.K.; Rajput, O.S.; Singh, D.; Kumar, S.; Kumar, A. Expression dynamics of genes encoding nitrate and ammonium assimilation enzymes in rice genotypes exposed to reproductive stage salinity stress. *Plant Physiol. Biochem.* **2021**, *165*, 161–172. [CrossRef]
20. Licht, S.; Halperin, L.; Kalina, M.; Zidman, M.; Halperin, N. Electrochemical potential tuned solar water splitting. *Chem. Commun.* **2003**, *24*, 3006–3007. [CrossRef]

21. Sanchez, D.H.; Lippold, F.; Redestig, H.; Hannah, M.A.; Erban, A.; Krämer, U.; Kopka, J.; Udvardi, M.K. Integrative functional genomics of salt acclimatization in the model legume Lotus japonicus. *Plant J.* **2008**, *53*, 973–987. [CrossRef]
22. Mostajeran, A.; Rahimi-Eichi, V. Effects of drought stress on growth and yield of rice (*Oryza sativa* L.) cultivars and accumulation of proline and soluble sugars in sheath and blades of their different ages leaves. *Agric. Environ. Sci.* **2009**, *5*, 264–272.
23. Sicher, R.C.; Timlin, D.; Bailey, B. Responses of growth and primary metabolism of water-stressed barley roots to rehydration. *J. Plant Physiol.* **2012**, *169*, 686–695. [CrossRef]
24. Xu, W.; Cui, K.; Xu, A.; Nie, L.; Huang, J.; Peng, S. Drought stress condition increases root to shoot ratio via alteration of carbohydrate partitioning and enzymatic activity in rice seedlings. *Acta Physiol. Plant.* **2015**, *37*, 1–11. [CrossRef]
25. Perlikowski, D.; Czyżniejewski, M.; Marczak, Ł.; Augustyniak, A.; Kosmala, A. Water deficit affects primary metabolism differently in two Lolium multiflorum/Festuca arundinacea introgression forms with a distinct capacity for photosynthesis and membrane regeneration. *Front. Plant Sci.* **2016**, *7*, 1063. [CrossRef] [PubMed]
26. Guo, X.; Peng, C.; Li, T.; Huang, J.; Song, H.; Zhu, Q.; Wang, M. The effects of drought and re-watering on non-structural carbohydrates of Pinus tabulaeformis seedlings. *Biology* **2021**, *10*, 281. [CrossRef] [PubMed]
27. Wu, D.; Cai, S.; Chen, M.; Ye, L.; Chen, Z.; Zhang, H.; Dai, F.; Wu, F.; Zhang, G. Tissue metabolic responses to salt stress in wild and cultivated barley. *PLoS ONE* **2013**, *8*, e55431. [CrossRef] [PubMed]
28. Li, P.-C.; Yang, X.-Y.; Wang, H.-M.; Ting, P.; Yang, J.-Y.; Wang, Y.-Y.; Yang, X.; Yang, Z.-F.; Xu, C.-W. Metabolic responses to combined water deficit and salt stress in maize primary roots. *J. Integr. Agric.* **2021**, *20*, 109–119. [CrossRef]
29. Patterson, J.H.; Newbigin, E.; Tester, M.; Bacic, A.; Roessner, U. Metabolic responses to salt stress of barley (*Hordeum vulgare* L.) cultivars, Sahara and Clipper, which differ in salinity tolerance. *J. Exp. Bot.* **2009**, *60*, 4089–4103.
30. Amiard, V.; Morvan-Bertrand, A.; Billard, J.-P.; Huault, C.; Keller, F.; Prud'homme, M.-P. Fructans, but not the sucrosyl-galactosides, raffinose and loliose, are affected by drought stress in perennial ryegrass. *Plant Physiol.* **2003**, *132*, 2218–2229. [CrossRef]
31. Roy-Macauley, H.; Zuily-Fodil, Y.; Kidric, M.; Thi, A.P.; de Silva, J.V. Effect of drought stress on proteolytic activities in Phaseolus and Vigna leaves from sensitive and resistant plants. *Physiol. Plant.* **1992**, *85*, 90–96. [CrossRef]
32. Ahmad, F.S.; Aminu, M.U.; Safwan, I.I. Response of Two Pearl Millet Cultivars (*Pennisetum glaucum*) to Water Stress in the University Garden at Gadau, Bauchi State University. *Int. J. Sci. Technoledge* **2016**, *4*, 111.
33. Sicher, R.C.; Barnaby, J.Y. Impact of carbon dioxide enrichment on the responses of maize leaf transcripts and metabolites to water stress. *Physiol. Plant.* **2012**, *144*, 238–253. [CrossRef]
34. Degenkolbe, T.; Do, P.T.; Kopka, J.; Zuther, E.; Hincha, D.K.; Köhl, K.I. Identification of drought tolerance markers in a diverse population of rice cultivars by expression and metabolite profiling. *PLoS ONE* **2013**, *8*, e63637. [CrossRef]
35. Yang, J.; Fleisher, D.H.; Sicher, R.C.; Kim, J.; Baligar, V.C.; Reddy, V.R. Effects of CO2 enrichment and drought pretreatment on metabolite responses to water stress and subsequent rehydration using potato tubers from plants grown in sunlit chambers. *J. Plant Physiol.* **2015**, *189*, 126–136. [CrossRef]
36. Melandri, G.; AbdElgawad, H.; Floková, K.; Jamar, D.C.; Asard, H.; Beemster, G.T.; Ruyter-Spira, C.; Bouwmeester, H.J. Drought tolerance in selected aerobic and upland rice varieties is driven by different metabolic and antioxidative responses. *Planta* **2021**, *254*, 13. [CrossRef]
37. Bandurska, H. Does proline accumulated in leaves of water deficit stressed barley plants confine cell membrane injuries? II. Proline accumulation during hardening and its involvement in reducing membrane injuries in leaves subjected to severe osmotic stress. *Acta Physiol. Plant.* **2001**, *23*, 483–490. [CrossRef]
38. Hayat, S.; Hayat, Q.; Alyemeni, M.N.; Wani, A.S.; Pichtel, J.; Ahmad, A. Role of proline under changing environments: A review. *Plant Signal. Behav.* **2012**, *7*, 1456–1466. [CrossRef]
39. Zhao, X.; Wang, W.; Zhang, F.; Deng, J.; Li, Z.; Fu, B. Comparative metabolite profiling of two rice genotypes with contrasting salt stress tolerance at the seedling stage. *PLoS ONE* **2014**, *9*, e108020. [CrossRef]
40. Ramesh, S.A.; Tyerman, S.D.; Xu, B.; Bose, J.; Kaur, S.; Conn, V.; Domingos, P.; Ullah, S.; Wege, S.; Shabala, S. GABA signalling modulates plant growth by directly regulating the activity of plant-specific anion transporters. *Nat. Commun.* **2015**, *6*, 7879. [CrossRef]
41. Aziz, A.; Martin-Tanguy, J.; Larher, F. Stress-induced changes in polyamine and tyramine levels can regulate proline accumulation in tomato leaf discs treated with sodium chloride. *Physiol. Plant.* **1998**, *104*, 195–202. [CrossRef]
42. Simon-Sarkadi, L.; Kocsy, G.; Varhegyi, A.; Galiba, G.; Ronde, J.A.d. Effect of drought stress at supraoptimal temperature on polyamine concentrations in transgenic soybean with increased proline levels. *Z. Für Nat. C* **2006**, *61*, 833–839. [CrossRef]
43. Yang, J.; Zhang, J.; Liu, K.; Wang, Z.; Liu, L. Involvement of polyamines in the drought resistance of rice. *J. Exp. Bot.* **2007**, *58*, 1545–1555. [CrossRef]
44. Ahangir, A.; Ghotbi-Ravandi, A.A.; Rezadoost, H.; Bernard, F. Drought tolerant maize cultivar accumulates putrescine in roots. *Rhizosphere* **2020**, *16*, 100260. [CrossRef]
45. Maiale, S.; Sánchez, D.H.; Guirado, A.; Vidal, A.; Ruiz, O.A. Spermine accumulation under salt stress. *J. Plant Physiol.* **2004**, *161*, 35–42. [CrossRef] [PubMed]
46. Do, P.T.; Degenkolbe, T.; Erban, A.; Heyer, A.G.; Kopka, J.; Köhl, K.I.; Hincha, D.K.; Zuther, E. Dissecting rice polyamine metabolism under controlled long-term drought stress. *PLoS ONE* **2013**, *8*, e60325. [CrossRef] [PubMed]

47. Santa-Cruz, A.; Acosta, M.; Pérez-Alfocea, F.; Bolarin, M.C. Changes in free polyamine levels induced by salt stress in leaves of cultivated and wild tomato species. *Physiol. Plant.* **1997**, *101*, 341–346. [CrossRef]
48. Bouchereau, A.; Aziz, A.; Larher, F.; Martin-Tanguy, J. Polyamines and environmental challenges: Recent development. *Plant Sci.* **1999**, *140*, 103–125. [CrossRef]

plants MDPI

Article

Physiological and Transcriptomic Analyses of the Effects of Exogenous Lauric Acid on Drought Resistance in Peach (*Prunus persica* (L.) Batsch)

Binbin Zhang, Hao Du, Sankui Yang, Xuelian Wu, Wenxin Liu, Jian Guo, Yuansong Xiao * and Futian Peng *

College of Horticulture Science and Engineering, Shandong Agricultural University, Taian 271018, China
* Correspondence: ysxiao@sdau.edu.cn (Y.X.); pft@sdau.edu.cn (F.P.); Tel.: +86-151-6387-3786 (Y.X.); +86-135-6382-1651 (F.P.)

Abstract: Peach (*Prunus persica* (L.) Batsch) is a fruit tree of economic and nutritional importance, but it is very sensitive to drought stress, which affects its growth to a great extent. Lauric acid (LA) is a fatty acid produced in plants and associated with the response to abiotic stress, but the underlying mechanism remains unclear. In this study, physiological analysis showed that 50 ppm LA pretreatment under drought stress could alleviate the growth of peach seedlings. LA inhibits the degradation of photosynthetic pigments and the closing of pores under drought stress, increasing the photosynthetic rate. LA also reduces the content of O_2^-, H_2O_2, and MDA under drought stress; our results were confirmed by Evans Blue, nitroblue tetrazolium (NBT), and DAB(3,3-diaminobenzidine) staining experiments. It may be that, by directly removing reactive oxygen species (ROS) and improving enzyme activity, i.e., catalase (CAT) activity, peroxidase (POD) activity, superoxide dismutase (SOD) activity, and ascorbate peroxidase (APX) activity, the damage caused by reactive oxygen species to peach seedlings is reduced. Peach seedlings treated with LA showed a significant increase in osmoregulatory substances compared with those subjected to drought stress, thereby regulating osmoregulatory balance and reducing damage. RNA-Seq analysis identified 1876 DEGs (differentially expressed genes) in untreated and LA-pretreated plants under drought stress. In-depth analysis of these DEGs showed that, under drought stress, LA regulates the expression of genes related to plant–pathogen interaction, phenylpropanoid biosynthesis, the MAPK signaling pathway, cyanoamino acid metabolism, and sesquiterpenoid and triterpenoid biosynthesis. In addition, LA may activate the Ca^{2+} signaling pathway by increasing the expressions of CNGC, CAM/CML, and CPDK family genes, thereby improving the drought resistance of peaches. In summary, via physiological and transcriptome analyses, the mechanism of action of LA in drought resistance has been revealed. Our research results provide new insights into the molecular regulatory mechanism of the LA-mediated drought resistance of peach trees.

Keywords: *Prunus persica* (L.) Batsch; lauric acid; drought stress; physiological indicators; transcriptome

Citation: Zhang, B.; Du, H.; Yang, S.; Wu, X.; Liu, W.; Guo, J.; Xiao, Y.; Peng, F. Physiological and Transcriptomic Analyses of the Effects of Exogenous Lauric Acid on Drought Resistance in Peach (*Prunus persica* (L.) Batsch). *Plants* **2023**, *12*, 1492. https://doi.org/10.3390/plants12071492

Academic Editors: Beata Prabucka, Mateusz Labudda, Marta Gietler, Justyna Fidler and Małgorzata Nykiel

Received: 20 February 2023
Revised: 21 March 2023
Accepted: 23 March 2023
Published: 29 March 2023

1. Introduction

In recent years, due to the increase in greenhouse gases around the world, global drought damage and water deprivation are rising, particularly in extremely arid and semi-arid regions [1–4]. Drought inhibits plants in various stages of normal growth and development, not only affecting the respiration and light interface effect, but also affecting osmotic regulation, protein interfaces, and material transportation, as well as some physiological processes, thus seriously influencing the survival of crops and their growth and yield [5–7]. Therefore, in order to promote plant growth, it is critical to investigate effective drought tolerance solutions [8]. It has been demonstrated that using biological modulators or organic acids to enhance plant growth and drought resistance is a successful and environmentally advantageous approach [9,10].

Many studies have confirmed that abiotic stress in plants is alleviated by the addition of plant growth regulators or organic acids, such as strigolactone, dopamine, salicylic acid (SA), jasmonic acid (JA), methyl jasmonate (MeJA), citric acid, and acetic acid, which can regulate the resistance of plants to abiotic stress [11–14]. For instance, by encouraging the production of JA and its signaling pathway in plants, exogenous acetic acid can enhance drought resistance in a range of plant species, including rice, maize, rapeseed, and wheat [9]. Recent research has demonstrated that acetic acid-induced modifications in plant physiology and metabolic composition increase the resistance of willow to drought [15]. Previous research has demonstrated that treating *Arabidopsis* with nicotinic acid (NA) increases the plant's tolerance to drought stress and encourages development [16]. In recent years, studies have shown that β-amino butyric acid (BABA) can improve the drought resistance of rice, cowpea, cherry, and other plants [17–19].

Fatty acids are a group of significant physiological chemicals that play a role in the energy conversion, membrane structure, and many of the signaling pathways of cells [20]. The medium-chain fatty acid lauric acid (C12) is more active than caproic acid (C6), caprylic acid (C8), and capric acid (C10) [21]. It was discovered to be the primary antibacterial and antiviral component in human breast milk [22]. As a result, lauric acid (LA) research primarily focuses on its antibacterial and bactericidal properties [23–25]. It has been demonstrated that a medium treated with LA may enhance the development of zygotic coconut (*Cocos nucifera* L.) embryos, while also greatly increasing plant growth. A radioactive LA metabolism experiment demonstrated that LA serves as a source of carbon for long-chain fatty acids, and that its principal byproduct is the phospholipid that comprises new cell membranes [26]. LA improved zygotic Syagrus coronata embryos during in vitro growth. Additionally, it was shown that plants treated with LA at the pre-acclimation stage had improved drought resistance via the controlling of stomatal closure, the preservation of leaf water content, and the maintenance of photosynthesis [27]. Through a study on the effects of drought stress on the growth and metabolism of loquat, it was found that among more than 100 leaf metabolites, 9 contributed to the metabolism of loquat under drought conditions, including LA [28].

In recent years, the peach (*Prunus persica*) industry has grown in the global economic market, and peach offers high economic, nutritional, and medicinal value [29]. However, drought seriously threatens peach yield and fruit quality during peach growth and development [30]. Therefore, the drought resistance mechanism of peach is being paid increasing attention. Exogenous LA has been applied to peaches in response to drought stress in agricultural production. So far, no studies have comprehensively explored the physiological and molecular mechanisms by which LA mediates peach drought resistance, and our understanding is still limited.

In this study, peach seedlings were subjected to drought stress, and LA was added at the same time to explore its alleviating effect. Additionally, a comparative transcriptome study was carried out on peaches that had been treated with LA versus those that had not, in order to further explore the molecular basis of LA-induced drought tolerance in peach seedlings. Our research should provide information about how the molecular control of peach drought resistance is mediated by LA, and should aid in the search for more potent ways to reduce the negative impacts of drought stress on peach output.

2. Results

2.1. Effects of LA on Growth and Membrane Permeability of P. persica under Drought

Figure 1A,B show that peach seedling development was considerably hampered by drought stress. Under drought conditions, peach seedlings' fresh weight, dry weight, and root activity declined by 38.55%, 55.37%, and 28.02%, respectively, in comparison to CK. The application of exogenous LA at a specific dosage during drought treatment was able to stimulate seedling growth, which increased by 21.99%, 23.94%, and 17.40% in comparison to drought treatment stages, respectively. The findings demonstrate that exogenous LA might reduce the impact of drought treatment on peach seedlings. Drought stress significantly decreased the leaf relative water content of peach seedlings. However, after pretreatment with 50 ppm LA, the leaf water status was maintained, and the relative water content of these leaves was higher than that of the control without LA (Figure 1C). Relative electrical leakage is an important physiological index reflecting the damage degree of plant leaves under stress, and the stability of the cell membrane. Here, the REL of the leaves of peach seedlings increased significantly under drought conditions, reaching 35.59%. Compared with plants without LA, the REL (relative electrolyte leakage) of plants treated with exogenous LA under drought stress conditions was lower, and the REL under drought stress was 15.28% with exogenous LA (Figure 1D).

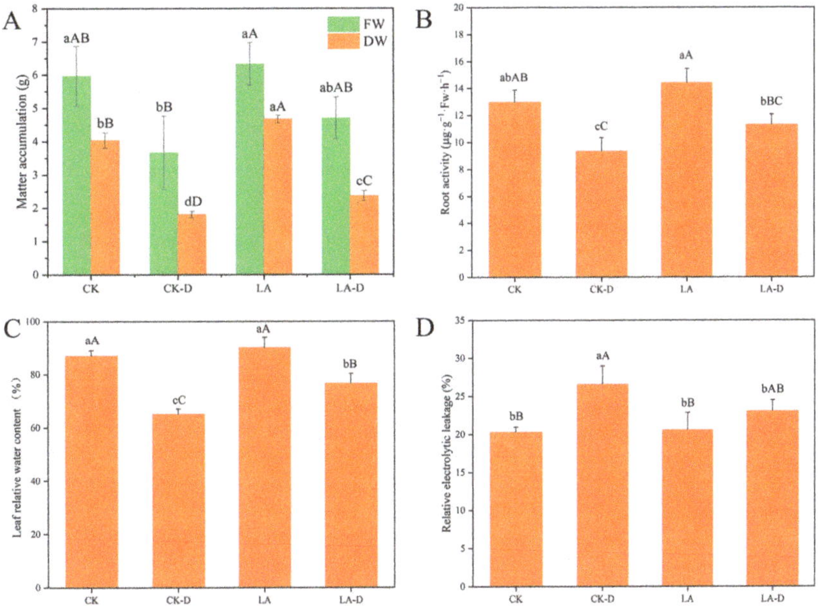

Figure 1. The effects of LA on *P. persica* growth under drought stress. Determination of dry and fresh weight of *P. persica* treated with LA under drought condition (**A**), determination of root activity (**B**). The effects of LA treatments on relative water content (**C**) and relative electrolyte leakage (**D**) in leaves under control and drought circumstances are shown in the graphs, while different small letter superscripts mean significant difference ($p < 0.05$), and different capital letter superscripts mean significant difference ($p < 0.01$). The same as below.

2.2. Characteristics of Photosynthetic Parameters and Photosynthetic Pigments of Leaves

As shown in Figure 2, drought stress significantly inhibited the net photosynthetic rate (Pn) of peach leaves. However, the exogenous application of 50 ppm LA increased the photosynthetic capacity of the plants (Figure 2A). The change trend of the stomatal conductance (Gs) value was basically consistent with Pn, showing that Gs decreased significantly under drought stress, while exogenous LA could maintain this at a relatively high level (Figure 2B). Similarly, the intercellular carbon dioxide concentration (Ci) and transpiration rate (Tr) were significantly decreased under drought stress, and exogenous LA alleviated the effects of drought on peach seedlings (Figure 2C,D). As shown in Figure 2, the contents of chlorophyll a (Chl a), chlorophyll b (Chl b), chlorophyll a + b (Chl a + b), and carotenoids (Car) in peach seedlings under drought stress were significantly lower than those under CK ($p < 0.05$), which values decreased by 22.37%, 16.22%, 20.85%, and 30.16% compared with CK treatment, respectively. These results indicate that drought stress could inhibit the accumulation of photosynthetic pigments in peach seedlings. When LA was applied in the drought treatment, the contents of Chl a, Chl b, Chl a + b, and

Car were increased by 10.85%, 8.83%, 8.35%, and 20.55%, respectively, indicating that LA significantly alleviated the inhibition of photosynthetic pigment synthesis or degradation in peach seedlings under drought stress. Thus, the photosynthesis of peach seedlings was effectively improved.

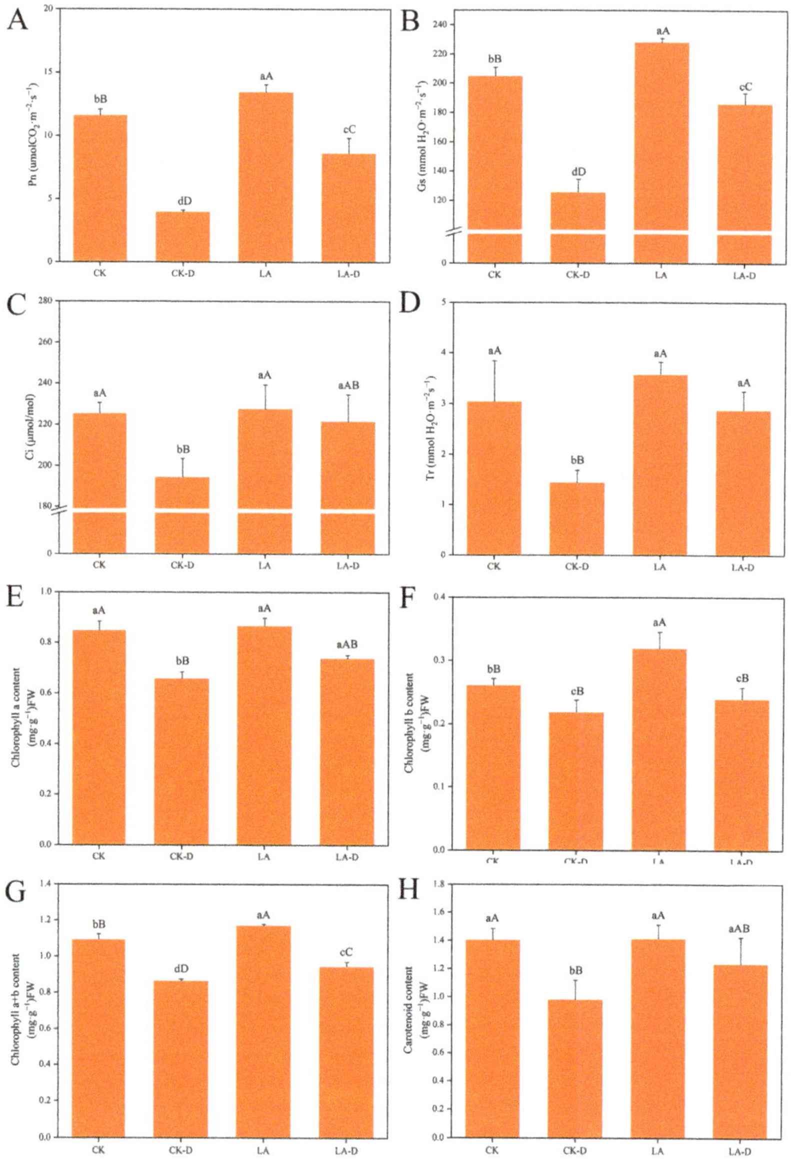

Figure 2. The effects of LA pretreatment on photosynthesis and chlorophyll content in drought-stressed *P. persica*. (**A**) net photosynthetic rate (Pn), (**B**) stomatal conductance (Gs), (**C**) intercellular carbon dioxide concentration (Ci), (**D**) transpiration rate (Tr), (**E**) chlorophyll a content (Chl a), (**F**) chlorophyll b content (Chl b), (**G**)total chlorophyll content (Chl a + b), (**H**) carotenoids (Car). Values represent means ± SD of three replicates. Different letters indicate significant differences according to Duncan's multiple range tests.

2.3. The Effects of LA on the Osmotic Regulation Substances of *P. persica* under Drought

The content of osmotic regulatory substances in plants reflects their stress resistance to a certain extent. Under drought stress, peach seedlings had considerably higher amounts of soluble sugar, soluble protein, proline, and free amino acids compared to normal circumstances (Figure 3). After drought treatment, the contents of soluble sugar, soluble protein, proline, and free amino acids in peach seedlings increased by 22.43%, 27.03%, 16.36%, and 52.39%, respectively, compared with CK. Compared with drought treatment, the contents of soluble sugar, soluble protein, proline, and free amino acids increased by 22.02%, 13.82%, 21.95%, and 9.46%, respectively, after applying 50 ppm LA.

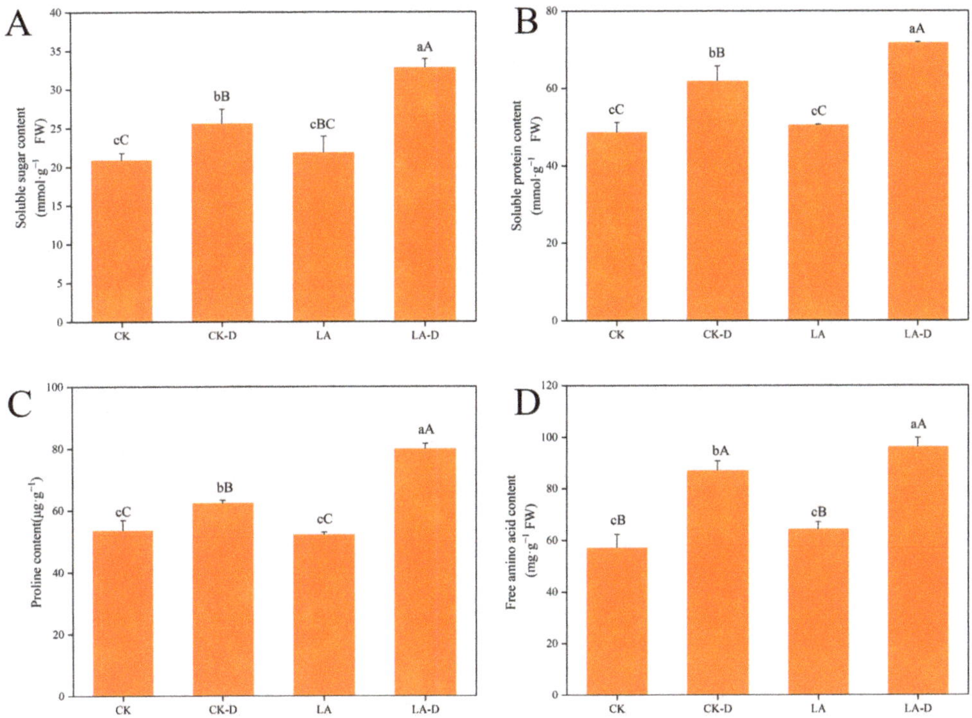

Figure 3. Effects of LA on osmotic regulation substances in *P. persica* under drought stress. (**A**) Soluble sugar content, (**B**) soluble protein, (**C**) proline content, (**D**) free amino acid content. Values represent means ± SD of three replicates. Different letters indicate significant differences according to Duncan's multiple range tests.

2.4. Effects of Cell Damage and ROS (Reactive Oxygen Species) Accumulation on Leaves of *P. persica*

Drought can lead to oxidative stress and cell damage. As a result, under various treatment settings, we found cell death and ROS buildup in peach seedling leaves. Cell death and ROS accumulation in peach seedling leaves were compared using Evans Blue, DAB(3,3-diaminobenzidine), and NBT (Nitroblue tetrazolium) staining. Figure 4A,C,E show that under drought stress conditions, the leaves show a darker color and a larger staining area, suggesting that the cells in the leaves are damaged, the number of dead cells has increased, and the generation of ROS in the leaves has greatly increased. The stained area of leaves treated with exogenous LA was minimal under drought stress, showing that exogenous LA might mitigate drought damage. Stress can lead to the accumulation of reactive oxygen species and the intensification of membrane lipid peroxidation in plant leaves, and the content of MDA (malondialdehyde) in plant leaves' cells can reflect the degree of oxidative damage. We examined the levels of O_2^- and H_2O_2 to investigate the influence of LA pretreatment on oxidative alterations (Figure 4B,D). Compared with control plants, drought stress significantly increased O_2^- and H_2O_2 contents in peach leaves by 45.33% and 32.47%, respectively. Compared with the drought-treated plants, the O_2^- and H_2O_2 contents in peach leaves after LA application decreased by 32.71% and 28.42%, respectively. According to Figure 4F, drought stress significantly increased the MDA content in the leaves of peach seedlings. The MDA content in the leaves increased by 214.48% after drought treatment compared with CK. After exogenous LA application, the MDA content decreased by 140.82% compared with under the drought treatment.

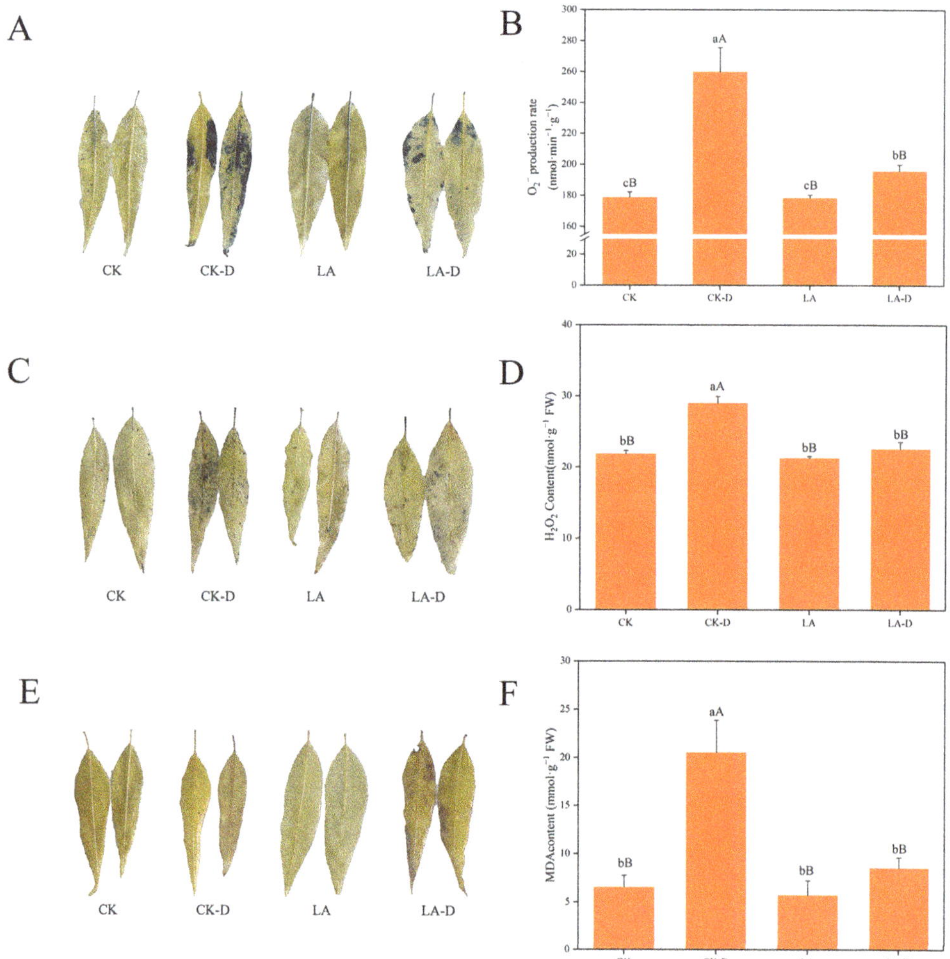

Figure 4. Effects of exogenous LA on cell death and ROS accumulation in leaves of *P. persica* under drought conditions. (**A**) Staining with Evans Blue, (**B**) effects of LA on O_2^- contents in leaves under drought, (**C**) staining with nitroblue tetrazolium (NBT), (**D**) effects of LA on hydrogen peroxide (H_2O_2) content in leaves under drought, (**E**) staining with 3,3-diaminobenzidine (DAB), (**F**) MDA content. Values represent means ± SD of three replicates. Different letters indicate significant differences according to Duncan's multiple range tests.

2.5. Exogenous LA Enhanced the Antioxidant Capacity of *P. persica* under Drought Stress

Catalase (CAT), peroxidase (POD), superoxide dismutase (SOD), and ascorbate peroxidase (APX) are key antioxidant enzymes related to a plant's restoration ability, which act by adjusting their activity to maintain the balance of active oxygen metabolism, thereby slowing the accumulation of active oxygen free radicals and membrane lipid peroxidation damage. As can be seen from Figure 5, compared with CK, the activities of CAT, POD, and SOD in peach leaves increased by 37.7%, 23.12%, and 13.69% after drought treatment, respectively. However, CAT activity decreased by 9.82%. Compared with the normal treatment, the activities of CAT, POD, SOD, and APX were increased by 10.58%, 24.35%, 13.56%, and 31.54%, respectively, in the LA-pretreated peach seedlings under drought stress. The results show that the exogenous application of LA increased the antioxidant enzyme activity of peach seedling leaves under drought stress, thereby removing reactive oxygen species and alleviating cell damage.

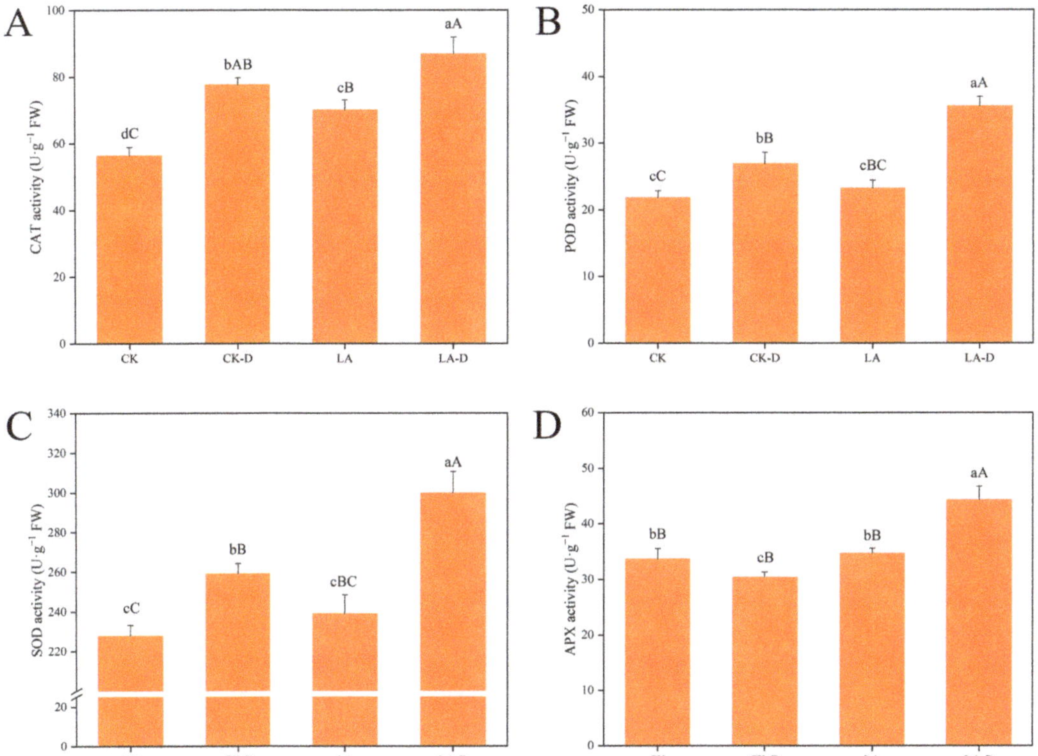

Figure 5. Effects of LA on antioxidant enzyme activity *P. persica* leaves under drought stress. (**A**) Catalase (CAT) activity, (**B**) peroxidase (POD) activity, (**C**) superoxide dismutase (SOD) activity, (**D**) ascorbate peroxidase (APX) activity. Values represent means ± SD of three replicates. Different letters indicate significant differences according to Duncan's multiple range tests.

2.6. Sequencing of mRNA and Alignment to Reference Genome

We conducted transcriptional profiling on the leaves treated as above, using plants subjected to drought or control circumstances for seven days in order to better understand the impact of exogenous LA on the drought resistance of *P. persica*. Three duplicates of each cDNA library were created, as indicated in Table 1. Three biological copies of each sample yielded an average original reading of 40 to 47 million RNA sequences. We acquired an average of 39–46 million clean reads after removing low-quality reads, with more than

97% of these reads having Q30-grade sequences. More than 89% of them were unique mappings, compared to fewer than 2.8% that were multiple mappings. Each sample's average GC content was almost 46%. Overall, the RNA sequence data quality justified further investigation.

Table 1. Summary of RNA-Seq reads mapped to the peach genome.

Sample	Raw Reads	Total Reads	Unique Mapped	Multiple Mapped Reads	Q30 (%)	GC (%)
CK1	44,207,050	43,662,964	39,265,717 (89.93%)	1,162,481 (2.66%)	97.31	46.28
CK2	40,366,010	39,694,068	35,823,821 (90.25%)	1,078,507 (2.72%)	97.44	46.3
CK3	43,599,416	43,065,770	38,840,909 (90.19%)	1,131,967 (2.63%)	97.38	45.98
CK-D1	46,361,522	45,764,628	41,264,769 (90.17%)	1,158,338 (2.53%)	97.47	45.83
CK-D2	44,368,670	43,916,542	39,814,963 (90.66%)	1,093,090 (2.49%)	97.81	45.97
CK-D3	42,321,410	41,890,470	37,653,846 (89.89%)	1,073,290 (2.56%)	97.2	45.68
LA1	45,264,872	44,661,824	40,217,866 (90.05%)	1,225,178 (2.74%)	97.54	46.35
LA2	45,406,182	44,989,512	40,658,962 (90.37%)	1,144,441 (2.54%)	97.57	46.26
LA3	42,587,194	41,938,716	37,738,248 (89.98%)	1,135,371 (2.71%)	97.19	46.1
LA-D1	44,720,060	44,057,232	39,717,836 (90.15%)	1,150,437 (2.61%)	97.62	45.93
LA-D2	43,577,364	43,201,694	38,994,746 (90.26%)	1,077,302 (2.49%)	97.61	46.14
LA-D3	43,898,624	43,239,670	38,731,686 (89.57%)	1,136,746 (2.63%)	97.26	45.67

2.7. Correlation of Gene Expression Level between Samples and Gene Expression Status

For each treatment of the three biological duplications, a correlation analysis was performed to determine the quality of the duplications (Pearson correlation coefficient was calculated using R language). Figure 6A shows that the three biological duplications of each treatment have strong correlation coefficients, which are all above 0.98, demonstrating that the three biological duplications are good and fulfill the statistical requirements. Principal component analysis (PCA) was performed using R software to assess sample repeatability and correlation (Figure 6B). The repeatability of the three biological replicates of each treatment is acceptable, and the differences between treatments are clear, indicating that the transcript results from this test are reliable. The standard has a |log2 (FPKM) ratio| ≥ 1, and the q value is ≤0.05. By comparing the gene expression intensity (FPKM) values between pairs of control (CK), drought stress treatment (CK-D), lauric acid treatment (LA), and lauric acid + drought stress treatment (LA-D), differentially expressed genes were identified. In CK-D vs. CK, LA-D vs. CK, LA-D vs. CK-D, LA-D vs. LA, LA-D vs. CK, and LA vs. CK-D, 5745, 4430, 1876, 5390, 3340, and 6936 DEGs (differentially expressed genes) were screened, including 2152, 1499, 1104, 1904, 2118, and 4135 up-regulated DEGs, while 2152, 2931, 772, 3486, 1222, and 2801 DEGs were down-regulated (Figure 6C,D).

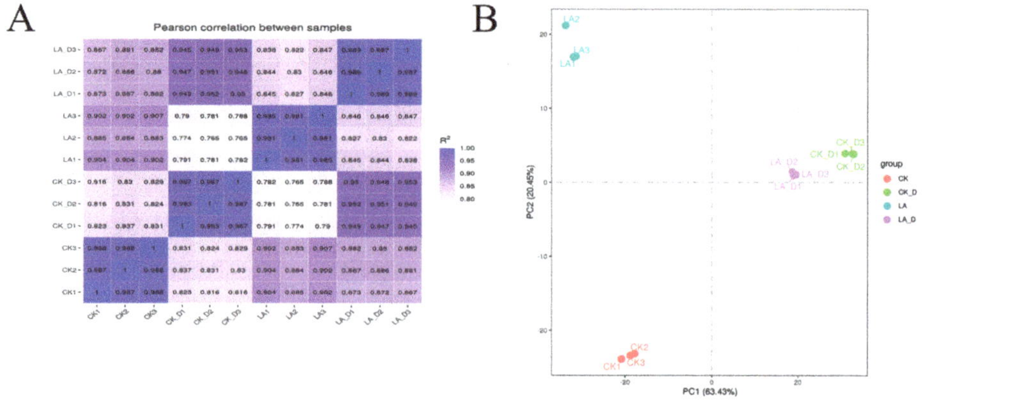

Figure 6. *Cont.*

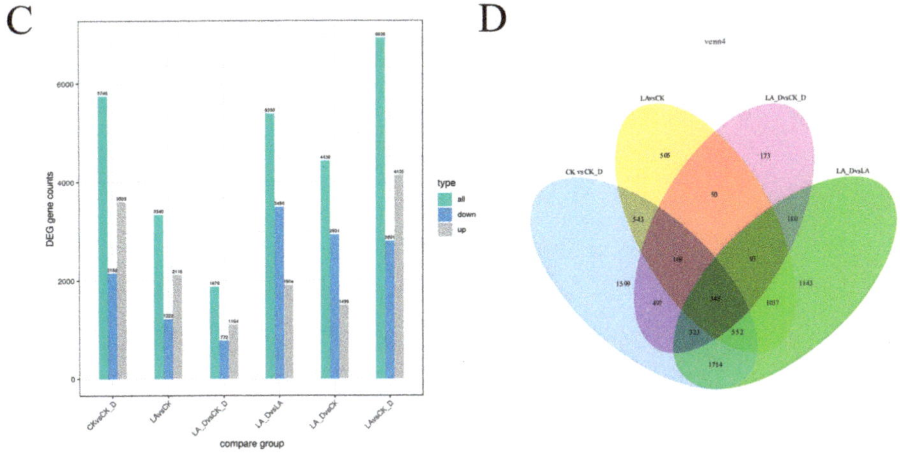

Figure 6. Correlation and differentially expressed genes among treatments at different sampling. (**A**) Heat map of the correlation coefficient between samples. (**B**) Principal component analysis (PCA) is also commonly used to assess inter-group differences and intra-group sample duplication. (**C**) Number of DEGs among the six comparison groups. (**D**) Venn diagram of differential gene expression analysis of DEGs.

2.8. GO Functional Analysis of DEGs

Functional enrichment analysis was carried out for DEGs in each group in order to determine the overall roles of these genes in various treatments. According to their GO annotations, the DEGs across the groups CK vs. CK-D, LA vs. CK, LA-D vs. CK-D, and LA vs. LA-D were split into three categories: biological process, cellular component, and molecular function.

The GO annotation results of DEGs between the CK and CK-D samples showed that 479 DEGs were annotated as biological processes (Figure 7A). In the biological process category, the minor categories with higher gene frequency and more significant GO annotation were carbohydrate metabolic process, response to oxidative stress, and photosynthesis. In total, 258 DEGs were annotated as cell components. Among the cell components, the minor categories with higher gene frequency and more significant GO annotations were extracellular region and cell wall. In total, 995 DEGs were annotated for molecular function. Among the molecular function categories, the minor categories with more significant gene frequencies than GO annotation were hydrolase activity, hydrolyzing O-glycosyl compounds, acting on glycosyl bonds, etc.

The GO annotation results show that 187 DEGs between CK and LA were annotated as biological processes (Figure 7B). Among the biological process categories, the minor categories with higher gene frequency and more significant GO annotation were defense response, response to biotic stimulus, and ion balance. In total, 117 DEGs were annotated as cell components. Among the major categories of cell components, the significant minor categories of GO annotation were extracellular region and cell periphery. In total, 487 DEGs were annotated for molecular function. Among the molecular function categories, the minor categories with higher gene frequency and more significant GO annotation were ADP binding, ion binding, etc.

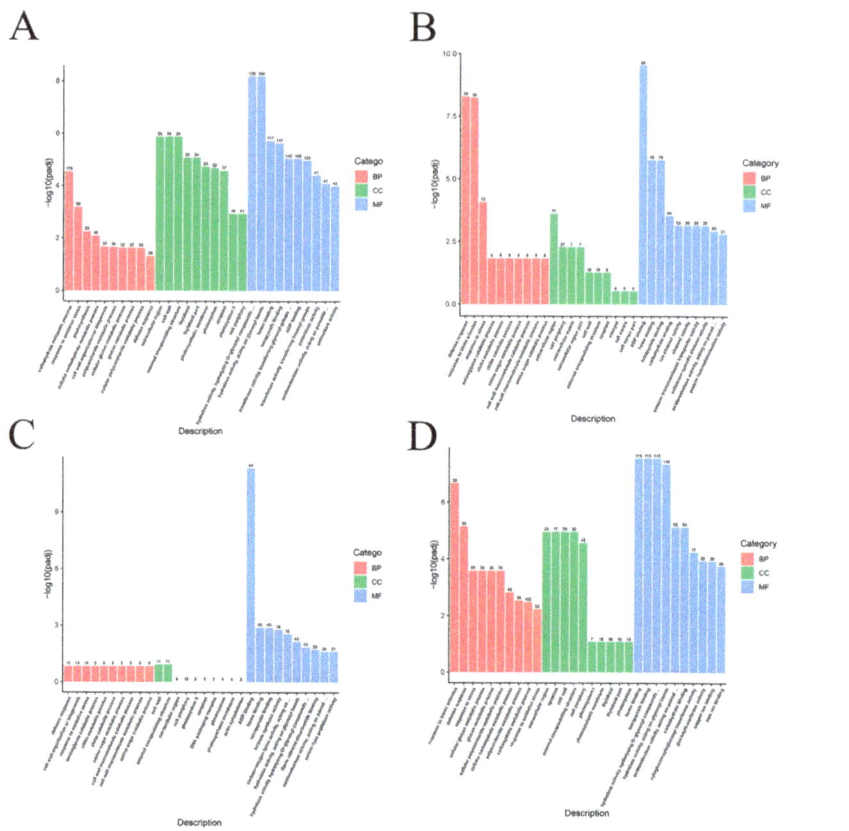

Figure 7. GO term enrichment analysis of DEGs in response to different stress treatments. (**A**) CK vs. CK-D; (**B**) LA vs. CK; (**C**) LA-D vs. CK-D; (**D**) LA-D vs. LA. The x-axis indicates the GO classifications, and the Y-axis indicates the number of genes in each classification. The top 30 GO terms with the most significant enrichment level are displayed in the order of *p*-value from small to large.

The GO annotation results of differentially expressed genes between LA-D and CK-D show that 75 DEGs were annotated as biological processes (Figure 7C). Among the biological process categories, the minor categories with higher gene frequency and more significant GO annotation were defense response, cell wall organization or biogenesis, and response to oxidative stress. In total, 58 DEGs were annotated as cell components. Among the major categories of cell components, the significant minor categories of GO annotation were cell wall and external encapsulating structure. In total, 336 DEGs were annotated for molecular function. Among the molecular function categories, the minor categories with higher gene frequency and more significant GO annotation were ADP binding, heme binding, etc.

The GO annotation results of differentially expressed genes between LA and LA-D show that 535 DEGs were annotated as biological processes (Figure 7D). Among the biological process categories, the minor categories with higher gene frequency and more significant GO annotation were response to biotic stimulus, defense response, and response to stress. In total, 213 DEGs were annotated as cell components. Among the major categories of cell components, the significant minor categories of GO annotation were extracellular region and apoplast. In total, 779 DEGs were annotated for molecular function. Among the molecular function categories, the minor categories with higher gene frequency and more significant GO annotation were heme binding, tetrapyrrole binding, etc.

2.9. KEGG Enrichment Analysis of DEGs

The biological processes carried out by various genes in vivo are coordinated. The pathways implicated in LA-mediated drought response in *P. persica* were further investigated using KEGG enrichment analysis, which was performed on the DEGs of each experimental group. The enrichment results are displayed in scatterplots (Figure 8).

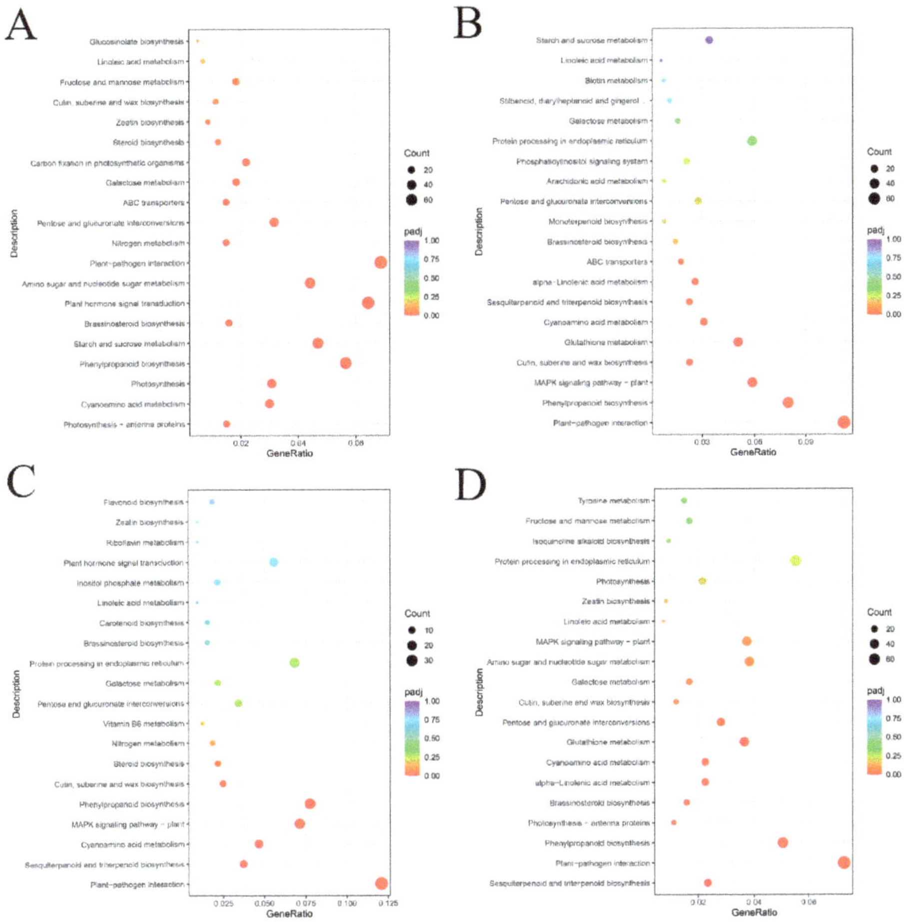

Figure 8. Kyoto Encyclopedia of Genes and Genomes (KEGG) pathway enrichment analysis of DEGs in response to different stress treatments. (**A**) CK vs. CK-D; (**B**) LA vs. CK; (**C**) LA-D vs. CK-D; (**D**) LA-D vs. LA. The X-axis shows the rich factor. Green represents a high q value, and red represents a low q value. The Y-axis shows the top 20 KEGG pathways. The bigger the size of the spot, the greater the number of DEGs enriched.

The KEGG enrichment analysis results of DEGs in the CK vs. CK-D treatment group are shown in Figure 8A. As can be seen, the DEGs between the CK and CK-D treatments were mostly concentrated in the following areas: plant–pathogen interaction, plant hormone signal transduction, phenylpropanoid biosynthesis, starch and sucrose metabolism, amino sugar and nucleotide sugar metabolism, and cyanoamino acid metabolism.

The KEGG enrichment analysis of the DEGs between the LA and CK treatment groups is illustrated in Figure 8B. It can be seen that the DEGs were mainly concentrated in plant–pathogen interaction; phenylpropanoid biosynthesis; MAPK signaling pathway; glutathione metabolism; cyanoamino acid metabolism; and cutin, suberine, and wax biosynthesis.

The KEGG enrichment analysis of the DEGs between the LA-D and CK-D treatment groups is shown in Figure 8C. As shown on the map, the DEGs were mainly concentrated in plant–pathogen interaction, phenylpropanoid biosynthesis, MAPK signaling pathway, cyanoamino acid metabolism, and sesquiterpenoid and triterpenoid biosynthesis.

The KEGG enrichment analysis of the DEGs between the LA-D and LA treatment groups is shown in Figure 8D. As shown on the map, the DEGs were mainly concentrated in plant–pathogen interaction, phenylpropanoid biosynthesis, MAPK signaling pathway, cyanoamino acid metabolism, and sesquiterpenoid and triterpenoid biosynthesis.

2.10. Hormone Signaling Pathway and Calcium (Ca^{2+}) Signaling Pathway Gene Expression Analysis

It is generally understood that plant hormones play an important role in stress responses by coordinating many signaling pathways. Under drought stress, LA pretreatment impacted the transcription levels of genes linked to auxin, ABA, and salicylic acid signal transduction pathways, implying that LA may influence *P. persica*'s drought adaptation by interacting with different hormones. LA induced changes in many genes related to drought resistance in peach seedlings. A total of 76 hormone-related proteins and genes were found in the transcriptome data of differentially expressed genes, including 29 ABA-related proteins and genes (25 up-regulated and 4 down-regulated), 5 SA-responsive proteins and genes (4 up-regulated and 1 down-regulated), and 6 ABA-related proteins and transcription factors (3 up-regulated and 3 down-regulated) (Figure 9).

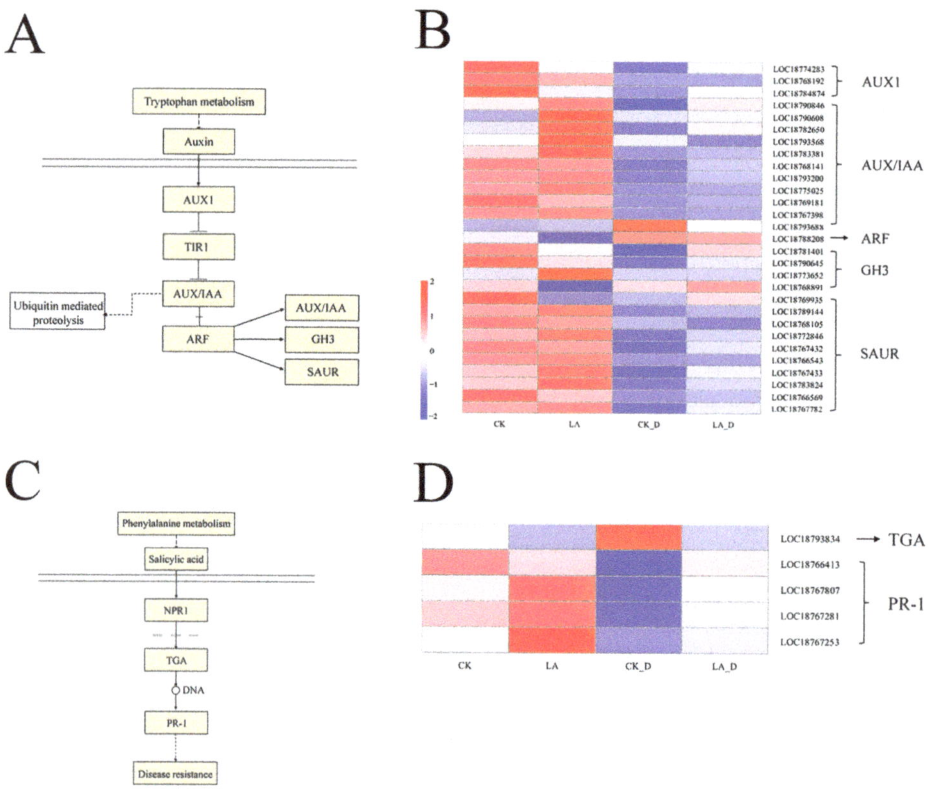

Figure 9. *Cont.*

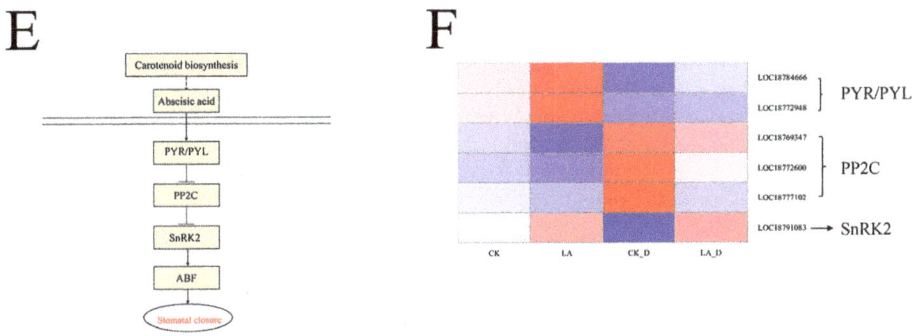

Figure 9. Effects of exogenous LA on hormone signaling in drought-stressed *P. persica*. Transduction of plant hormone signals in response to exogenous melatonin administration (**A**,**C**,**E**). Gene expression profiles for the signaling pathway for plant hormones (**B**,**D**,**F**). The expression levels were assessed by log2-transformed FPKM values.

Notably, 21 up- or down-regulated DEGs were associated with the calcium (Ca^{2+}) signaling pathway (Figure 10), indicating the integration of Ca^{2+} signaling mechanisms with LA-regulated responses to drought stress in peach seedlings. More specifically, there were seven, six and one DEGs belonging to the CNGC (cyclic nucleotide gated channel) family, CaM/CML (calmodulin/calmodulin-like protein) family, and CDPK (Ca^{2+} dependent protein kinase) family, respectively (Figure 10B). At the same time, a total of seven DEGS were changed in the FLS22 signaling pathway, including five flagellin receptors (FLS2), one promotor of mitogen-activated protein kinases (MEKK1), and one RBOH protein (RBOH).

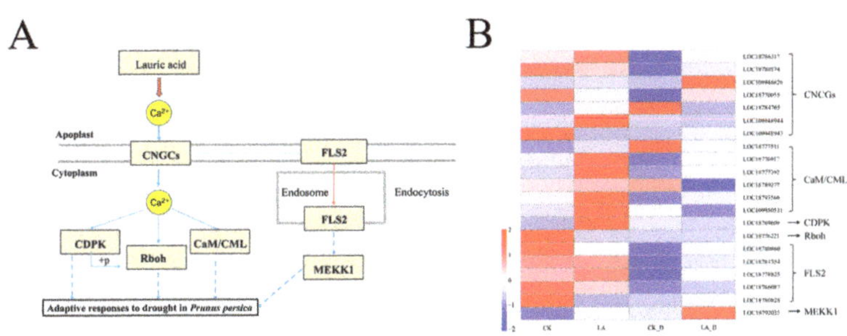

Figure 10. The Ca^{2+} signaling pathway induced by lauric acid in *P. persica* during drought stress (**A**) and a heat map of expression levels of CNGC, CAM/CML, CDPK, Rboh, FLS2, and MEKK1 (**B**). The expression levels were assessed by log2-transformed FPKM values.

3. Discussion

Plants face a variety of stress conditions in their growth and development stages, which will have different impacts and reduce crop yield [31,32]. Drought is the single most destructive environmental stress, and it causes more serious crop yield losses than any other stress [33]. Drought seriously affects the growth and development of plants, greatly reducing the crop growth rate and biomass accumulation [34]. The main consequences of drought are reductions in the rate of plant cell division and expansion, reduced leaf area, inhibitions in the rate of stem elongation and root proliferation, disrupted stomatal behavior, reductions in the nutrient relationship between water and crop yield, and reductions in water use efficiency [35–37]. When faced with drought stress, plants often adjust through a series of complex physiological and biochemical reactions [38].

Lauric acid (LA) is a kind of saturated fatty acid; the research on this compound mainly focuses on its antibacterial properties in relation to animals and plants [39,40].

It has also been found that LA exhibits certain biological activity, regulating nematode avoidance [41] and actively regulating the growth of soybean (*Glycine max* L.) [23]. By studying the changes in metabolic substances in loquat (*Eriobotrya japonica* Lindl.) under drought stress, it was found that LA accumulated significantly. It was speculated that LA might have similar functions to decanoic acid and abscisic acid (ABA) in loquat leaves, regulating the opening and closing of stomata to enhance drought resistance [28]. In this study, we treated *P. persica* subjected to drought stress with exogenous LA to investigate how LA alleviated the consequential damage. In addition, we identified the potential mitigation mechanism of exogenous LA in relation to drought stress through transcriptome analysis and the comparison of treatment groups. It has been reported that drought stress will also significantly reduce the growth of many important crops [42]. In this study, drought stress also significantly inhibited the fresh and dry weights and root activities of peach, as has been found in many studies of other plants, such as tomato [5], maize [35], willow [15], apple [43], and peach [30]. In this study, the fresh and dry weights and root activity following LA treatment were relatively higher than those following CK under drought stress, implying the effective alleviation of the adverse factors caused by drought.

The relative water content (RWC) of leaves may be the most appropriate measure of the physiological consequences of plant water status [44]. As shown in Figure 1, under drought stress, the RWC of LA-treated leaves was higher than that of untreated plants. This is consistent with the research results of [27], who reported that LA has the ability to maintain leaf water content under water deficit stress. Under normal conditions, the relative electrolyte leakage (REL) level of LA-treated plants did not change significantly compared with that of untreated plants. Under the condition of water shortage, the REL of LA-treated plants decreased by 15.32% compared with that of untreated plants (Figure 1). REL is an indicator of membrane damage. From our results, we can clearly see that LA plays an important role in preventing the cell membrane damage caused by water deficit stress [43].

Photosynthetic pigments are crucial for the absorption, transmission, and conversion of light energy during photosynthesis [17,43,45]. The content of photosynthetic pigment directly affects the photosynthesis of plants. Most studies have shown that the photosynthetic pigment content decreases with the increase in drought stress [46], but we found that LA treatment significantly inhibited this decline. Our results show that under certain drought stress conditions, applying the appropriate concentration of LA could help maintain higher chlorophyll and carotenoid contents in peach leaves. The increase in photosynthetic pigment content will ensure better light energy absorption and conversion efficiency in peach leaves. Photosynthesis is the basic function of plant growth, and it can be used to reflect the growth potential and drought resistance of plants [12,27,47]. Previous studies have shown that drought stress significantly reduces leaves' photosynthetic rate, while LA treatment increases the photosynthetic rate [27]. We found that the net photosynthetic rate (Pn), stomatal conductance (Gs), intercellular carbon dioxide concentration (Ci), and transpiration rate (Tr) decreased to different degrees under drought stress. LA treatment can effectively alleviate drought damage, such that the Pn, Gs, Tr, and Ci of peach leaves are maintained at a higher level (Figure 2). These results indicate that LA could effectively maintain high levels of photosynthesis under certain degrees of drought stress.

Plants display corresponding adjustments and regulation mechanisms when under drought stress, which is of great significance in osmotic regulation [8,17,35]. Plants actively accumulate osmotic-regulating substances, increase cell fluid concentration, reduce cell water potential, and increase their water absorption capacity [48,49]. This study shows that the contents of osmotic adjustment substances of plants under drought stress, such as soluble sugar, soluble protein, proline, and free amino acids, increased to a certain extent, which indicates that peach seedlings can increase their ability to resist drought in this way (Figure 3). After applying LA at an appropriate concentration, this regulation mechanism was increased, thus slowing the damage caused by drought stress in peach seedlings.

When plants are subjected to stress, they produce a large amount of ROS, aggravate the degree of membrane lipid peroxidation, produce MDA, and destroy the integrity of cell membranes [50–52]. MDA content can reflect the degree of damage to cell membrane integrity [53]. When plants suffer from drought stress, the contents of H_2O_2, O_2^-, and MDA increase significantly [15,54]. In this study, after drought stress, these contents in the LA peach seedlings were slightly lower than in untreated plants, indicating that the protoplasm membrane system of LA-treated plants was less damaged and had better tolerance to drought (Figure 4). The leaves subjected to drought treatment were stained with cell death and reactive oxygen species chemicals, and the degree of staining was the greatest, indicating that the accumulation of H_2O_2 and O_2^- in untreated plants was significantly higher than that in LA-treated plants, and the untreated plants suffered the most severe oxidative stress. Under drought stress, LA treatment can enhance the capacity for scavenging reactive oxygen species in peach leaf cells, and promote the reduction in H_2O_2 and O_2^- accumulation, thus reducing cell damage or cell death, and enhancing plant stress resistance. The contents of ROS that accumulate in plants are low, the degree of oxidative damage to cells is light, and the integrity of cell membranes is good, which also enables the plant to maintain good photosynthesis [55]. Plants eliminate excessive active oxygen and reduce its toxicity by increasing the activity of their antioxidant enzymes. SOD, POD, CAT, and APX are important antioxidant enzymes in plants [18,19,34]. Under drought stress, the activities of SOD, POD, CAT, and APX in peach seedling leaves were significantly higher than those in the control plants (Figure 5). This indicates that plants can improve their drought resistance by increasing the activities of four of their antioxidant enzymes, eliminating excess ROS in their bodies, alleviating drought damage, and improving drought resistance [11,51]. After treatment with LA, the activities of four enzymes in the leaves of peach seedlings were further improved, indicating that LA can enhance the activities of antioxidant enzymes, thereby improving the antioxidant and anti-aging capacities of peach seedlings, and alleviating the toxic effects of drought stress. It was found that exogenous LA could reduce the oxidative damage of peach leaves and improve the activity of antioxidant enzymes to alleviate the damage caused by drought stress.

The secondary effects of drought stress are complex. The combination of RNA-Seq sequencing and DEG can be used to clarify the role of exogenous substances in directly or indirectly regulating plant stress resistance [32,43]. In order to better understand the mechanism of LA in drought-resistant plants, we used RNA-Seq technology to study the molecular mechanism of exogenous LA in relation to regulating peach drought resistance. This is the first study using transcriptome methods to explore LA and regulate plant stress resistance. Because the aims of this study include exploring the effects of exogenous LA on the molecular regulation mechanism of peach seedlings under drought stress, we focus on LA-D vs. CK-D. We identified 1876 DEGs between LA-D and CK-D: 1104 up-regulated and 772 down-regulated. GO analysis shows that most genes are involved in biological process categories, cell components, and molecular function.

As signaling molecules, plant endogenous hormones participate in the regulation of plant growth and development, play an important role in plant response to abiotic stress, and coordinate different signaling pathways [56,57]. In this study, the LA pretreatment affected the transcription levels of genes related to auxin, JA (salicylic acid), and ABA (abscisic acid) signal transduction pathways under drought stress, suggesting that LA may regulate the drought adjustment of peach seedlings through interactions with various hormones. After analysis, the majority of DEGs were allocated to the auxin pathway. Most of the genes related to auxin signal transduction were up-regulated, including 3 AUX1 (Auxin1) genes (2 up-regulated and 1 down-regulated), 12 AUX/IAA (Auxin/indole-3-acetic acid) genes (11 up-regulated and 1 down-regulated), 1 ABF (ABA response element binding factor) gene, 4 GH3 (Gretchen Hagen 3) genes (3 up-regulated and 1 down-regulated), and all 10 SAUR (small auxin-up RNA) genes. The critical role of auxin signaling genes in response to drought stress has also been demonstrated in other species. Auxin is a regulator of plant drought resistance, acting by regulating the AUX/IAA affinity of the auxin co-receptor, and it activates the transcriptional regulation of ARF-mediated early auxin response genes [58,59]. In addition to the ARF gene family, the auxin response genes include AUX/IAA, GH3, and SAUR. These three gene families are collectively referred to as auxin early response genes. Refs. [60,61] reported that the Aux/IAA protein OsIAA20, which is up-regulated under abiotic stress conditions, has a positive effect on the resistance to abiotic stress, enhances the growth of rice at different stages of development, reduces water loss, and increases pore closure. The authors of [62] reported the GH3 gene in three cotton varieties, and found it plays an important role in plants' adjustment to drought and salt stress conditions. Ref. [63] showed that the overexpression of TaSAUR75 enhanced drought tolerance in Arabidopsis, and H2O2 accumulation in transgenic plants was lower than that in control plants.

SA, as a plant hormone, is an immune signal that causes systemic acquired resistance (SAR). The TGA (transcription factor TGA) and PR-1 (pathogenesis-related protein 1) gene families play an important role in the metabolism of plants in response to biological and abiotic stress [64–66]. It has been found that drought stress causes the up-regulation of all SlPR-1 genes, by up to 50 times, and our results show that the SlPR-1 gene played an active role in drought response. We found that under drought stress, LA treatment significantly up-regulates the related PR-1 gene.

The ABA signaling pathway is activated and regulated under drought stress and plays a key role in the regulation of plant resilience [67]. The four key genes of the ABA receptor, PYR/PYL, PP2C (protein phosphatases type 2C), SnRK2 (SNF1-related protein kinase 2), and downstream transport factor ABF (ABA-responsive element binding factors), are involved in the regulation of ABA signals [68]. They transmit ABA signals under the induction of ABA and stress, abscisic acid binds to PYR/PYL/ABA receptors, and the ABA receptor complex inhibits PP2C phosphatase activity, resulting in the PP2C-mediated inhibition of the release of SnRK2. Activated SnRK2 can phosphorylate downstream transcription factors and ion channels, and activate ABA signaling pathways and stress response processes, inducing various reactions such as increased ROS and pore closure [69,70]. In this study, LA caused changes in many genes associated with the ABA signaling pathway (Figure 9E,F).

The second messenger Ca^{2+} is widely involved in many types of signal transduction in plants [43]. Plants respond to many stimuli, such as the external environment and hormones, by starting gene expression through changes in the concentration of free Ca^{2+} in the cytoplasm, and ultimately improving stress adjustment [71–73]. In the current study, plant–pathogen interaction is one of the most significantly changed pathways (Figure 8C), and DEGs mapped to this pathway are mainly concentrated in Ca^{2+} pathway signals. Studies have shown that the Ca^{2+} signaling pathway is activated under drought stress, and this can improve the drought resistance of plants [57,74,75]. In addition, it has been found that osmotic stress induces an increase in exoplast Ca^{2+}, which leads to the activation of CNGC (cyclic nucleotide-gated channel), thus causing responses in CaM/CML (Calmodulin/Calmodulin-like protein) and CDPK (calcium-dependent protein kinases) [43,57]. Increasing numbers of CaMs/CML family members have been identified and proven to be related to drought resistance [76]. The CGNCs in plants are a group of non-selective cation channels, which can promote the uptake of Ca^{2+} and thus mediate the acquisition of resistance [77]. Therefore, LA may improve the drought resistance of peach seedlings by activating the expressions of CNGC, CAM/CML, and CDPK family genes. In plants, oxidative stress has been proven to be related to CaM, which is a Ca^{2+}-dependent activator [43]. Previous studies have found that SOD is a CaM-dependent enzyme [78]. Therefore, LA may increase the expression of the CaM/CML gene, activate the Ca^{2+} signaling pathway, and thus regulate antioxidant enzyme activity. The respiratory burst oxidase homolog (RBOH, also known as NADPH oxidase) is one of the key enzymes by which plants produce ROS. When cells are stimulated by external stress, they produce Ca^{2+}, which is activated by the combination of RBOH's EF structure [79]. Therefore, LA may regulate the activity of antioxidant enzymes through the regulation of the RBOH gene.

Flagellin-sensitive 2 (FLS2) is an important immunokinase receptor, which can effectively recognize the 22-amino acid peptide of plant–pathogen conserved peptide flg22 [80]. In recent years, important signal peptides and receptors that regulate plant response to drought have been discovered [81,82]. Based on our results, we propose that under LA treatment, FLS2 is involved in the perception of drought stress (Figure 10). Further verification of this potential function of FLS2 will provide new insights into the mechanism by which plants respond to drought stress.

In this study, using DEGs, multiple lncRNAs (long non-coding RNA) were predicted to explore the drought response induced by LA treatment. However, due to the lack of annotated information, it is difficult to determine the function of lncRNA in drought adjustment. However, the study of lncRNA in drought stress has become a hot topic [83,84]. We believe that the expression profile of lncRNA in response to drought is useful and valuable for the molecular breeding of peaches. At the same time, this study is the first to describe the response of lncRNA to drought in peaches under LA treatment.

4. Materials and Methods

4.1. Experimental Materials

Experiments were carried out at the experimental base of Shandong Agricultural University in 2022, located in Tai'an City, Shandong Province, China (36°17′7459″ N, 117°16′7712″ E). We selected uniform and plump peach seeds that had been steeped in a 400 ppm gibberellin solution for 24 h, before being put in seedling trays. We selected seedlings of comparable sizes that were free of disease and insect pests to plant in the basin when they reached approximately 5 cm in height. Each pot was square in shape (7 cm × 7 cm). The media in the pot comprised a 2:1 volume ratio blend of garden soil and vermiculite. Each pot of culture media weighed 200 g. The seedlings were managed on a regular basis.

4.2. Experimental Design

We selected 160 potted peach seedlings that had uniform development and were around 10 cm high. Among them, 80 pots of peach seedlings received 10 days of irrigation with 50 ppm LA (Shanghai macklin Biochemical Co., Ltd., Shanghai, China). Reagent concentrations were determined from the results of earlier preparatory studies. The results show that LA had no negative effect on the growth and development of peach seedlings. Then, the seedlings in each group were divided into two groups: conventional watering treatment (CK, LA) and natural drought treatment for 10 days (CK-D, LA-D), with 40 pots in each group. After 10 days of treatment, the leaves were carefully washed with tap water, frozen in liquid nitrogen, and stored at −80 °C for later use.

4.3. Determination of Plant Growth, RWC, and REL of Leaves

At the end of the drought treatment, peach seedlings were removed whole and cleaned, and fresh (FW) and dry (DW) weights were determined after oven-drying at 60 °C to a constant weight.

The phenyltetrazolium chloride (TTC) method was used for determination. The root tips were cut and weighed at 0.5 g. The sample was mixed with 0.4% TTC and pH 7.0 phosphate buffer in equal amounts up to 10 mL, sealed, and reacted in the dark at 37 °C for 4 h. The reaction was terminated by adding 1 mol·L^{-1} sulfuric acid up to 2 mL. The sample was blotted and 4 mL of ethyl acetate and quartz sand was added, before the sample was ground well. The extract was transferred to a test tube and the residue was rinsed with ethyl acetate, and finally fixed with ethyl acetate up to 10 mL. Colorimetric assessment was undertaken at 485 nm. Three plants were chosen for each treatment, and two leaves were removed from each plant and promptly placed in an aluminum box of known weight, and the fresh weight (Wf) was determined. The leaves were submerged in distilled water for approximately 1 h before being removed and weighed to determine the saturated fresh weight (Wt) of the sample. The dry weight was then obtained by drying to constant weight

(Wd). Relative water content (RWC) was determined using the following formula [84]: (Wf − Wd)/(Wt − Wd) × 100%.

The relative electrolyte leakage (REL) was measured with reference to [85], whose method was slightly modified. We punched 10 tiny discs from ripe leaves and placed them in a 20 mL centrifuge tube with 10 mL of deionized water. After 4 h at room temperature, we used a Raymag DDS-307 conductivity meter to test the solution's conductivity, designated as S1; at the same time, we measured the conductivity of deionized water, denoted as S0. The centrifuge tube was then put in a 100 °C water bath for 20 min. After cooling, it was shaken well and we measured the conductivity S2. To represent the relative permeability of the plasma membrane, we calculated the relative electrolyte leakage using the following formula: REL (%) = (S1 − S0)/(S2 − S0) × 100%.

4.4. Determination of Photosynthetic Parameters and Leaf Photosynthetic Pigments

On the 14th day after drought stress, photosynthesis was monitored from 10:00 to 11:30 am. We measured the following parameters with a CIRAS-3 portable photosynthetic system (PP Systems, Massachusetts, USA): net photosynthetic rate (Pn), stomatal conductance (Gs), intercellular CO_2 concentration (Ci), and transpiration rate (Tr).

Samples (0.2 g) were taken from fresh, clean peach leaves, and extracted for 24 h in 95% ethanol solution. We then used a Pharma-Spec UC-2450 ultraviolet spectrophotometer (Shimadzu, Kyoto, Japan) to measure the OD665, OD649, and OD470 of the extract. In order to calculate leaf chlorophyll and carotenoid contents, we used the methods of [43].

4.5. Determination of Osmotic Regulation Substances

The soluble sugar content was determined via the anthrone colorimetric method [47], and the absorbance at 620 nm was measured. The proline content was determined via the ninhydrin method [86], and the absorbance value of the toluene layer was measured using a spectrophotometer at 520 nm. Soluble protein content was determined via Coomassie brilliant blue staining [46]; we recorded the absorbance at 595 nm, and assessed the soluble protein concentration using the standard curve. The method for measuring the total amount of free amino acids was based on [87], with slight modifications, and we added 60% ethanol to 5 mL and tested the absorbance at 570 nm.

4.6. Histochemical Evaluation of Oxidative Damage and Cell Death

The peach leaves were stained with Evans Blue to measure cell death [55]. 3,3-diaminobenzidine (DAB) and nitroblue tetrazolium (NBT) histochemical staining was performed to observe the color development of superoxide anion and hydrogen peroxide, as described in the method of [88] with a slight modification.

4.7. Determination of Leaf Reactive Oxygen and Lipid Peroxidation

We measured the hydrogen peroxide contents (H_2O_2) of leaves using the method of [55]. Briefly, leaf samples (0.1 g) were placed in sterilized centrifuge tubes and then ground with liquid nitrogen. After centrifugation at 6000 rpm for 150 s, the samples were mixed with 1.5 mL of 0.1% trichloroacetic acid (TCA) and placed on ice. The samples were then centrifuged again at 12,000 rpm for 15 min at 4 °C. A 0.5 mL sample of supernatant was then collected and mixed with 0.5 mL of phosphate-buffered saline (PBS) and 1 mL potassium iodide (KI). The resulting solution was shaken well and held at 28 °C for 1 h, after which we measured the absorbance at 390 nm.

We measured the O_2^- contents of leaves using the method of [55]. Briefly, we chopped 1 g samples of peach leaves and added 3 mL of phosphate buffer (pH = 7.8), placed them in an ice bath and ground them, and then centrifuged the samples at 4000× *g* for 15 min. The supernatant was collected and mixed with 0.1 mL of 10 mM hydroxylamine hydrochloride solution and incubated at 25 °C for 20 min. We then added 1 mL of 17 mM p-aminobenzene sulfonic acid and 1 mL of 7 mM a-naphthylamine solution and incubated the mixture at 25 °C for 20 min. After this, we added an equal volume of chloroform to extract the pigment and centrifuged the mixture at 10,000 rpm for 3 min. The pink extract was collected in order to measure the OD530.

Malondialdehyde (MDA), a biomarker of lipid peroxidation caused by oxidative stress, was measured using the method described by [46], with some modifications.

4.8. Determination of Leaf CAT, POD, SOD, and APX Activities

We weighed a mixed sample of 0.5 g, performed liquid nitrogen grinding, and added 4 mL of phosphate buffer at a concentration of 0.05 M pH 7.8 (0.3% EDTA with 0.1 mM, Triton-X100 and 4% of polyvinylpyrrolidone); the sample was then ground. Next, the sample was placed in a centrifugal tube, flushed twice with 6 mL buffer at 4 °C, then centrifuged at 12,000 rpm for 20 min; we then collected the supernatant and stored it at 4 °C until use.

Superoxide dismutase (SOD) activity was determined according to the method of [47], with some modifications. Peroxidase (POD) activity was determined according to the method of [55], with some modifications. Catalase (CAT) activity was determined according to the method of [47], with slight modifications.

4.9. RNA Preparation, Library Construction, and Sequencing

Total RNA was used as the input material for the RNA sample preparations. Briefly, mRNA was purified from total RNA using poly-T oligo-attached magnetic beads. Fragmentation was carried out using divalent cations at an elevated temperature in First Strand Synthesis Reaction Buffer (5X). First-strand cDNA was synthesized using random hexamer primer and M-MuLV Reverse Transcriptase (RNase H-). Second-strand cDNA synthesis was subsequently performed using DNA Polymerase I and RNase H. The remaining overhangs were converted into blunt ends via exonuclease/polymerase activities. After the adenylation of the 3′ ends of DNA fragments, adaptors with a hairpin loop structure were ligated to prepare the sample for hybridization. In order to preferentially select cDNA fragments of 370~420 bp in length, the library fragments were purified with the AMPure XP system (Beckman Coulter, Beverly, MA, USA). Then, PCR was performed with Phusion High-Fidelity DNA polymerase, Universal PCR primers and Index (X) Primer. Lastly, the PCR products were purified (AMPure XP system) and the library quality was assessed on the Agilent Bioanalyzer 2100 system.

4.10. Differential Expression Analysis

The differential expression analysis of four conditions/groups (three biological replicates per condition) was performed using the DESeq2 R package (1.20.0). DESeq2 provides statistical routines for determining differential expressions in digital gene expression data using a model based on the negative binomial distribution. The resulting *p*-values were adjusted using Benjamini and Hochberg's approach for controlling the false discovery rate. Genes with an adjusted *p*-value ≤ 0.05 found by DESeq2 were assigned as differentially expressed. Prior to differential gene expression analysis, for each sequenced library, the read counts were adjusted using the edgeR program package through one scaling normalized factor. Differential expression analysis of two conditions was performed using the edgeR R package (3.22.5). The *p*-values were adjusted using the Benjamini and Hochberg method. A corrected *p*-value of 0.05 and absolute fold change of 2 were set as the threshold for significantly differential expression.

4.11. GO and KEGG Enrichment Analysis of Differentially Expressed Genes

The gene ontology (GO) enrichment analysis of differentially expressed genes was implemented using the clusterProfiler R package, in which gene length bias was corrected. GO terms with corrected *p*-values less than 0.05 were considered significantly enriched by differentially expressed genes. KEGG is a database resource used for understanding high-level functions and utilities of the biological system, such as the cell, the organism, and the ecosystem, via molecular-level information, usually through large-scale molecular datasets generated by genome sequencing and other high-throughput experimental technologies (http://www.genome.jp/kegg/ (accessed on 1 June 2022)). We used the clusterProfiler R package to test the statistical enrichment of differentially expressed genes in the KEGG pathway.

4.12. Statistical Analysis

SPSS 17.0 (IBM, New York, NY, USA) statistical analysis software was used to perform one-way ANOVA and the Duncan multiple comparison test. Statistical significance, designated as the lettered results in the charts below, was determined at 5% ($p < 0.05$). All chart data are shown as mean $\pm$ error bars' deviation.

5. Conclusions

Through physiological and biochemical research and transcriptome data analysis, the effects of exogenous LA treatment on peach tree seedlings under drought conditions were studied. The results show the following: (i) LA treatment enhanced the growth of peach tree seedlings under drought stress and the relative water contents of leaves. (ii) The LA treatment of peach tree seedlings under drought stress improved the photosynthetic ability, chlorophyll content, and antioxidant enzyme activity, and reduced the H_2O_2, O_2^- and MDA levels and conductivity, which may be related to the addition of LA. (iii) Compared with the untreated drought stress seedlings, a total of 1876 significant DEGs were identified, of which 1104 were up-regulated and 772 were down-regulated. In-depth analysis of the DEGs has shown that the LA-mediated drought response may involve the complex coordination of multiple plant hormone signals and synthesis pathways, as well as the activation of Ca^{2+} signals. At the same time, the lncRNAs involved in LA treatment and drought response were identified. This study provides a new method by which to reduce the impact of drought stress on plants and increase plant yield.

Author Contributions: F.P. designed the experiments; B.Z. wrote the manuscript; B.Z. and J.G. analyzed most of the experiments; B.Z. and H.D. helped obtain the experimental data; S.Y. and H.D. performed the experiments; X.W. and W.L. revised the English composition; F.P. and Y.X. provided all financial and critical support. All authors discussed the results and commented on the manuscript. All authors have read and agreed to the published version of the manuscript.

Funding: This work was supported by the Rural Revitalization Science and Technology Innovation Action Plan Project of Shandong: 2021TZXD013; the National Modern Agro-industry Technology Research System Fund: Grant No. CARS-30-2-02.

Data Availability Statement: The data presented in this study are available on request from the corresponding author.

Conflicts of Interest: The authors declare no conflict of interest.

References

1. Chiang, F.; Mazdiyasni, O.; AghaKouchak, A. Evidence of anthropogenic impacts on global drought frequency, duration, and intensity. *Nat. Commun.* **2021**, *12*, 1–10. [CrossRef] [PubMed]
2. Mora, C.; Spirandelli, D.; Franklin, E.C.; Lynham, J.; Kantar, M.B.; Miles, W.; Smith, C.Z.; Freel, K.; Moy, J.; Louis, L.V.; et al. Broad threat to humanity from cumulative climate hazards intensified by greenhouse gas emissions. *Nat. Clim. Chang.* **2018**, *8*, 1062–1071. [CrossRef]
3. Mukherjee, S.; Mishra, A.; Trenberth, K.E. Climate Change and Drought: A Perspective on Drought Indices. *Curr. Clim. Chang. Rep.* **2018**, *4*, 145–163. [CrossRef]
4. Zhao, C.; Yang, M.; Wu, X.; Wang, Y.; Zhang, R. Physiological and transcriptomic analyses of the effects of exogenous melatonin on drought tolerance in maize (*Zea mays* L.). *Plant Physiol. Biochem.* **2021**, *168*, 128–142. [CrossRef]
5. Altaf, M.A.; Shahid, R.; Ren, M.-X.; Naz, S.; Altaf, M.M.; Khan, L.U.; Tiwari, R.K.; Lal, M.K.; Shahid, M.A.; Kumar, R.; et al. Melatonin Improves Drought Stress Tolerance of Tomato by Modulating Plant Growth, Root Architecture, Photosynthesis, and Antioxidant Defense System. *Antioxidants* **2022**, *11*, 309. [CrossRef]
6. Osmolovskaya, N.; Shumilina, J.; Kim, A.; Didio, A.; Grishina, T.; Bilova, T.; Keltsieva, O.A.; Zhukov, V.; Tikhonovich, I.; Tarakhovskaya, E.; et al. Methodology of Drought Stress Research: Experimental Setup and Physiological Characterization. *Int. J. Mol. Sci.* **2018**, *19*, 4089. [CrossRef]

7. Zheng, H.; Zhang, X.; Ma, W.; Song, J.; Rahman, S.U.; Wang, J.; Zhang, Y. Morphological and physiological responses to cyclic drought in two contrasting genotypes of Catalpa bungei. *Environ. Exp. Bot.* **2017**, *138*, 77–87. [CrossRef]
8. Desoky, E.-S.; Mansour, E.; Ali, M.; Yasin, M.; Abdul-Hamid, M.; Rady, M.; Ali, E. Exogenously Used 24-Epibrassinolide Promotes Drought Tolerance in Maize Hybrids by Improving Plant and Water Productivity in an Arid Environment. *Plants* **2021**, *10*, 354. [CrossRef]
9. Kim, J.-M.; To, T.K.; Matsui, A.; Tanoi, K.; Kobayashi, N.I.; Matsuda, F.; Habu, Y.; Ogawa, D.; Sakamoto, T.; Matsunaga, S. Acetate-mediated novel survival strategy against drought in plants. *Nat. Plants* **2017**, *3*, 1–7.
10. Zhang, H.; Sun, X.; Dai, M. Improving crop drought resistance with plant growth regulators and rhizobacteria: Mechanisms, applications, and perspectives. *Plant Commun.* **2021**, *3*, 100228. [CrossRef]
11. Li, M.; Gao, S.; Luo, J.; Cai, Z.; Zhang, S. Mitigation effects of exogenous acetic acid on drought stress in Cunninghamia lanceolata. *Plant Soil* **2022**, 1–16. [CrossRef]
12. Sedaghat, M.; Tahmasebi-Sarvestani, Z.; Emam, Y.; Mokhtassi-Bidgoli, A. Physiological and antioxidant responses of winter wheat cultivars to strigolactone and salicylic acid in drought. *Plant Physiol. Biochem.* **2017**, *119*, 59–69. [CrossRef] [PubMed]
13. Wang, D.; Chen, Q.; Chen, W.; Guo, Q.; Xia, Y.; Wang, S.; Jing, D.; Liang, G. Physiological and transcription analyses reveal the regulatory mechanism of melatonin in inducing drought resistance in loquat (*Eriobotrya japonica* Lindl.) seedlings. *Environ. Exp. Bot.* **2020**, *181*, 104291. [CrossRef]
14. Xie, H.; Bai, G.; Lu, P.; Li, H.; Fei, M.; Xiao, B.; Chen, X.; Tong, Z.; Wang, Z.; Yang, D. Exogenous citric acid enhances drought tolerance in tobacco (*Nicotiana tabacum*). *Plant Biol.* **2021**, *24*, 333–343. [CrossRef]
15. Kong, X.; Guo, Z.; Yao, Y.; Xia, L.; Liu, R.; Song, H.; Zhang, S. Acetic acid alters rhizosphere microbes and metabolic composition to improve willows drought resistance. *Sci. Total. Environ.* **2022**, *844*, 157132. [CrossRef]
16. Ahmad, Z.; Bashir, K.; Matsui, A.; Tanaka, M.; Sasaki, R.; Oikawa, A.; Hirai, M.Y.; Chaomurilege; Zu, Y.; Kawai-Yamada, M.; et al. Overexpression of nicotinamidase 3 (NIC3) gene and the exogenous application of nicotinic acid (NA) enhance drought tolerance and increase biomass in Arabidopsis. *Plant Mol. Biol.* **2021**, *107*, 63–84. [CrossRef]
17. Javadi, T.; Rohollahi, D.; Ghaderi, N.; Nazari, F. Mitigating the adverse effects of drought stress on the morpho-physiological traits and anti-oxidative enzyme activities of Prunus avium through β-amino butyric acid drenching. *Sci. Hortic.* **2017**, *218*, 156–163. [CrossRef]
18. Jisha, K.C.; Puthur, J.T. Seed priming with BABA (β-amino butyric acid): A cost-effective method of abiotic stress tolerance in *Vigna radiata* (L.) Wilczek. *Protoplasma* **2015**, *253*, 277–289. [CrossRef]
19. Jisha, K.C.; Puthur, J.T. Seed Priming with Beta-Amino Butyric Acid Improves Abiotic Stress Tolerance in Rice Seedlings. *Rice Sci.* **2016**, *23*, 242–254. [CrossRef]
20. Liu, S.; Ruan, W.; Li, J.; Xu, H.; Wang, J.; Gao, Y.; Wang, J. Biological Control of Phytopathogenic Fungi by Fatty Acids. *Mycopathologia* **2008**, *166*, 93–102. [CrossRef]
21. Dayrit, F.M. The Properties of Lauric Acid and Their Significance in Coconut Oil. *J. Am. Oil Chem. Soc.* **2014**, *92*, 1–15. [CrossRef]
22. Lieberman, S.; Enig, M.G.; Preuss, H.G. A Review of Monolaurin and Lauric Acid:Natural Virucidal and Bactericidal Agents. *Altern. Complement. Ther.* **2006**, *12*, 310–314. [CrossRef]
23. Liang, C.; Gao, W.; Ge, T.; Tan, X.; Wang, J.; Liu, H.; Wang, Y.; Han, C.; Xu, Q.; Wang, Q. Lauric Acid Is a Potent Biological Control Agent That Damages the Cell Membrane of Phytophthora sojae. *Front. Microbiol.* **2021**, *12*, 666761. [CrossRef] [PubMed]
24. Muniyan, R.; Gurunathan, J. Lauric acid and myristic acid from *Allium sativum* inhibit the growth of *Mycobacterium tuberculosis* H37Ra: In silico analysis reveals possible binding to protein kinase B. *Pharm. Biol.* **2016**, *54*, 2814–2821. [CrossRef] [PubMed]
25. Solano, R.J.; Sierra, C.A.; Murillo, M. Antifungal activity of LDPE/lauric acid films against Colletotrichum tamarilloi. *Food Packag. Shelf Life* **2020**, *24*, 100495. [CrossRef]
26. López-Villalobos, A.; Dodds, P.F.; Hornung, R. Lauric acid improves the growth of zygotic coconut (*Cocos nucifera* L.) embryos in vitro. *Plant Cell Tissue Organ Cult. (PCTOC)* **2011**, *106*, 317–327. [CrossRef]
27. Medeiros, M.J.; Oliveira, D.S.; Oliveira, M.T.; Willadino, L.; Houllou, L.; Santos, M.G. Ecophysiological, anatomical and biochemical aspects of in vitro culture of zygotic Syagrus coronata embryos and of young plants under drought stress. *Trees* **2015**, *29*, 1219–1233. [CrossRef]
28. Gugliuzza, G.; Talluto, G.; Martinelli, F.; Farina, V.; Bianco, R.L. Water Deficit Affects the Growth and Leaf Metabolite Composition of Young Loquat Plants. *Plants* **2020**, *9*, 274. [CrossRef]
29. Wang, X.; Gao, Y.; Wang, Q.; Chen, M.; Ye, X.; Li, D.; Chen, X.; Li, L.; Gao, D. 24-Epibrassinolide-alleviated drought stress damage influences antioxidant enzymes and autophagy changes in peach (*Prunus persicae* L.) leaves. *Plant Physiol. Biochem.* **2018**, *135*, 30–40. [CrossRef]
30. Haider, M.S.; Kurjogi, M.M.; Khalil-ur-Rehman, M.; Pervez, T.; Songtao, J.; Fiaz, M.; Jogaiah, S.; Wang, C.; Fang, J. Drought stress revealed physiological, biochemical and gene-expressional variations in 'Yoshihime'peach (*Prunus Persica* L) cultivar. *J. Plant Interact.* **2018**, *13*, 83–90. [CrossRef]
31. Ashraf, M.A.; Akbar, A.; Askari, S.H.; Iqbal, M.; Rasheed, R.; Hussain, I. Recent advances in abiotic stress tolerance of plants through chemical priming: An overview. In *Advances in Seed Priming*; Rakshit, A., Singh, H.B., Eds.; Springer: Singapore, 2018; pp. 51–79, ISBN 9789811300325.

32. Bai, M.; Zeng, W.; Chen, F.; Ji, X.; Zhuang, Z.; Jin, B.; Wang, J.; Jia, L.; Peng, Y. Transcriptome expression profiles reveal response mechanisms to drought and drought-stress mitigation mechanisms by exogenous glycine betaine in maize. *Biotechnol. Lett.* **2022**, *44*, 367–386. [CrossRef] [PubMed]
33. Yang, X.; Lu, M.; Wang, Y.; Wang, Y.; Liu, Z.; Chen, S. Response Mechanism of Plants to Drought Stress. *Horticulturae* **2021**, *7*, 50. [CrossRef]
34. Khaleghi, A.; Naderi, R.; Brunetti, C.; Maserti, B.E.; Salami, S.A.; Babalar, M. Morphological, physiochemical and antioxidant responses of Maclura pomifera to drought stress. *Sci. Rep.* **2019**, *9*, 1–12. [CrossRef]
35. Anjum, S.A.; Ashraf, U.; Tanveer, M.; Khan, I.; Hussain, S.; Shahzad, B.; Zohaib, A.; Abbas, F.; Saleem, M.F.; Ali, I.; et al. Drought Induced Changes in Growth, Osmolyte Accumulation and Antioxidant Metabolism of Three Maize Hybrids. *Front. Plant Sci.* **2017**, *8*, 69. [CrossRef] [PubMed]
36. Bhusal, N.; Lee, M.; Han, A.R.; Han, A.; Kim, H.S. Responses to drought stress in Prunus sargentii and Larix kaempferi seedlings using morphological and physiological parameters. *For. Ecol. Manag.* **2020**, *465*, 118099. [CrossRef]
37. Misra, V.; Solomon, S.; Mall, A.; Prajapati, C.; Hashem, A.; Abd_Allah, E.F.; Ansari, M.I. Morphological assessment of water stressed sugarcane: A comparison of waterlogged and drought affected crop. *Saudi J. Biol. Sci.* **2020**, *27*, 1228–1236. [CrossRef]
38. Shan, X.; Li, Y.; Jiang, Y.; Jiang, Z.; Hao, W.; Yuan, Y. Transcriptome Profile Analysis of Maize Seedlings in Response to High-salinity, Drought and Cold Stresses by Deep Sequencing. *Plant Mol. Biol. Rep.* **2013**, *31*, 1485–1491. [CrossRef]
39. Wu, Y.; Zhang, H.; Zhang, R.; Cao, G.; Li, Q.; Zhang, B.; Wang, Y.; Yang, C. Serum metabolome and gut microbiome alterations in broiler chickens supplemented with lauric acid. *Poult. Sci.* **2021**, *100*, 101315. [CrossRef]
40. Cai, Q.; Zhang, Y.; Fang, X.; Lin, S.; He, Z.; Peng, S.; Liu, W. Improving Anti-listeria Activity of Thymol Emulsions by Adding Lauric Acid. *Front. Nutr.* **2022**, *9*, 859293. [CrossRef]
41. Dong, L.; Li, X.; Huang, L.; Gao, Y.; Zhong, L.; Zheng, Y.; Zuo, Y. Lauric acid in crown daisy root exudate potently regulates root-knot nematode chemotaxis and disrupts Mi-flp-18 expression to block infection. *J. Exp. Bot.* **2013**, *65*, 131–141. [CrossRef]
42. Xu, L.; Wang, A.; Wang, J.; Wei, Q.; Zhang, W. Piriformospora indica confers drought tolerance on *Zea mays* L. through enhancement of antioxidant activity and expression of drought-related genes. *Crop. J.* **2017**, *5*, 251–258. [CrossRef]
43. Gao, T.; Zhang, Z.; Liu, X.; Wu, Q.; Chen, Q.; Liu, Q.; van Nocker, S.; Ma, F.; Li, C. Physiological and transcriptome analyses of the effects of exogenous dopamine on drought tolerance in apple. *Plant Physiol. Biochem.* **2020**, *148*, 260–272. [CrossRef] [PubMed]
44. Jungklang, J.; Saengnil, K.; Uthaibutra, J. Effects of water-deficit stress and paclobutrazol on growth, relative water content, electrolyte leakage, proline content and some antioxidant changes in Curcuma alismatifolia Gagnep. cv. Chiang Mai Pink. *Saudi J. Biol. Sci.* **2017**, *24*, 1505–1512. [CrossRef]
45. Song, Y.; Li, J.; Liu, M.; Meng, Z.; Liu, K.; Sui, N. Nitrogen increases drought tolerance in maize seedlings. *Funct. Plant Biol.* **2019**, *46*, 350–359. [CrossRef]
46. Han, D.; Tu, S.; Dai, Z.; Huang, W.; Jia, W.; Xu, Z.; Shao, H. Comparison of selenite and selenate in alleviation of drought stress in *Nicotiana tabacum* L. *Chemosphere* **2021**, *287*, 132136. [CrossRef] [PubMed]
47. Li, Y.; Lv, Y.; Lian, M.; Peng, F.; Xiao, Y. Effects of combined glycine and urea fertilizer application on the photosynthesis, sucrose metabolism, and fruit development of peach. *Sci. Hortic.* **2021**, *289*, 110504. [CrossRef]
48. Per, T.S.; Khan, N.A.; Reddy, P.S.; Masood, A.; Hasanuzzaman, M.; Khan, M.I.R.; Anjum, N.A. Approaches in modulating proline metabolism in plants for salt and drought stress tolerance: Phytohormones, mineral nutrients and transgenics. *Plant Physiol. Biochem.* **2017**, *115*, 126–140. [CrossRef]
49. Vardhini, B.V.; Anjum, N.A. Brassinosteroids make plant life easier under abiotic stresses mainly by modulating major components of antioxidant defense system. *Front. Environ. Sci.* **2015**, *2*, 67. [CrossRef]
50. Golldack, D.; Li, C.; Mohan, H.; Probst, N. Tolerance to drought and salt stress in plants: Unraveling the signaling networks. *Front. Plant Sci.* **2014**, *5*, 151. [CrossRef]
51. Khoyerdi, F.F.; Shamshiri, M.H.; Estaji, A. Changes in some physiological and osmotic parameters of several pistachio genotypes under drought stress. *Sci. Hortic.* **2016**, *198*, 44–51. [CrossRef]
52. Laxa, M.; Liebthal, M.; Telman, W.; Chibani, K.; Dietz, K.-J. The Role of the Plant Antioxidant System in Drought Tolerance. *Antioxidants* **2019**, *8*, 94. [CrossRef]
53. Haider, M.S.; Kurjogi, M.M.; Khalil-Ur-Rehman, M.; Fiaz, M.; Pervaiz, T.; Jiu, S.; Haifeng, J.; Chen, W.; Fang, J. Grapevine immune signaling network in response to drought stress as revealed by transcriptomic analysis. *Plant Physiol. Biochem.* **2017**, *121*, 187–195. [CrossRef]
54. Silva, E.N.; Silveira, J.A.; Ribeiro, R.V.; Vieira, S.A. Photoprotective function of energy dissipation by thermal processes and photorespiratory mechanisms in Jatropha curcas plants during different intensities of drought and after recovery. *Environ. Exp. Bot.* **2015**, *110*, 36–45. [CrossRef]
55. Xiao, Y.; Wu, X.; Sun, M.; Peng, F. Hydrogen Sulfide Alleviates Waterlogging-Induced Damage in Peach Seedlings via Enhancing Antioxidative System and Inhibiting Ethylene Synthesis. *Front. Plant Sci.* **2020**, *11*, 696. [CrossRef] [PubMed]
56. Gao, H.; Yu, W.; Yang, X.; Liang, J.; Sun, X.; Sun, M.; Xiao, Y.; Peng, F. Silicon enhances the drought resistance of peach seedlings by regulating hormone, amino acid, and sugar metabolism. *BMC Plant Biol.* **2022**, *22*, 1–17. [CrossRef]
57. Liu, Q.; Feng, Z.; Xu, W.; Vetukuri, R.R.; Xu, X. Exogenous melatonin-stimulated transcriptomic alterations of Davidia involucrata seedlings under drought stress. *Trees* **2021**, *35*, 1025–1038. [CrossRef]

58. Jung, H.; Lee, D.-K.; Do Choi, Y.; Kim, J.-K. OsIAA6, a member of the rice Aux/IAA gene family, is involved in drought tolerance and tiller outgrowth. *Plant Sci.* **2015**, *236*, 304–312. [CrossRef]
59. Wang, X.; Li, Q.; Xie, J.; Huang, M.; Cai, J.; Zhou, Q.; Dai, T.; Jiang, D. Abscisic acid and jasmonic acid are involved in drought priming-induced tolerance to drought in wheat. *Crop. J.* **2020**, *9*, 120–132. [CrossRef]
60. Luo, J.; Zhou, J.-J.; Zhang, J.-Z. Aux/IAA Gene Family in Plants: Molecular Structure, Regulation, and Function. *Int. J. Mol. Sci.* **2018**, *19*, 259. [CrossRef]
61. Zhang, A.; Yang, X.; Lu, J.; Song, F.; Sun, J.; Wang, C.; Lian, J.; Zhao, L.; Zhao, B. OsIAA20, an Aux/IAA protein, mediates abiotic stress tolerance in rice through an ABA pathway. *Plant Sci.* **2021**, *308*, 110903. [CrossRef] [PubMed]
62. Kirungu, J.N.; Magwanga, R.O.; Lu, P.; Cai, X.; Zhou, Z.; Wang, X.; Peng, R.; Wang, K.; Liu, F. Functional characterization of Gh_A08G1120 (GH3.5) gene reveal their significant role in enhancing drought and salt stress tolerance in cotton. *BMC Genet.* **2019**, *20*, 1–17. [CrossRef]
63. Guo, Y.; Jiang, Q.; Hu, Z.; Sun, X.; Fan, S.; Zhang, H. Function of the auxin-responsive gene TaSAUR75 under salt and drought stress. *Crop. J.* **2018**, *6*, 181–190. [CrossRef]
64. Li, B.; Liu, Y.; Cui, X.-Y.; Fu, J.-D.; Zhou, Y.-B.; Zheng, W.-J.; Lan, J.-H.; Jin, L.-G.; Chen, M.; Ma, Y.-Z.; et al. Genome-Wide Characterization and Expression Analysis of Soybean TGA Transcription Factors Identified a Novel TGA Gene Involved in Drought and Salt Tolerance. *Front. Plant Sci.* **2019**, *10*, 549. [CrossRef]
65. Park, S.-H.; Lee, B.-R.; La, V.; Mamun, A.; Bae, D.-W.; Kim, T.-H. Drought Intensity-Responsive Salicylic Acid and Abscisic Acid Crosstalk with the Sugar Signaling and Metabolic Pathway in *Brassica napus*. *Plants* **2021**, *10*, 610. [CrossRef]
66. Akbudak, M.A.; Yildiz, S.; Filiz, E. Pathogenesis related protein-1 (PR-1) genes in tomato (*Solanum lycopersicum* L.): Bioinformatics analyses and expression profiles in response to drought stress. *Genomics* **2020**, *112*, 4089–4099. [CrossRef] [PubMed]
67. Castillo, M.-C.; Lozano-Juste, J.; González-Guzmán, M.; Rodriguez, L.; Rodriguez, P.L.; León, J. Inactivation of PYR/PYL/RCAR ABA receptors by tyrosine nitration may enable rapid inhibition of ABA signaling by nitric oxide in plants. *Sci. Signal.* **2015**, *8*, ra89. [CrossRef]
68. Rodriguez, P.L.; Lozano-Juste, J.; Albert, A. *PYR/PYL/RCAR ABA Receptors, Advances in Botanical Research*; Elsevier: Amsterdam, The Netherlands, 2019; pp. 51–82.
69. Gonzalez-Guzman, M.; Pizzio, G.A.; Antoni, R.; Vera-Sirera, F.; Merilo, E.; Bassel, G.W.; Fernández, M.A.; Holdsworth, M.J.; Perez-Amador, M.A.; Kollist, H.; et al. Arabidopsis PYR/PYL/RCAR Receptors Play a Major Role in Quantitative Regulation of Stomatal Aperture and Transcriptional Response to Abscisic Acid. *Plant Cell* **2012**, *24*, 2483–2496. [CrossRef]
70. Zhao, Y.; Zhang, Z.; Gao, J.; Wang, P.; Hu, T.; Wang, Z.; Hou, Y.-J.; Wan, Y.; Liu, W.; Xie, S.; et al. Arabidopsis Duodecuple Mutant of PYL ABA Receptors Reveals PYL Repression of ABA-Independent SnRK2 Activity. *Cell Rep.* **2018**, *23*, 3340–3351.e5. [CrossRef] [PubMed]
71. Ahmad, P.; Abd_Allah, E.F.; Alyemeni, M.N.; Wijaya, L.; Alam, P.; Bhardwaj, R.; Siddique, K.H.M. Exogenous application of calcium to 24-epibrassinosteroid pre-treated tomato seedlings mitigates NaCl toxicity by modifying ascorbate–glutathione cycle and secondary metabolites. *Sci. Rep.* **2018**, *8*, 1–15. [CrossRef]
72. Li, L.-B.; Yu, D.-W.; Zhao, F.-L.; Pang, C.-Y.; Song, M.-Z.; Wei, H.-L.; Fan, S.-L.; Yu, S.-X. Genome-wide analysis of the calcium-dependent protein kinase gene family in Gossypium raimondii. *J. Integr. Agric.* **2015**, *14*, 29–41. [CrossRef]
73. Roy, P.R.; Arif, T.U.; Polash, M.A.S.; Hossen, Z.; Hossain, M.A. Physiological mechanisms of exogenous calcium on alleviating salinity-induced stress in rice (*Oryza sativa* L.). *Physiol. Mol. Biol. Plants* **2019**, *25*, 611–624. [CrossRef] [PubMed]
74. Jha, S.K.; Sharma, M.; Pandey, G.K. Role of cyclic nucleotide gated channels in stress management in plants. *Curr. Genom.* **2016**, *17*, 315–329. [CrossRef] [PubMed]
75. Mahmood, T.; Khalid, S.; Abdullah, M.; Ahmed, Z.; Shah, M.K.N.; Ghafoor, A.; Du, X. Insights into Drought Stress Signaling in Plants and the Molecular Genetic Basis of Cotton Drought Tolerance. *Cells* **2019**, *9*, 105. [CrossRef] [PubMed]
76. Mittal, S.; Mallikarjuna, M.G.; Rao, A.R.; Jain, P.A.; Dash, P.K.; Thirunavukkarasu, N. Comparative Analysis of CDPK Family in Maize, Arabidopsis, Rice, and Sorghum Revealed Potential Targets for Drought Tolerance Improvement. *Front. Chem.* **2017**, *5*, 115. [CrossRef] [PubMed]
77. Nawaz, Z.; Kakar, K.U.; Ullah, R.; Yu, S.; Zhang, J.; Shu, Q.-Y.; Ren, X.-L. Genome-wide identification, evolution and expression analysis of cyclic nucleotide-gated channels in tobacco (*Nicotiana tabacum* L.). *Genomics* **2019**, *111*, 142–158. [CrossRef]
78. Valivand, M.; Amooaghaie, R.; Ahadi, A. Interplay between hydrogen sulfide and calcium/calmodulin enhances systemic acquired acclimation and antioxidative defense against nickel toxicity in zucchini. *Environ. Exp. Bot.* **2018**, *158*, 40–50. [CrossRef]
79. Rodrigues, O.; Shan, L. Stomata in a state of emergency: H_2O_2 is the target locked. *Trends Plant Sci.* **2021**, *27*, 274–286. [CrossRef]
80. Liu, Z.; Guo, C.; Wu, R.; Hu, Y.; Zhou, Y.; Wang, J.; Yu, X.; Zhang, Y.; Bawa, G.; Sun, X. FLS2–RBOHD–PIF4 Module Regulates Plant Response to Drought and Salt Stress. *Int. J. Mol. Sci.* **2022**, *23*, 1080. [CrossRef]
81. Qi, C.-H.; Zhao, X.-Y.; Jiang, H.; Liu, H.-T.; Wang, Y.-X.; Hu, D.-G.; Hao, Y.-J. Molecular cloning and functional identification of an apple flagellin receptor MdFLS2 gene. *J. Integr. Agric.* **2018**, *17*, 2694–2703. [CrossRef]
82. Dong, S.-M.; Xiao, L.; Li, Z.-B.; Shen, J.; Yan, H.-B.; Li, S.-X.; Liao, W.-B.; Peng, M. A novel long non-coding RNA, DIR, increases drought tolerance in cassava by modifying stress-related gene expression. *J. Integr. Agric.* **2022**, *21*, 2588–2602. [CrossRef]
83. Wan, L.; Li, Y.; Li, S.; Li, X. Transcriptomic Profiling Revealed Genes Involved in Response to Drought Stress in Alfalfa. *J. Plant Growth Regul.* **2021**, *41*, 92–112. [CrossRef]

84. Wang, W.-N.; Min, Z.; Wu, J.-R.; Liu, B.-C.; Xu, X.-L.; Fang, Y.-L.; Ju, Y.-L. Physiological and transcriptomic analysis of Cabernet Sauvginon (*Vitis vinifera* L.) reveals the alleviating effect of exogenous strigolactones on the response of grapevine to drought stress. *Plant Physiol. Biochem.* **2021**, *167*, 400–409. [CrossRef] [PubMed]
85. Liu, M.; Ju, Y.; Min, Z.; Fang, Y.; Meng, J. Transcriptome analysis of grape leaves reveals insights into response to heat acclimation. *Sci. Hortic.* **2020**, *272*, 109554. [CrossRef]
86. Mehdizadeh, L.; Farsaraei, S.; Moghaddam, M. Biochar application modified growth and physiological parameters of Ocimum ciliatum L. and reduced human risk assessment under cadmium stress. *J. Hazard. Mater.* **2020**, *409*, 124954. [CrossRef] [PubMed]
87. Vardharajula, S.; Zulfikar Ali, S.; Grover, M.; Reddy, G.; Bandi, V. Drought-tolerant plant growth promoting Bacillus spp.: Effect on growth, osmolytes, and antioxidant status of maize under drought stress. *J. Plant Interact.* **2011**, *6*, 1–14. [CrossRef]
88. Hu, D.-G.; Ma, Q.-J.; Sun, C.-H.; Sun, M.-H.; You, C.-X.; Hao, Y.-J. Overexpression of MdSOS2L1, a CIPK protein kinase, increases the antioxidant metabolites to enhance salt tolerance in apple and tomato. *Physiol. Plant.* **2015**, *156*, 201–214. [CrossRef]

plants

MDPI

Article

ZmLBD5 Increases Drought Sensitivity by Suppressing ROS Accumulation in Arabidopsis

Jing Xiong [1,†], Weixiao Zhang [1,†], Dan Zheng [1], Hao Xiong [1], Xuanjun Feng [1,2], Xuemei Zhang [1], Qingjun Wang [1], Fengkai Wu [1], Jie Xu [1] and Yanli Lu [1,2,*]

[1] Maize Research Institute, Sichuan Agricultural University, Wenjiang 611130, China; xiongjingsmile@163.com (J.X.); zhangweixiao136@163.com (W.Z.); zhengdan610065@163.com (D.Z.); xiongphd@icloud.com (H.X.); xuanjunfeng@sicau.edu.cn (X.F.); 18283581522@163.com (X.Z.); wdqdjm@126.com (Q.W.); wfk0909@163.com (F.W.); jiexu28@gmail.com (J.X.)

[2] State Key Laboratory of Crop Gene Exploration and Utilization in Southwest China, Wenjiang 611130, China

* Correspondence: yanli.lu82@hotmail.com; Tel./Fax: +86-028-8629-3108

† These authors contributed equally to this work.

Abstract: Drought stress is known to significantly limit crop growth and productivity. Lateral organ boundary domain (LBD) transcription factors—particularly class-I members—play essential roles in plant development and biotic stress. However, little information is available on class-II *LBD* genes related to abiotic stress in maize. Here, we cloned a maize class-II LBD transcription factor, *ZmLBD5,* and identified its function in drought stress. Transient expression, transactivation, and dimerization assays demonstrated that *ZmLBD5* was localized in the nucleus, without transactivation, and could form a homodimer or heterodimer. Promoter analysis demonstrated that multiple drought-stress-related and ABA response cis-acting elements are present in the promoter region of *ZmLBD5*. Overexpression of *ZmLBD5* in Arabidopsis promotes plant growth under normal conditions, and suppresses drought tolerance under drought conditions. Furthermore, the overexpression of *ZmLBD5* increased the water loss rate, stomatal number, and stomatal apertures. DAB and NBT staining demonstrated that the reactive oxygen species (ROS) decreased in *ZmLBD5*-overexpressed Arabidopsis. A physiological index assay also revealed that SOD and POD activities in *ZmLBD5*-overexpressed Arabidopsis were higher than those in wild-type Arabidopsis. These results revealed the role of *ZmLBD5* in drought stress by regulating ROS levels.

Keywords: LBD; drought stress; ROS; stomata; maize

check for updates

Citation: Xiong, J.; Zhang, W.; Zheng, D.; Xiong, H.; Feng, X.; Zhang, X.; Wang, Q.; Wu, F.; Xu, J.; Lu, Y. *ZmLBD5* Increases Drought Sensitivity by Suppressing ROS Accumulation in Arabidopsis. *Plants* **2022**, *11*, 1382. https://doi.org/10.3390/plants11101382

Academic Editors: Małgorzata Nykiel, Beata Prabucka, Mateusz Labudda, Marta Gietler and Justyna Fidler

Received: 22 April 2022
Accepted: 18 May 2022
Published: 23 May 2022

1. Introduction

Drought tolerance is a complex trait that involves a series of adaptive changes in the molecular, cellular, physiological, and morphological levels [1]. Reactive oxygen species (ROS) accumulated under drought stress are versatile in plant development and environmental stress responses [2,3]. ROS act as secondary messengers in stress signaling pathways by triggering defensive/adaptive responses to stress at low-to-moderate concentrations, such as stomatal closure, deposition of lignin and cellulose, and modulation of protein activity and gene expression [3,4]. In addition, ABA stimulates the production of H_2O_2 through NADPH oxidase in guard cells [4]. In rice, *Abscisic acid, Stress and Ripening5 (ASR5)* is known to potentiate ABA biosynthesis, the expression of peroxidase 24 precursor, H_2O_2 accumulation, and stomatal closure, as well as the osmotic and drought tolerance of the seedlings [5]. However, ROS causes growth retardation and eventual cell death once a threshold of ROS concentration is reached [2]. Therefore, ROS detoxification is essential for cell survival, metabolism, and development. Recent studies reveal that increased expression of ROS-scavenging-related genes can improve plants' drought tolerance. Yang et al. found that drought-tolerant maize seedlings have higher antioxidant activities and, consequently, accumulate fewer ROS than sensitive genotypes when exposed to water deficit [6]. Overexpression of *OsLG3* significantly improves rice's tolerance to drought stress by triggering the ROS scavenging system, whereas suppression of *OsLG3* results in decreased ROS scavenging activity and increased drought susceptibility [7].

Lateral organ boundary domain (LBD) proteins, defined by a conserved lateral organ boundary (LOB) domain, belong to the plant-specific transcription factor family [8]. The characteristic LOB domain comprises a C-block containing four cysteine residues (CX2CX6CX3C) required for DNA binding, a Gly-Ala-Ser (GAS) block, and a leucine zipper-like coiled-coil motif (LX6LX3LX6L) responsible for protein dimerization [8–10]. The variable C-terminal region of LBD confers transcriptional control of downstream gene expression [10]. Based on the conserved regions, most LBD genes belong to class-I, which is characterized by a complete leucine-zipper-like domain, while all members of class-II have an incomplete or no leucine zipper-like coiled-coil motif [8]. Several LBD proteins form homo-interactions and hetero-interactions, such as LBD10, LBD27, LBD18, and LBD33 in Arabidopsis [11,12], and RTCS (rootless concerning crown and seminal roots), RTCL (RTCS-like), IG1, and RS2 in maize [9,13]. According to whole-genome sequencing, Arabidopsis [8], rice [14], maize [15], barley [16], tomato [17], grape [18], Eucalyptus [19], and potato [20] have been identified as harboring 43, 35, 44, 24, 46, 49, 47, and 43 LBD genes, respectively, of which 7, 5, 7, 5, 6, 7, 8, and 8 are LBD class-II family members, respectively. There were significantly fewer LBD class-II members than LBD class-I members. The expression patterns of LBD genes were diverse, revealing the diversity of LBD function [8,14].

Most studies have revealed the function of class-I members, including organ development, plant regeneration, photomorphogenesis, and environmental cue responses. *AtASL4* was first discovered to regulate leaf development in Arabidopsis [8]. AtASL4 interacts with AtAS1 to bind the cis-element of the *KNAT1* promoter, inhibits the expression of *KNAT1*, and promotes the development of leaf primordia and inflorescence [21]. In maize, *IG1 (indeterminate gametophyte1)* interacts with *RS (AtAS1 homolog)* to regulate the development of female gametophytes and the number of tassel branches [13,22]. At*LBD16*, *OsCRL1*, and *ZmRTCS* are involved in root development downstream of the auxin signal transduction pathway [9,23–25]. In trees, *LBD* genes also promote stem thickening by accelerating the cell division activity of vascular cambium cells during secondary growth [26–28]. In addition to participating in plant growth and development, *LBD* genes are involved in plants' responses to external biotic and abiotic stresses [10]. In Arabidopsis, the root-specific LBD gene *AtLBD20* inhibits the defense genes *THI2* (*thionin 2.1*) and *VSP2* (*volatile storage protein 2*) via *COI1 (coronatine-insensitive 1)/MYC2*-mediated jasmonic acid (JA) signaling, thereby preventing the damage of the root-invading fungal pathogen *Fusarium oxysporum* in plants. *AtLBD14* regulates the branching of lateral roots through the ABA signaling pathway [29,30]. *AtLBD15* directly binds to the promoter of *AtABI4* (*ABSCISIC ACID INSENSITIVE4*) to activate its expression, resulting in stomatal closure, reduced water loss rate, and enhanced drought tolerance in *AtLBD15*-overexpressed plants [31]. In rice, *OsLBD12-1* directly interacts with the *OsAGO10* promoter to inhibit its expression, resulting in growth retardation, leaf distortion, anther abnormality, and SAM reduction, and *LBD12-1* has a stronger effect on *AGO10* under salt stress [32].

In contrast, reports about class-II members remain limited. The class-II *LBD* genes characterized thus far are mainly involved in metabolism, such as anthocyanin biosynthesis and nitrogen metabolism [33–35]. In this study, the role of *ZmLBD5*—a class-II member—in drought response was investigated. Overexpression of *ZmLBD5* in Arabidopsis caused the drought-sensitive phenotype by suppressing ROS accumulation, increasing the stomatal aperture and water loss. Our results suggest that *ZmLBD5* mediates the response of maize seedlings to drought by regulating H_2O_2 homeostasis, and is expected to be used in genetically modified crops.

2. Results

2.1. ZmLBD5 Was Induced by Osmotic Stress in Maize

In the present study, we cloned the class-II LBD gene *ZmLBD5* from inbred B73 maize. The full-length CDS of *ZmLBD5* is 942 bp, and encodes a polypeptide of 313 amino acid residues with a predicted molecular mass of 33.96 kD and a pI value of 6.61. Sequence alignment revealed that *ZmLBD5* contained a typical DNA-binding domain, CX2CX6CX3C, whereas the GAS block and LX6LX3LX6L coiled-coil motif were incomplete, allowing for a distinction between the class-I and class-II members of the LBD family (Figure 1A).

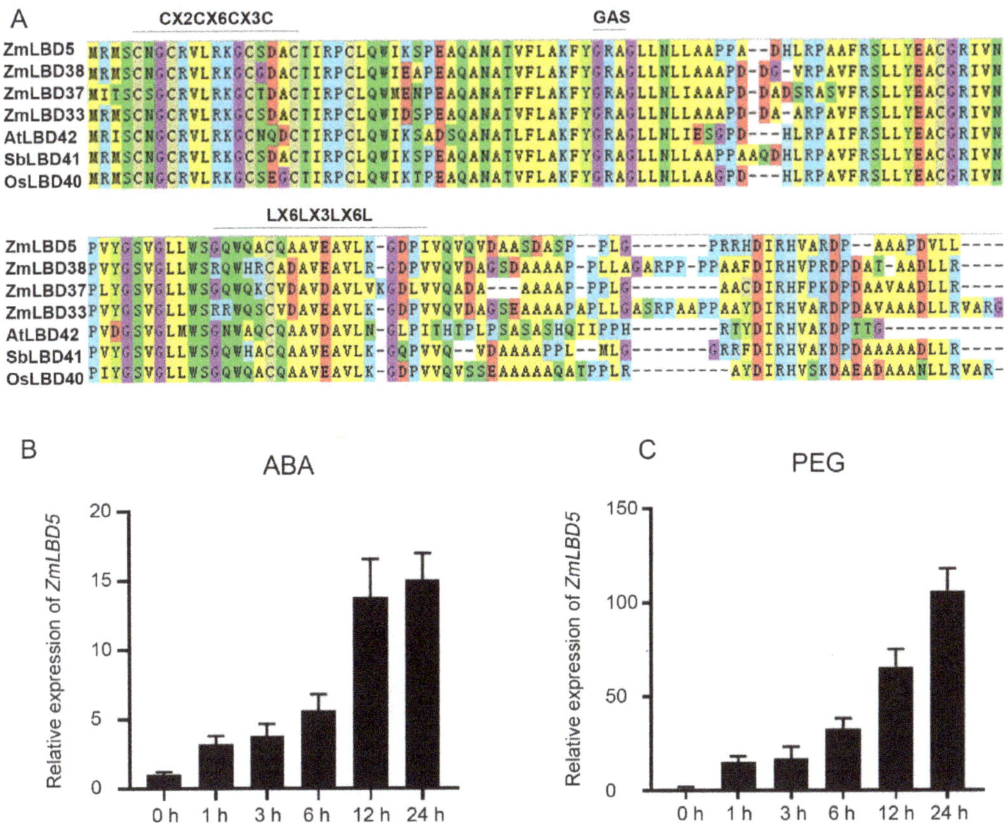

Figure 1. Sequence and expression pattern analysis of *ZmLBD5*: (**A**) LOB domain sequence alignment of ZmLBD5 and LBD members from other plant species. The class-I LBD members had typical CX2CX6CX3C, GAS, and LX6LX3LX6L domains. The same position in class II is marked. (**B**,**C**) The expression of *ZmLBD5* upon ABA and drought treatment in maize. Total RNA was isolated from 3-leaf seedlings grown without (0 h) or with 10 μM ABA and 20% PEG6000 treatment. Transcript levels of *ZmLBD5* were determined by qPCR, using *Zme1F1α* and *Zm18S* as reference genes. Fold change was calculated by $2^{-\Delta\Delta t}$. All bars represent means ± SD (n = 3).

Given that gene expression levels are regulated by promoters, we examined the *ZmLBD5* promoter region (approximately 2000 bp upstream of the first codon). Several stress-response-related cis-acting elements were present in the *ZmLBD5* promoter, including seven ABREs (ABA-responsive elements), three DREs (dehydration-responsive elements), two LTREs (low-temperature-responsive elements), one MBS (MYB-binding site, involved in drought-inducibility), eight MYBRSs (MYB recognition sites), and other light-response elements (Table 1). Therefore, the response of *ZmLBD5* to drought stress was investigated by RT-qPCR using drought- and ABA- treated maize plants. The results showed that the expression of *ZmLBD5* was strongly induced by ABA and drought stress (Figure 1B,C). These results suggest that *ZmLBD5* plays a prominent role in the response to drought stress in maize.

Table 1. Cis-elements in the promoter region (~2 kb) of ZmLBD5.

Site Name	Sequence	Position	Strand	Function
ABRE	ACGTG	−1984	-	Abscisic acid responsiveness
ABRE	CGTACGTGCA	−1730	-	Abscisic acid responsiveness
ABRE	CACGTG	−1596	+	Abscisic acid responsiveness
ABRE	ACGTG	−1595	+	Abscisic acid responsiveness
ABRE	ACGTG	−1528	+	Abscisic acid responsiveness
ABRE	ACGTG	−71	-	Abscisic acid responsiveness
ABRE	CCACGTGG	−1597	+	Abscisic acid responsiveness
DRE	GCCGAC	−1896	-	Dehydration-responsive element
DRE	GCCGAC	−1495	-	Dehydration-responsive element
DRE	ACCGAGA	−38	+	Dehydration-responsive element
LTR	CCGAAA	−1635	+	Low-temperature responsiveness
LTR	CCGAAA	−262	+	Low-temperature responsiveness
MBS	CAACTG	−597	-	MYB-binding site involved in drought-inducibility
MYBRS	CAACCA	−1566	-	MYB recognition site
MYBRS	CAACTG	−597	-	MYB recognition site
MYBRS	TAACCA	−593	-	MYB recognition site
MYBRS	CAACCA	−518	+	MYB recognition site
MYBRS	CAACCA	−100	+	MYB recognition site
MYBRS	CAACCA	−96	+	MYB recognition site
MYBRS	CCGTTG	−1844	+	MYB recognition site
MYBRS	TAACCA	−593	-	MYB recognition site
CCAAT-box	CAACGG	−1844	-	MYBHv1-binding site
ARE	AAACCA	−1631	+	Anaerobic induction
ARE	AAACCA	−259	+	Anaerobic induction
G-box	CACGTC	−1984	+	Light responsiveness
G-box	GCCACGTGGA	−1598	+	Light responsiveness
G-box	CACGTG	−1596	+	Light responsiveness
G-Box	CACGTG	−1596	+	Light responsiveness
G-Box	CACGTT	−1529	-	Light responsiveness
G-box	CACGTC	−71	+	Light responsiveness

"+" represents sense strand, and "-" represents the antisense strand.

2.2. *ZmLBD5 Is Localized in the Nucleus, and Could Form Dimers*

Understanding the subcellular localization of gene expression products is important for the functional analysis of genes. To determine the subcellular localization of ZmLBD5, ZmLBD5-GFP was transiently expressed in tobacco leaf cells and maize protoplasts under the control of the cauliflower mosaic virus (CaMV) 35S promoter. The strong green fluorescence signal of GFP was mainly distributed in the nucleus and the cytoplasm, whereas the green fluorescence signal of ZmLBD5-GFP was observed in the nucleus, which completely overlapped with the red fluorescence signal of the nuclear localization signal (Figure 2A,B). In addition, GFP fluorescence was observed in the nuclei of the root cells in ZmLBD5-GFP-overexpressed Arabidopsis (Figure 2C).

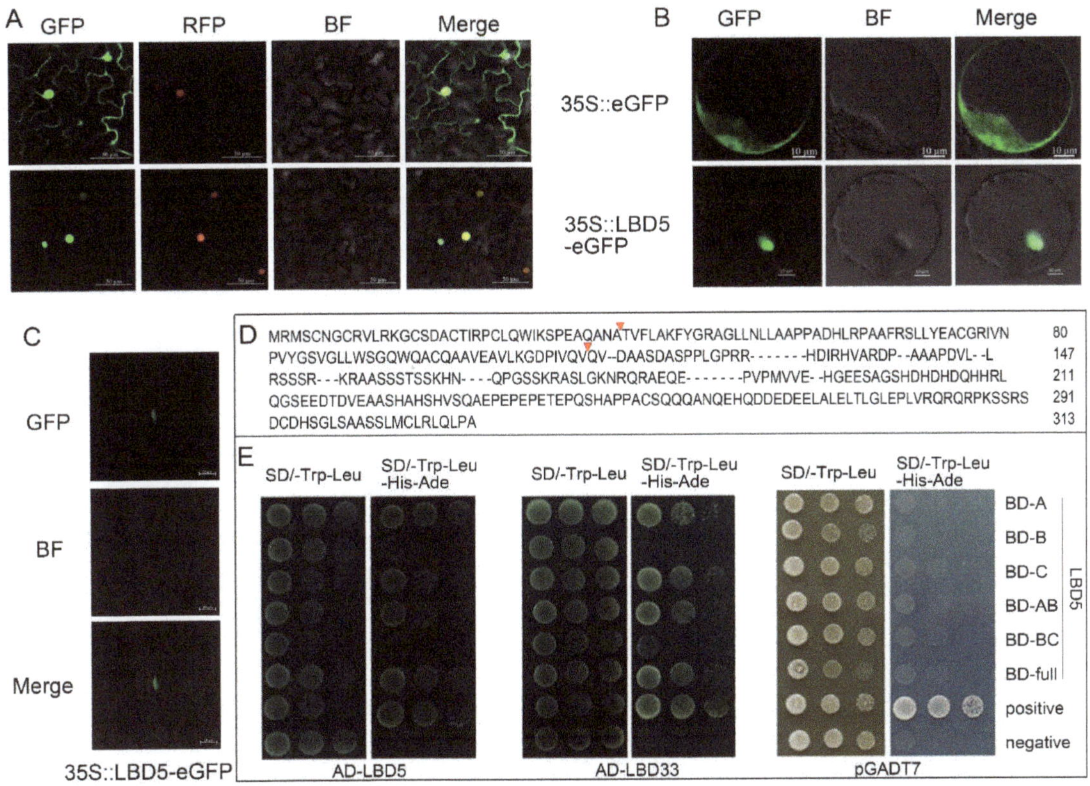

Figure 2. Subcellular localization and dimer-forming ability of ZmLBD5: (**A–C**) Subcellular localization of ZmLBD5 in tobacco leaves, maize protoplasts, and 35S::ZmLBD5-eGFP transgenic Arabidopsis, respectively (bar = 50, 10, and 20 μm, respectively). (**D**) Three fragments of ZmLBD5 (**A–C**) were divided and the red triangle indicated the position of truncation. (**E**) The ability of ZmLBD5 to form dimers in the yeast strain Y2H Gold. The Y2H Gold strains containing target plasmids were diluted and cultured on no-selection synthetic dropout (SD) media without tryptophan and leucine (SD/-T-L), or on selection SD media without tryptophan, leucine, histidine, and adenine (SD/-T-L-H-A). Photos were taken 3 days after inoculation for the plates. Fragments A, B, and C represent CX2CX6CX3C, GAS, and LX6LX3LX6L, and various C-terminal domains, respectively.

Given that the GAS block and LX6LX3LX6L coiled-coil motif of class-I members are essential for protein dimerization, and class-II members are characterized by a lack of or an incomplete domain, the ability of ZmLBD5 to dimerize was tested. The full length of ZmLBD5 and its five truncated peptide fragments (A, B, C, AB, and BC) were tested for the interaction with ZmLBD5 itself and another LBD member, ZmLBD33. Fragments A, B, and C represent the N-terminal C-block (CX2CX6CX3C), the GAS and LX6LX3LX6L coiled-coil motifs, and the C-terminal domain, respectively (Figure 2D). Although it was difficult to determine the essential region for the interaction, homo- and heterodimerization were clearly detected through yeast two-hybrid screening (Figure 2E).

2.3. Overexpression of ZmLBD5 Decreased Drought Tolerance in Transgenic Arabidopsis

ZmLBD5 was overexpressed in Arabidopsis to observe its function. Eleven transgenic lines were generated, and the three homozygous lines with the highest expression levels (OE3, OE10, and OE19) were selected for subsequent experiments (Figure 3A). Transgenic lines and the wild-type seeds were exposed to different concentrations of mannitol (0, 200,

250, and 300 mM). The germination rates of *ZmLBD5*-overexpressed plants were comparable to that of the wild type, and it was significantly delayed along with the increase in mannitol concentration (Figure 3B–E). The cotyledon greening rate of the overexpressed lines was significantly lower than that of the wild type under 250 mM and 300 mM mannitol stresses 3 days after germination (Figure 3F,G).

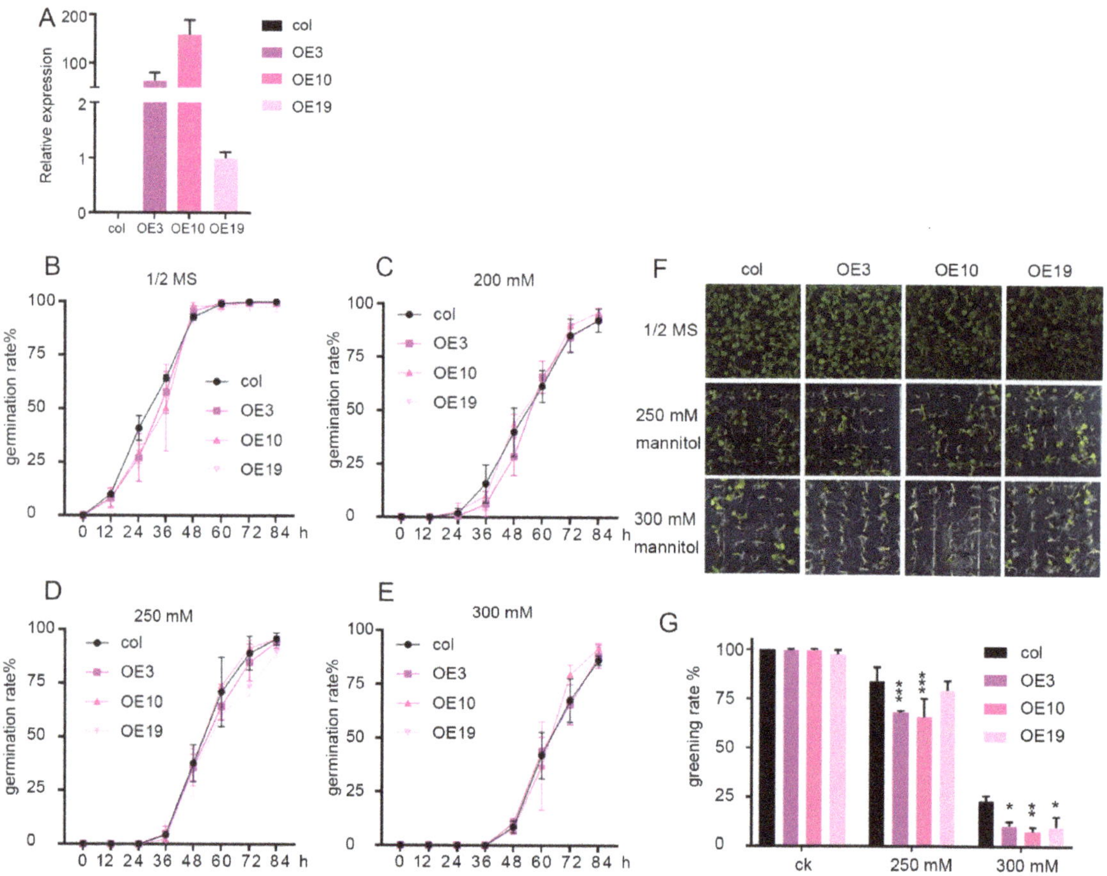

Figure 3. The germination and greening rate of *ZmLBD5* transgenic Arabidopsis with mannitol treatment: (**A**) RT-qPCR analysis of *ZmLBD5* expression in the overexpression lines and wild type. (**B–E**) Statistical analysis of *ZmLBD5* transgenic lines' and wild-type's seed germination rates under 0 mM, 200 mM, 250 mM, and 300 mM mannitol, respectively. (**F**) The phenotypes of *ZmLBD5* transgenic and wild-type Arabidopsis lines treated with 0 mM, 250 mM, and 300 mM mannitol. (**G**) Greening rate of *ZmLBD5* transgenic and wild-type Arabidopsis under 0 mM, 250 mM, and 300 mM mannitol stress. Significance was calculated by one-way ANOVA. * $p < 0.05$; ** $p < 0.01$; *** $p < 0.001$. All bars represent means ± SD, ($n \geq 3$).

To further characterize the responses of the wild-type and *ZmLBD5*-overexpressed plants to osmotic stress, 5-day-old seedlings were exposed to different concentrations of mannitol (0, 200, 250, and 300 mM) for 7 days. Unexpectedly, there was no significant difference between the wild-type and the transgenic seedlings, except for line OE19 (Figure 4).

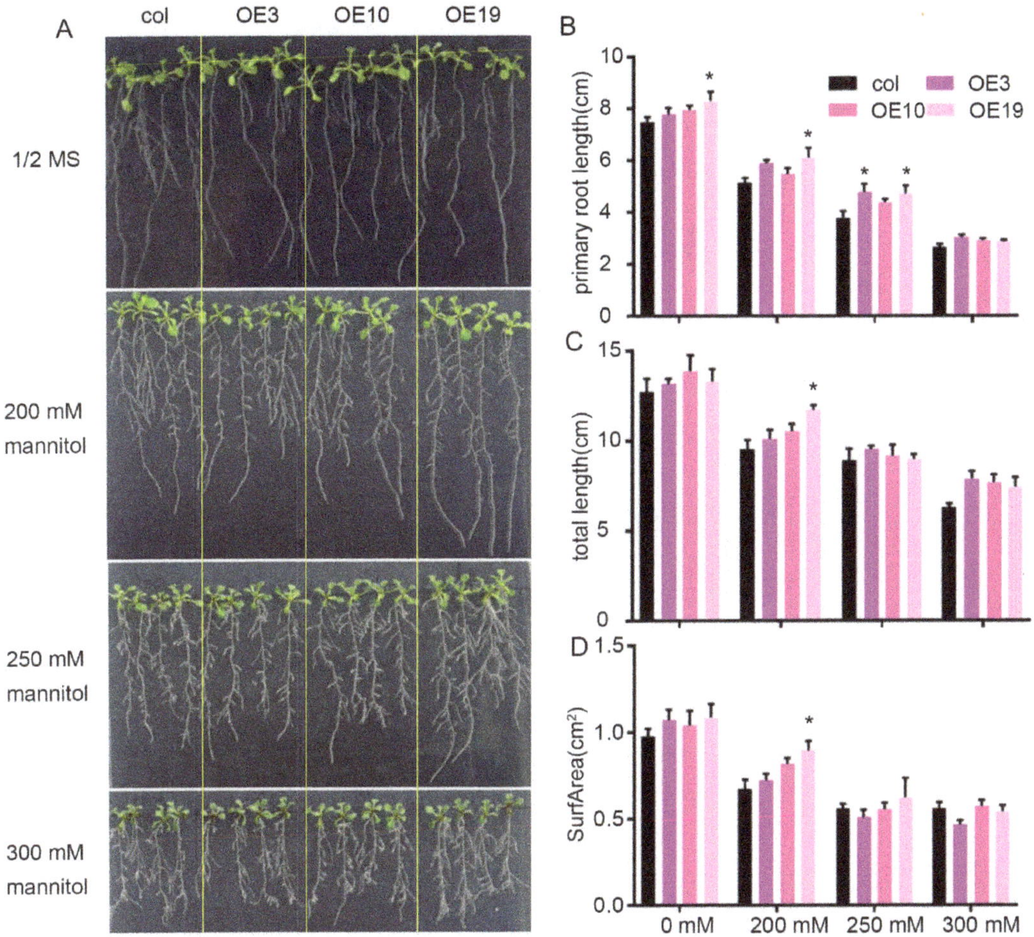

Figure 4. The phenotype of *ZmLBD5* transgenic Arabidopsis with mannitol treatment: (**A**) Seedlings of wild-type and *ZmLBD5* transgenic lines grown on 1/2 MS medium with 0 mM, 200 mM, 250 mM, and 300 mM mannitol. (**B–D**) Statistical analysis of primary root length, total root length, and root surface area of wild-type and *ZmLBD5* transgenic seedlings grown on 1/2 MS medium with or without mannitol treatment. Mean values and standard errors (bars) are shown from 12 independent seedlings. Asterisks on bar represent the difference compared with wild type is significant (* $p < 0.05$).

To further understand the role of *ZmLBD5* under drought stress, 7-day-old plants were transplanted into the soil and grown for one month. Then, plants were exposed to drought stress for 10 days. Three days after rewatering, *ZmLBD5*-overexpressed plants displayed a higher survival rate than that of the wild type (Figure 5C,D). Under well-watered conditions, the rosette leaf areas of *ZmLBD5*-overexpressed Arabidopsis were significantly larger than those of the wild type (Figure 5A), and the fresh weight increased in *ZmLBD5*-overexpressed seedlings (Figure 5B). These results indicated that *ZmLBD5* promotes seedling growth under normal conditions, and increases drought sensitivity under drought stress.

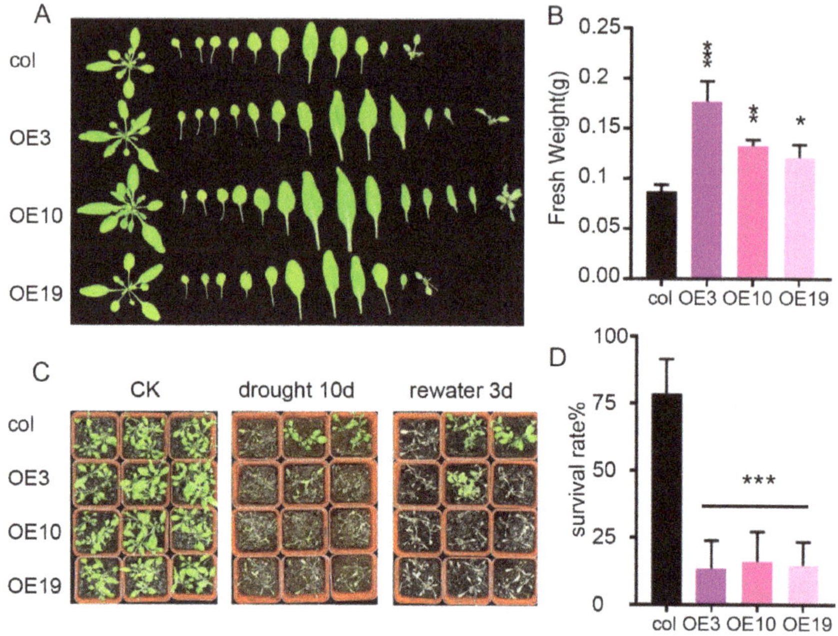

Figure 5. Phenotype and survival rate of *ZmLBD5* transgenic seedlings under normal or drought conditions in soil: (**A**,**B**) Phenotype and fresh weight of *ZmLBD5* transgenic seedlings under normal conditions in soil. (**C**,**D**) Survival rate of *ZmLBD5* transgenic seedlings under drought conditions in soil. Significance was analyzed by one-way ANOVA. * $p < 0.05$; ** $p < 0.01$; *** $p < 0.001$. All bars represent means ± SD ($n \geq 12$).

2.4. *ZmLBD5* Increased the Water Loss Rate by Enhancing the Stomatal Density and Aperture

We measured the rate of water loss from detached leaves to investigate why *ZmLBD5*-overexpressed seedlings displayed drought sensitivity. The results showed that detached leaves of *ZmLBD5* transgenic seedlings lost water at a greater rate than that of the wild-type plants after 1 h of dehydration (Figure 6A), indicating that *ZmLBD5*-overexpressed seedlings ran out of soil water more rapidly than the wild-type seedlings, leading to earlier wilting. Given that water evaporated mainly through the stomata, the stomatal number and apertures on abaxial leaves were analyzed. The stomatal number and stomatal aperture of *ZmLBD5*-overexpressed plants were both greater than those of the wild-type plants (Figure 6B–D). Thus, the overexpression of *ZmLBD5* enhanced drought sensitivity by increasing the stomatal number and apertures in Arabidopsis under drought conditions.

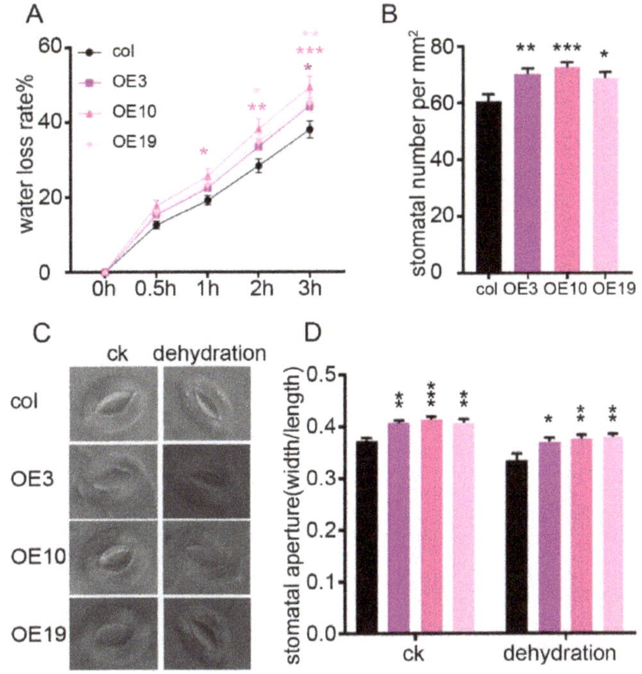

Figure 6. Water loss rate, stomatal number, and apertures of leaves in *ZmLBD5* transgenic seedlings: (**A**) Water loss rate of leaves in *ZmLBD5* transgenic seedlings. (**B**) Stomatal number of the fourth leaves in *ZmLBD5* transgenic seedlings. Each sample had at least 12 seedlings. (**C**,**D**) Stomatal aperture of fourth leaves after detachment for 1 h in *ZmLBD5* transgenic seedlings ($n > 30$ for each sample). Significance was analyzed by one-way ANOVA. * $p < 0.05$; ** $p < 0.01$; *** $p < 0.001$. All bars represent means ± SD ($n \geq 12$).

2.5. Overexpression of ZmLBD5 Improved Antioxidant Enzyme Activity and Blocked ROS Accumulation in Arabidopsis

ROS are important molecular signaling and cytotoxic substances in plants' response to drought stress. The accumulation of H_2O_2 is important for stomatal closure [4,5,36]. Therefore, H_2O_2 and superoxide anions were investigated using DAB and NBT staining, and H_2O_2 content was quantitatively measured by the potassium iodide method. The accumulation of H_2O_2 and superoxide anions in the leaves of *ZmLBD5*-overexpressed seedlings was significantly lower than that of the wild-type seedlings (Figure 7A,B,D). Many antioxidant enzymes—such as POD, SOD, and catalase (CAT)—are associated with ROS levels and the tolerance of plants to abiotic stress [3,37]. SOD activity was higher in *ZmLBD5*-overexpressed seedlings than that of the wild-type seedlings under both normal and drought conditions (Figure 7C). The activities of POD and CAT were not different between the wild-type plants and the transgenic plants, except for OE3 (Figure 7E,F).

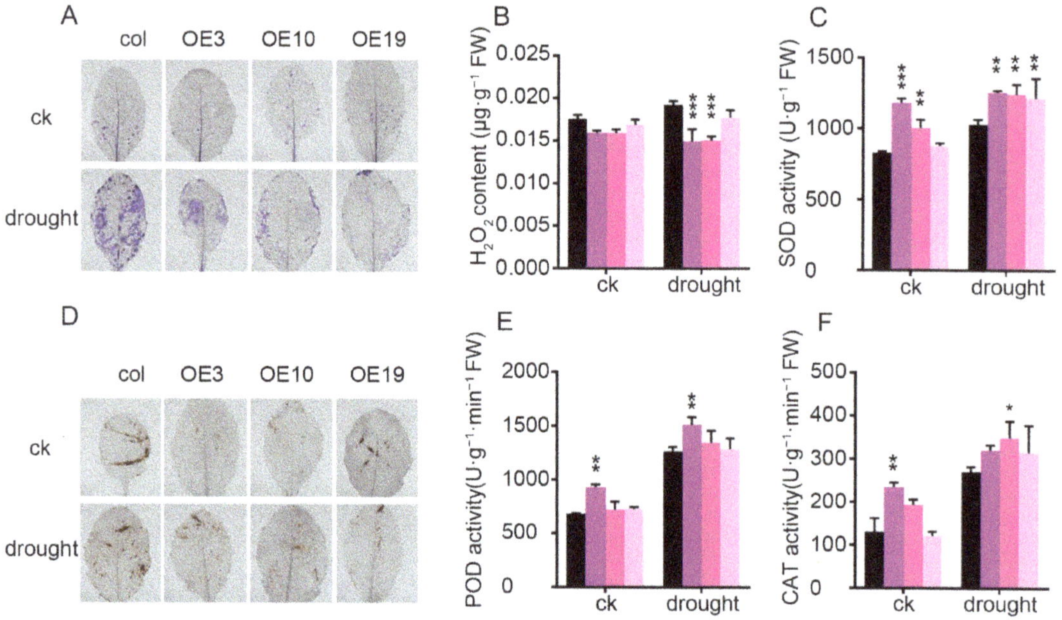

Figure 7. Reactive oxygen species staining and physiological indices in *ZmLBD5* transgenic Arabidopsis: (**A**,**D**) NBT and DAB staining of leaves for H_2O_2 from *ZmLBD5* transgenic seedlings and wild-type plants under normal conditions and drought stress (3-week-old seedlings were subjected to drought for 7 days). (**B**) H_2O_2 content in leaves from *ZmLBD5* transgenic seedlings and wild-type plants under normal conditions and drought stress (withholding water for 7 days). (**C**) SOD activity, (**E**) POD activity, and (**F**) CAT activity in leaves from *ZmLBD5* transgenic seedlings and wild-type plants under normal and drought conditions (withholding water for 7 days). Significance was analyzed by one-way ANOVA. * $p < 0.05$; ** $p < 0.01$; *** $p < 0.001$. All bars represent means ± SD ($n = 6$).

2.6. *ZmLBD5 Negatively Regulates Drought-Related Genes' Expression in Transgenic Arabidopsis*

From the above findings, *ZmLBD5* is a negative regulator of plant in drought tolerance. Therefore, we analyzed the expression of several widely reported drought-related genes (*PP2CA, RD17, RD26, DREB2A, RD29A, and RD29B*) in *ZmLBD5*-overexpressed plants and wild-type plants. Under normal conditions, the expression levels of these genes were similar between the transgenic plants and the wild-type plants, except for *PP2CA* (Figure 8). Under drought conditions, the expression levels of these tested genes were remarkably increased in both the transgenic plants and the wild-type plants. However, the expression levels were lower in the transgenic plants than those in the wild-type plants (Figure 8), indicating that *ZmLBD5* suppressed drought-related genes' expression under drought stress.

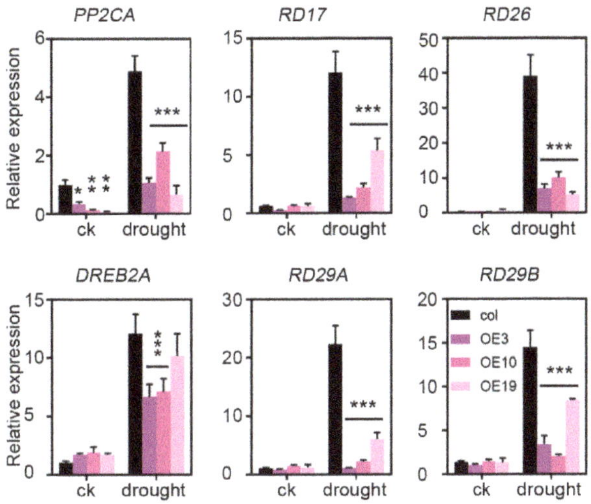

Figure 8. Expression levels of drought-stress-related genes in *ZmLBD5* transgenic Arabidopsis: Total RNA was isolated from 15-day-old seedlings grown without (CK) or with 250 mM mannitol treatment for 7 days. Transcript levels of *PP2CA*, *RD17*, *RD26*, *DREB2A*, *RD29A*, and *RD29B* in the transgenic lines and wild type were determined by qPCR, using *AtACTIN8* and *AtUBQ10* as reference genes. Fold change was calculated by $2^{-\Delta\Delta t}$. Significance was analyzed by one-way ANOVA. * $p < 0.05$; ** $p < 0.01$; *** $p < 0.001$. All bars represent means ± SD ($n = 3$).

3. Discussion

The LBD genes are plant-specific transcription factors. According to their GAS and leucine zipper domains, LBD genes are divided into class-I and class-II members. With the publication of genomic data on different species, the distribution, gene structure, and expression pattern of the LBD gene family involved in plants' development and stress response have been displayed [14,15,38–40]. Most class-I LBD genes are involved in root growth, leaf extension, pollen development, plant regeneration, photomorphogenesis, pathogen response, and secondary cell wall development [10,33,35]. However, studies of class-II members are scarce, and these are involved in anthocyanin synthesis, nitrogen metabolism, root development, auxin response, and GA response [33,34,41,42]. In this study, *ZmLBD5* was reported to negatively regulate drought tolerance by decreasing ROS levels and suppressing stomatal closure.

LBD proteins generally function by forming homodimers or heterodimers with themselves or other proteins [43]. In Arabidopsis, the dimerization of AtLBD16 and AtLBD18 is critical for lateral root formation [44]. AtLBD10 interacts with AtLBD27 to regulate pollen development [11]. In maize, RTCS and RTCL form heterodimers to affect the initial crown root generation [9]. The difference between class-I and class-II LBD proteins is in the GAS and leucine zipper domains, which are proposed to be necessary for protein–protein dimerization [10,35]. In this study, ZmLBD5, without the complete GAS and leucine zipper domains, could form homodimers and heterodimers in the same way as class-I members, implying that the GAS and leucine zipper domains may be not essential for dimerization.

Previous studies have shown that the functions of the class-I LBDs are mainly related to organ determination, including embryo, root, leaf, and inflorescence development [13,45–47]. In Arabidopsis, *AtASL4* is expressed at the boundary between the developing leaf primordia and the shoot-tip meristem to regulate leaf development [45]. *AtAS2 (At LBD6)* represses cell proliferation in the adaxial domain, and is critical for the development of properly expanded leaves [46,47]. *IG1*, an LBD gene, affects the formation of the leaf's ligular region by inhibiting the expression of *KNOX* while also inhibiting the development of female gametophytes in *ig1* mutants, and limiting the number of male ear branches in maize [13]. In addition, the class-I member *AtLBD15* promotes drought tolerance by increasing stomatal closure and reducing the water loss rate [31]. *OsLBD12-1* reduces SAM size by directly binding to the promoter region, and strongly represses the expression of *AGO10* under salt conditions [32]. *ZmLBD5,* a class-II LBD gene, also regulates organ development and response to drought stress in transgenic Arabidopsis, indicating the functional overlap between class-I and class-II members. The function of *ZmLBD5* is similar to that of the class-I members of the LBD gene family, thereby portraying substantial evidence for the study of class-II members in plant growth and the regulation of drought.

Drought stress brings about the production of ROS. The excessive accumulation of ROS causes damage to plant cells; however, they also act as signaling molecules that participate in the regulation of stomatal number and apertures. Therefore, ROS have been widely reported to regulate drought tolerance in various studies [5–7,48]. Many studies have shown that ABA stimulates the production of H_2O_2 through NADPH oxidase in guard cells, and H_2O_2 is an important signaling molecule that activates calcium channels in the plasma membrane, thereby mediating ABA-induced stomatal closure [36,49,50]. In rice, H_2O_2 accumulated in guard cells in *dst* (drought- and salt-tolerant transcription factor) mutants was able to promote stomatal closure, reduce water loss, and improve drought and salt tolerance. *DST* also inhibited stomatal closure by directly regulating the expression of the H_2O_2 homeostasis-related gene peroxidase 24 predictor [5,48]. In this study, the activities of SOD and POD were remarkably higher, and the ROS level was significantly lower in *ZmLBD5*-overexpressed Arabidopsis than that in the wild type. This indicates that *ZmLBD5* regulates the response of the seedlings to drought through the regulation of H_2O_2 signal molecules on the stomatal aperture—not the antioxidant pathway. Accordingly, *ZmLBD5* overexpression increased water loss rate by suppressing stomatal closure, and resulted in the drought-sensitive phenotype.

4. Conclusions

In summary, this study demonstrated *ZmLBD5* as a negative regulator of drought tolerance. Overexpression of *ZmLBD5* increased the stomatal aperture and water loss rate by suppressing ROS accumulation. Furthermore, the enhancement of SOD and POD activities in *ZmLBD5*-overexpressed Arabidopsis revealed the role of *ZmLBD5* in drought stress by regulating ROS levels. The study of *ZmLBD5* promoted our understanding of the function of the class-II LBD genes in maize.

5. Materials and Methods

5.1. Plant Materials and Growth Conditions

ZmLBD5 transgenic lines and wild-type (ecotype: Col-0) seeds were surface-sterilized with 3% $NaClO_3$ for 8 min, and washed five times with sterile water. The sterilized Arabidopsis seeds were plated on half of the Murashige and Skoog (1/2 MS) medium and stored at 4 °C for 72 h in darkness. Then, they were transferred into a growth incubator (22 °C, 16 h light/8 h dark) for germination and growth. Seven days later, the seedlings were transplanted into soil and grown in a greenhouse (22 °C, 16 h light/8 h dark). Tissues were harvested from the seedlings for further study.

For drought stress and ABA treatment, two-leaf-stage seedlings were transferred to Hoagland nutrient solution in a greenhouse with a 14 h light/10 h dark photoperiod at 28 °C, and grown to the three-leaf stage. Then, seedlings were subjected to polyethylene glycol 6000 (PEG6000) (20% *w/v*) and ABA (10 μM). The roots were collected after 0, 1, 3, 6, 12, and 24 h of treatment. The harvested samples were frozen immediately in liquid nitrogen and used for RNA isolation.

5.2. Sequence Analysis

The *ZmLBD5* cDNA was obtained from MaizeGDB (https://www.maizegdb.org/) (accessed on 16 February 2022). Homologous sequences of ZmLBD5 were retrieved from the Phytozome database (https://phytozome-next.jgi.doe.gov/) (accessed on 16 February 2022), and sequence alignment was performed using ClustalX. Plant CARE (http://bioinformatics.psb.ugent.be/webtools/plantcare/html/) (accessed on 1 March 2022) was used to analyze the promoter sequences of different abiotic-stress-related cis-elements.

5.3. Subcellular Localization

The full-length CDS of *ZmLBD5* was inserted into the binary vector pCAMBIA2300-eGFP to generate the pCAMBIA2300-ZmLBD5-eGFP vector. The constructed vector was introduced into Arabidopsis and tobacco leaves using agrobacterium-mediated methods. In addition, the plasmid was transformed into maize protoplasts via PEG-mediated methods. GFP fluorescence was investigated using a laser confocal microscope (LSM800, Zeiss, Germany). The primers used here are listed in Supplementary Table S1.

5.4. RNA Extraction and Quantitative RT-qPCR Analysis

Total RNA was extracted from Arabidopsis or maize (inbred line: B73) seedlings according to the manufacturer's protocol of the Plant Total RNA Isolation Kit (FOREGENE, Re-05014), and treated with DNase I (Trans, GD201-01) at 37 °C for 30 min to eliminate genomic DNA contamination. The PrimeScript RT reagent kit with a gDNA eraser (Takara, RR047A) was used to synthesize first-strand cDNA as real-time PCR templates with gene-specific primers. Furthermore, we performed qPCR amplification on a Bio-Rad CFX96 PCR instrument according to the SYBR Green Fast qPCR Mix Kit instructions (ABclonal, RM21203). *AtACTIN8* and *AtUBQ10* were used as internal reference genes for Arabidopsis. *ZmeF1α* and *Zm18S* were used as internal reference genes for maize. The mean values and standard deviations were estimated using $2^{-\Delta\Delta CT}$ from the data of three biological experiments. All of the primers used in the experiments are listed in Supplementary Table S1.

5.5. Generation of Transgenic Plants and Phenotypic Analysis

The coding sequence of *ZmLBD5* was cloned into the pCAMBIA2300-eGFP vector to generate the 35S::ZmLBD5-eGFP vector. The product was introduced into the wild-type Col-0 using the agrobacterium-mediated floral-dip method. Homozygous plants were screened under 10 μg/mL neomycin (G418) conditions, and three high-expression lines were used for further study. All of the primers used in the experiments are listed in Supplementary Table S1.

Seeds were sterilized and sowed on 1/2 MS medium containing 0 mM, 200 mM, 250 mM, and 300 mM mannitol and grown in a greenhouse. Germination rates were recorded every 12 h. Seeds were recorded as germinated when the radicles protruded from the seed coat. After germination for 5 d, the greening rate of the seedlings was recorded. Each sample contained three biological replicates. Statistical analysis was performed with one-way ANOVA, and the mean value each transgenic line was compared with the wild type.

Five-day-old seedlings vertically grew on 1/2 MS medium containing 0 mM, 200 mM, 250 mM, and 300 mM mannitol for one week. Primary root length was measured using ImageJ software. Root lengths and surface areas were collected and analyzed using an Epson 11,000 × l root scanner and WinRHIZO pro2013. All experiments were performed in triplicate. Each biological replicate contained at least 12 seedlings.

Seven-day-old plants were transplanted into the soil under short-day conditions to grow for three weeks. Subsequently, water was withheld for approximately 10 days, and photographs were taken. After re-watering for 3 days, the survival rates were investigated.

5.6. Water Loss Measurement

To assay the water loss rate, 12 one-month-old seedlings of each sample were detached on the laboratory bench and weighed at different time points. The experiment was replicated three times. ANOVA was used to assess the differences between the wild-type and transgenic plants.

5.7. Stomatal Density and Stomatal Aperture

The fourth expanded rosette leaves of Arabidopsis were detached on a laboratory bench for one hour. Leaves were placed into the Carnot fixed solution (absolute ethanol: glacial acetic acid = 3:1) for 24 h, followed by dehydration with 30%, 50%, 70%, 80%, 85%, 90%, 95%, and 100% alcohol for 30 min, dehydration with 100% alcohol again, and placement into a transparent solution (trichloroacetaldehyde: water: glycerol = 8:3:1). Stomatal apertures were observed under microscopes, and the ratio of the stomatal length to width was recorded using ImageJ software. At least 30 stomata of each sample per replicate were measured, and three replicates were performed.

5.8. ROS Measurements

Histochemical assays for reactive oxygen species (ROS) accumulation were performed using DAB and NBT staining. The detached leaves of Arabidopsis seedlings were treated with DAB staining solution (0.1 g/mL DAB, PH 3.8) through a vacuum pump for 30 min, and placed in the dark at room temperature for 10 h, soaked in decolorizing solution (acetic acid: glycerol: ethanol = 1:1:3) in 95 °C boiling water for 5 min, stored in 95% ethanol, and observed under a stereomicroscope. Superoxide anion accumulation was detected using NBT staining. The detached leaves of Arabidopsis were immediately immersed in 50 mM phosphate buffer (pH 7.5) containing 0.1 g/mL NBT at room temperature for 8 h in the dark. The decolorization method was similar to that of DAB staining. Each line contained at least 10 different seedlings, and representative images are shown.

Quantitative measurement of H_2O_2 concentration was performed using the potassium iodide method [51]. Briefly, 100 mg leaf samples were placed in liquid nitrogen and ground into powder. Furthermore, 1 mL of precooled 0.1% trichloroacetic acid (TCA) solution was immediately added and mixed with the samples. After centrifugation (10,000× *g*, 4 °C, 10 min), equal volumes of PBS buffer were added to the 500 μL supernatant, and then 1 mL of 1 M potassium iodide solution was added, and the mixture was shaken with 150 rpm at 30 °C for 1 h. Absorbance was measured at 390 nm. In addition, 300 μmol/L H_2O_2 was used to obtain a standard curve. Each experiment was performed in six replicates.

The activities of antioxidant enzymes (SOD, POD, and CAT) were measured following the aforementioned protocols [52–54]. The units of the antioxidant enzyme activities were defined as follows: a unit of SOD activity is the quantity of enzyme required to cause 50% inhibition of the photochemical reduction of NBT per minute at 560 nm [54]; a unit of POD activity is the amount of enzyme required to cause a 0.01 increase in the absorbance of H_2O_2 per minute at 470 nm [52,53]; and a unit of CAT activity is the amount of enzyme required to cause a 0.01 decrease in the absorbance per minute at 240 nm [53].

Supplementary Materials: The following supporting information can be downloaded at: https://www.mdpi.com/article/10.3390/plants11101382/s1, Table S1: The primers of RT-PCR and vector construction.

Author Contributions: Conceptualization, X.F. and Y.L.; Methodology, J.X. (Jing Xiong), W.Z., D.Z. and H.X.; Data Curation, J.X. (Jing Xiong), W.Z. and X.Z.; Writing—original draft preparation, J.X. (Jing Xiong) and X.F.; Writing—review and editing, Q.W., F.W., J.X. (Jie Xu) and Y.L. All authors have read and agreed to the published version of the manuscript.

Funding: This research was funded by the key Research Program of the Department of Science and Technology of Sichuan province, China (2021YFH0053, 2020YFH0116), and the National Natural Science Foundation of China (31901557, 32072074).

Conflicts of Interest: The authors declare no conflict of interest.

References

1. Blum, A. Drought resistance, water-use efficiency, and yield potential—are they compatible, dissonant, or mutually exclusive? *Aust. J. Agric. Res.* **2005**, *56*, 1159–1168. [CrossRef]
2. Apel, K.; Hirt, H. Reactive oxygen species: Metabolism, oxidative stress, and signal transduction. *Annu. Rev. Plant Biol.* **2004**, *55*, 373–399. [CrossRef]

3. Miller, G.; Suzuki, N.; Ciftci-Yilmaz, S.; Mittler, R. Reactive oxygen species homeostasis and signalling during drought and salinity stresses. *Plant Cell Environ.* **2010**, *33*, 453–467. [CrossRef]
4. Marino, D.; Dunand, C.; Puppo, A.; Pauly, N. A burst of plant NADPH oxidases. *Trends Plant Sci.* **2012**, *17*, 9–15. [CrossRef]
5. Li, J.; Li, Y.; Yin, Z.; Jiang, J.; Zhang, M.; Guo, X.; Ye, Z.; Zhao, Y.; Xiong, H.; Zhang, Z.; et al. OsASR5 enhances drought tolerance through a stomatal closure pathway associated with ABA and H_2O_2 signalling in rice. *Plant Biotechnol. J.* **2017**, *15*, 183–196. [CrossRef]
6. Yang, L.; Jake, F.; Hui, W.; Xinzhi, N.; Pingsheng, J.; Robert, L.; Robert, K.; Brian, S.; Baozhu, G. Stress sensitivity is associated with differential accumulation of reactive oxygen and nitrogen species in maize genotypes with contrasting levels of drought tolerance. *Int. J. Mol. Sci.* **2015**, *16*, 24791–24819. [CrossRef]
7. Xiong, H.; Yu, J.; Miao, J.; Li, J.; Zhang, H.; Wang, X.; Liu, P.; Zhao, Y.; Jiang, C.; Yin, Z.; et al. Natural variation in OsLG3 increases drought tolerance in rice by inducing ROS scavenging. *Plant Physiol.* **2018**, *178*, 451–467. [CrossRef]
8. Shuai, B.; Reynaga-Peña, C.G.; Springer, P.S. The lateral organ boundaries gene defines a novel, plant-specific gene family. *Plant Physiol.* **2002**, *129*, 747–761. [CrossRef]
9. Majer, C.; Xu, C.; Berendzen, K.W.; Hochholdinger, F. Molecular interactions of rootless concerning crown and semial roots, a LOB domain protein regulating shoot-borne root initiation in maize (*Zea mays* L.). *Philosphical Trans. Rayal Soc. B* **2012**, *367*, 1542–1551. [CrossRef]
10. Xu, C.; Luo, F.; Hochholdinger, F. LOB domain proteins: Beyond lateral organ boundaries. *Trends Plant Sci.* **2016**, *21*, 159–167. [CrossRef]
11. Kim, M.J.; Kim, M.; Lee, M.R.; Park, S.K.; Kim, J. Lateral organ boundaries domain (LBD) 10 interacts with sidecar pollen/LBD27 to control pollen development in Arabidopsis. *Plant J.* **2015**, *81*, 794–809. [CrossRef]
12. Berckmans, B.; Vassileva, V.; Schmid, S.P.C.; Maes, S.; Veylder, L.D. Auxin-dependent cell cycle reactivation through transcriptional regulation of Arabidopsis E2Fa by lateral organ boundary proteins. *Plant Cell* **2011**, *23*, 3671–3683. [CrossRef]
13. Evans, M.M.S. The indeterminate gametophyte1 gene of maize encodes a LOB domain protein required for embryo Sac and leaf development. *Plant Cell* **2007**, *19*, 46–62. [CrossRef]
14. Yang, Y.; Yu, X.; Wu, P. Comparison and evolution analysis of two rice subspecies lateral organ boundaries domain gene family and their evolutionary characterization from Arabidopsis. *Mol. Phylogenetics Evol.* **2006**, *39*, 248–262. [CrossRef]
15. Zhang, Y.M.; Zhang, S.Z.; Zheng, C.C. Genomewide analysis of lateral organ boundaries domain gene family in zea mays. *J. Genet.* **2014**, *93*, 79–91. [CrossRef]
16. Guo, B.J.; Wang, J.; Lin, S.; Tian, Z.; Zhou, K.; Luan, H.Y.; Lyu, C.; Zhang, X.Z.; Xu, R.G. A genome-wide analysis of the asymmetric leaves2/lateral organ boundaries(AS2/LOB) gene family in barley (*Hordeum vulgare* L.). *Biomed. Biotechnol.* **2016**, *17*, 763–774. [CrossRef]
17. Gupta, K.; Gupta, S. Molecular and in silico characterization of tomato LBD transcription factors reveals their role in fruit development and stress responses. *Plant Gene* **2021**, *27*, 100309. [CrossRef]
18. Grimplet, J.; Pimentel, D.; Agudelo-Romero, P.; Martinez-Zapater, J.M.; Fortes, A.M. The lateral organ boundaries domain gene family in grapevine: Genome-wide characterization and expression analyses during developmental processes and stress responses. *Sci. Rep.* **2017**, *7*, 15968. [CrossRef]
19. Lu, Q.; Shao, F.; Macmillan, C.; Wilson, I.W.; van der Merwe, K.; Hussey, S.G.; Myburg, A.A.; Dong, X.; Qiu, D. Genomewide analysis of the lateral organ boundaries domain gene family in Eucalyptus grandis reveals members that differentially impact secondary growth. *Plant Biotechnol. J.* **2018**, *16*, 124–136. [CrossRef]
20. Liu, H.; Cao, M.; Chen, X.; Ye, M.; Zhao, P.; Nan, Y.; Li, W.; Zhang, C.; Kong, L.; Kong, N.; et al. Genome-wide analysis of the lateral organ boundaries domain (LBD) gene family in Solanum tuberosum. *Int. J. Mol. Sci.* **2019**, *20*, 5360. [CrossRef]
21. Guo, M.; Thomas, J.; Collins, G.; Timmermans, M.C.P. Direct repression of KNOX loci by the asymmetric leaves1 complex of Arabidopsis. *Plant Cell* **2008**, *20*, 48–58. [CrossRef]
22. Bortiri, E.; Chuck, G.; Vollbrecht, E.; Rocheford, T.; Martienssen, R.; Hake, S. ramosa2 encodes a lateral organ boundary domain protein that determines the fate of stem cells in branch meristems of maize. *Plant Cell* **2006**, *18*, 574–585. [CrossRef]
23. Taramino, G.; Sauer, M.; Stauffer, J.L., Jr.; Multani, D.; Niu, X.; Sakai, H.; Hochholdinger, F. The maize (Zea mays L.) RTCS gene encodes a LOB domain protein that is a key regulator of embryonic seminal and post-embryonic shoot-borne root initiation. *Plant J.* **2007**, *50*, 649–659. [CrossRef]
24. Inukai, Y.; Sakamoto, T.; Ueguchi-Tanaka, M.; Shibata, Y.; Gomi, K.; Umemura, I.; Hasegawa, Y.; Ashikari, M.; Kitano, H.; Matsuoka, M. Crown rootless1, which is essential for crown root formation in rice, is a target of an auxin response factor in auxin signaling. *Plant Cell* **2005**, *17*, 1387–1396. [CrossRef]
25. Goh, T.; Toyokura, K.; Yamaguchi, N.; Okamoto, Y.; Uehara, T.; Kaneko, S.; Takebayashi, Y.; Kasahara, H.; Ikeyama, Y.; Okushima, Y.; et al. Lateral root initiation requires the sequential induction of transcription factors LBD16 and PUCHI in Arabidopsis thaliana. *New Phytol.* **2019**, *224*, 749–760. [CrossRef]
26. Ye, L.; Wang, X.; Lyu, M.; Siligato, R.; Eswaran, G.; Vainio, L.; Blomster, T.; Zhang, J.; Mähönen, A.P. Cytokinins initiate secondary growth in the Arabidopsis root through a set of LBD genes. *Curr. Biol.* **2021**, *31*, 3365–3373.e3367. [CrossRef]
27. Bdeir, R.; Busov, V.; Yordanov, Y.; Gailing, O. Gene dosage effects and signatures of purifying selection in lateral organ boundaries domain (LBD) genes LBD1 and LBD18. *Plant Syst. Evol.* **2016**, *302*, 433–445. [CrossRef]

28. Yordanov, Y.S.; Busov, V. Boundary genes in regulation and evolution of secondary growth. *Plant Signal Behav.* **2011**, *6*, 688–690. [CrossRef]
29. Jeon, E.; Young Kang, N.; Cho, C.; Joon Seo, P.; Chung Suh, M.; Kim, J. LBD14/ASL17 positively regulates lateral root formation and is involved in ABA response for root architecture in Arabidopsis. *Plant Cell Physiol.* **2017**, *58*, 2190–2201. [CrossRef]
30. Jeon, B.W.; Kim, J. Role of LBD14 during ABA-mediated control of root system architecture in Arabidopsis. *Plant Signal Behav.* **2018**, *13*, e1507405. [CrossRef]
31. Guo, Z.; Xu, H.; Lei, Q.; Du, J.; Li, C.; Wang, C.; Yang, Y.; Yang, Y.; Sun, X. The Arabidopsis transcription factor LBD15 mediates ABA signaling and tolerance of water-deficit stress by regulating ABI4 expression. *Plant J.* **2020**, *104*, 510–521. [CrossRef]
32. Ma, W.; Wu, F.; Sheng, P.; Wang, X.; Zhang, Z.; Zhou, K.; Zhang, H.; Hu, J.; Lin, Q.; Cheng, Z.; et al. The LBD12-1 transcription factor suppresses apical meristem size by repressing argonaute 10 expression. *Plant Physiol.* **2017**, *173*, 801–811. [CrossRef]
33. Rubin, G.; Tohge, T.; Matsuda, F.; Saito, K.; Scheible, W.R. Members of the LBD family of transcription factors repress anthocyanin synthesis and affect additional nitrogen responses in Arabidopsis. *Plant Cell* **2009**, *21*, 3567–3584. [CrossRef]
34. Albinsky, D.; Kusano, M.; Higuchi, M.; Hayashi, N.; Kobayashi, M.; Fukushima, A.; Mori, M.; Ichikawa, T.; Matsui, K.; Kuroda, H.; et al. Metabolomic screening applied to rice FOX Arabidopsis lines leads to the identification of a gene-changing nitrogen metabolism. *Mol. Plant* **2010**, *3*, 125–142. [CrossRef]
35. Majer, C.; Hochholdinger, F. Defining the boundaries: Structure and function of LOB domain proteins. *Trends Plant Sci.* **2011**, *16*, 47–52. [CrossRef]
36. Wang, P.; Song, C.P. Guard-cell signalling for hydrogen peroxide and abscisic acid. *New Phytol.* **2008**, *178*, 703–718. [CrossRef]
37. Miller, G. Reactive oxygen signaling and abiotic stress. *Physiol. Plant.* **2008**, *133*, 481–489. [CrossRef]
38. Zhang, Y.; Li, Z.; Ma, B.; Hou, Q.; Wan, X. Phylogeny and functions of LOB domain proteins in plants. *Intern. J. Mol. Sci.* **2020**, *21*, 2278. [CrossRef]
39. Zhang, X.; He, Y.; He, W.; Su, H.; Wang, Y.; Hong, G.; Xu, P. Structural and functional insights into the LBD family involved in abiotic stress and flavonoid synthases in Camellia sinensis. *Sci. Rep.* **2019**, *9*, 15651. [CrossRef]
40. Xu, J.; Hu, P.; Tao, Y.; Song, P.; Gao, H.; Guan, Y. Genome-wide identification and characterization of the Lateral Organ Boundaries Domain (LBD) gene family in polyploid wheat and related species. *PeerJ* **2021**, *9*, e11811. [CrossRef]
41. Zentella, R.; Zhang, Z.L.; Park, M.; Thomas, S.G.; Endo, A.; Murase, K.; Fleet, C.M.; Jikumaru, Y.; Nambara, E.; Kamiya, Y.; et al. Global analysis of della direct targets in early gibberellin signaling in Arabidopsis. *Plant Cell* **2007**, *19*, 3037–3057. [CrossRef]
42. Ariel, F.; Diet, A.; Verdenaud, M.; Gruber, V.; Frugier, F.; Chan, R.; Crespi, M. Environmental regulation of lateral root emergence in Medicago truncatula requires the HD-Zip I transcription factor HB1. *Plant Cell* **2010**, *22*, 2171–2183. [CrossRef]
43. Lee, H.W.; Kim, M.J.; Park, M.Y.; Han, K.H.; Kim, J. The conserved proline residue in the LOB domain of LBD18 is critical for DNA-binding and biological function. *Mol. Plant* **2013**, *6*, 1722–1725. [CrossRef]
44. Lee, H.W.; Kang, N.Y.; Pandey, S.K.; Cho, C.; Lee, S.H.; Kim, J. Dimerization in LBD16 and LBD18 transcription factors is critical for lateral root formation. *Plant Physiol.* **2017**, *174*, 301–311. [CrossRef]
45. Thatcher, L.F.; Kazan, K.; Manners, J.M. Lateral organ boundaries domain transcription factors: New roles in plant defense. *Plant Signal. Behav.* **2012**, *7*, 1702–1704. [CrossRef]
46. Semiarti, E.; Ueno, Y.; Tsukaya, H.; Iwakawa, H.; Machida, C.; Machida, Y. The ASYMMETRIC LEAVES2 gene of Arabidopsis thaliana regulates formation of a symmetric lamina, establishment of venation and repression of meristem-related homeobox genes in leaves. *Development* **2001**, *128*, 1771–1783. [CrossRef]
47. Iwakawa, H.; Iwasaki, M.; Kojima, S.; Ueno, Y.; Soma, T.; Tanaka, H.; Semiarti, E.; Machida, Y.; Machida, C. Expression of the ASYMMETRIC LEAVES2 gene in the adaxial domain of Arabidopsis leaves represses cell proliferation in this domain and is critical for the development of properly expanded leaves. *Plant J.* **2007**, *51*, 173–184. [CrossRef]
48. Huang, X.Y.; Chao, D.Y.; Gao, J.P.; Zhu, M.Z.; Shi, M.; Lin, H.X. A previously unknown zinc finger protein, DST, regulates drought and salt tolerance in rice via stomatal aperture control. *Genes Dev.* **2009**, *23*, 1805–1817. [CrossRef]
49. Wang, X.; Zhang, J.; Song, J.; Huang, M.; Cai, J.; Zhou, Q.; Dai, T.; Jiang, D. Abscisic acid and hydrogen peroxide are involved in drought priming-induced drought tolerance in wheat (*Triticum aestivum* L.). *Plant Biol.* **2020**, *22*, 1113–1122. [CrossRef]
50. Zhang, X.; Zhang, L.; Dong, F.; Gao, J.; Galbraith, D.W.; Song, C.-P. Hydrogen peroxide is involved in abscisic acid-induced stomatal closure in Vicia faba. *Plant Physiol.* **2001**, *126*, 1438–1448. [CrossRef]
51. Sergiev, I.; Alexieva, V.; Karanov, E.N.; Karanov, E.; Sergiev, L.M.; Karanova, E.; Alexieva, V. Effect of spermine, atrazine and combination between them on some endogenous protective systems and stress markers in plants. *Comptes Rendus Acad. Bulg. Sci.* **1997**, *51*, 121–124.
52. Otter, P.A. Apoplastic peroxidases and lignification in needles of Norway Spruce (*Picea abies* L.). *Plant Physiol.* **1994**, *106*, 53–60.
53. Chance, M.A. The assay of catalases and peroxidases. *Methods Biochem. Anal.* **1954**, *1*, 357–424.
54. Fridovich, B.A. Superoxide dismutase: Improved assays and an assay applicable to acrylamide gels. *Anal. Biochem.* **1971**, *44*, 276–287.

plants

MDPI

Article

Physiological and Transcriptome Analyses of Photosynthesis in Three Mulberry Cultivars within Two Propagation Methods (Cutting and Grafting) under Waterlogging Stress

Yong Li [1], Jin Huang [1], Cui Yu [1], Rongli Mo [1], Zhixian Zhu [1], Zhaoxia Dong [1], Xingming Hu [1], Chuxiong Zhuang [2,*] and Wen Deng [1,*]

1 Cash Crops Research Institute, Hubei Academy of Agricultural Sciences, Wuhan 430064, China
2 College of Life Sciences, South China Agricultural University, Guangzhou 510642, China
* Correspondence: zhuangcx@scau.edu.cn (C.Z.); dengwen@hbaas.com (W.D.)

Abstract: Mulberry is a valuable woody plant with significant economic importance. It can be propagated through two main methods: cutting and grafting. Waterlogging can have a major impact on mulberry growth and can significantly reduce production. In this study, we examined gene expression patterns and photosynthetic responses in three waterlogged mulberry cultivars propagated through cutting and grafting. Compared to the control group, waterlogging treatments reduced levels of chlorophyll, soluble protein, soluble sugars, proline, and malondialdehyde (MDA). Additionally, the treatments significantly decreased the activities of ascorbate peroxidase (APX), peroxidase (POD), and catalase (CAT) in all three cultivars, except for superoxide dismutase (SOD). Waterlogging treatments also affected the rate of photosynthesis (Pn), stomatal conductance (Gs), and transpiration rate (Tr) in all three cultivars. However, no significant difference in physiological response was observed between the cutting and grafting groups. Gene expression patterns in the mulberry changed dramatically after waterlogging stress and varied between the two propagation methods. A total of 10,394 genes showed significant changes in expression levels, with the number of differentially expressed genes (DEGs) varying between comparison groups. GO and KEGG analysis revealed important DEGs, including photosynthesis-related genes that were significantly downregulated after waterlogging treatment. Notably, these genes were upregulated at day 10 in the cutting group compared to the grafting group. In particular, genes involved in carbon fixation were significantly upregulated in the cutting group. Finally, cutting propagation methods displayed better recovery capacity from waterlogging stress than grafting. This study provides valuable information for improving mulberry genetics in breeding programs.

Keywords: mulberry; waterlogging; photosynthesis; gene regulation

Citation: Li, Y.; Huang, J.; Yu, C.; Mo, R.; Zhu, Z.; Dong, Z.; Hu, X.; Zhuang, C.; Deng, W. Physiological and Transcriptome Analyses of Photosynthesis in Three Mulberry Cultivars within Two Propagation Methods (Cutting and Grafting) under Waterlogging Stress. *Plants* **2023**, *12*, 2066. https://doi.org/10.3390/plants12112066

Academic Editors: Małgorzata Nykiel, Mateusz Labudda, Beata Prabucka, Marta Gietler and Justyna Fidler

Received: 11 April 2023
Revised: 10 May 2023
Accepted: 19 May 2023
Published: 23 May 2023

1. Introduction

Waterlogging stress is a major hindrance to plant growth and can result in significant yield losses in many plants [1–3]. Waterlogging creates hypoxic conditions due to the slow diffusion of molecular oxygen in water, leading to various morphological and cellular acclimation responses [4,5]. Studies on hypoxia in crops have shown that it causes rapid changes in gene expression and cellular metabolism [6,7]. While hypoxia affects energy metabolism and root zone hypoxia is a key component of waterlogging stress, field observations have also shown reduced growth rates, photosynthesis rates, and stomatal conductivity in waterlogged plants [8]. It has been established that waterlogging stress is more complex than mere altered energy metabolism in plants. In response to waterlogging stress, plants exhibit both short-term and long-term adaptations. In the short-term, plants alter their physiological processes by reducing stomatal conductance and net photosynthetic rate [9]. In the long-term, plants regulate the expression of hundreds of relevant genes, consequently modifying individual phenotypes and some morphological and anatomical features [10–12]. One adaptation strategy is to change structural arrangement of stem cells and cell organelles [13,14]. The hypertrophy of lenticels and formation of aerenchyma facilitate plant oxygenation and contribute to maintaining root aerobic respiration and water uptake under waterlogging conditions [15]. Waterlogging-adapted species are characterized by higher density of xylem vessels, larger size of vessels, and a greater number of secondary meristem cells compared with waterlogging-sensitive species [14].

As photosynthesis is affected by waterlogging stress, which in turn affects plant production, it is crucial to understand the physiological and genetic regulation of waterlogged plants [16–19]. In plants, genetically regulated hormones such as ethylene, abscisic acid, and gibberellic acid play important roles in responding to environmental stress [20,21]. The transcriptional and physiological interplay in response to waterlogging stress may provide key insights into plant waterlogging tolerance [22,23]. However, the level and degree of tolerance may vary by species or abiotic pressures and requires further study. Investigating regulatory pathways and interactions in crops such as wheat, rice, soybeans, and mulberry will provide valuable resources for enhancing economically and agriculturally important traits in waterlogged areas.

Mulberry (*Morus* spp.) is a perennial woody plant with significant economic importance. Its nutritious fruits are widely used in pharmaceutical and traditional Chinese medicine [24,25]. Notably, over 60% of the total cost of cocoon production is spent on mulberry production alone [26]. The primary goal of improving mulberry germplasm is to develop new cultivars with high leaf yield, fruit quality, pest resistance, and tolerance to various abiotic stresses such as drought, waterlogging, and salt. In agricultural practice, grafting and cutting are two important propagation methods for mulberry [27,28]. However, little information is available on the waterlogging tolerance of mulberry propagated through these methods. While physiological responses at the cellular level have been studied in many crops [24,29], the physiological and genetic regulation of photosynthesis in response to waterlogging stress in mulberry remains unclear, particularly across different cultivars. In order to investigate the differences in waterlogging tolerance and response mechanisms among different mulberry cultivars and between different propagation methods (grafting and cutting) under waterlogging stress, and to provide scientific recommendations for mulberry breeding, we conducted this study. In this study, we examined the photosynthetic responses to waterlogging stress in three widely grown mulberry fruit cultivars (AY, SG, and ZZ) from the Yangtze River basin and used RNA-seq to analyze gene expression patterns under waterlogging stress. We also compared the waterlogging stress tolerance of mulberry propagated through cutting and grafting methods to identify the ideal propagation method for waterlogging tolerance. This study provides valuable information for improving mulberry varieties in waterlogged areas.

2. Results

2.1. Comparisons of Osmotic Regulatory Substances between Cutting and Grafting Groups in Three Cultivars

After waterlogging treatments, the levels of chlorophyll, soluble protein, soluble sugars, proline, and MDA in the waterlogged groups were significantly lower than those in the control groups in all three mulberry cultivars (Figure 1). Overall, levels of soluble protein, soluble sugar, proline, and MDA gradually decreased over 20 days. Notably, waterlogging stress reduced the levels of these osmotic regulatory substances. Levels of osmotically-regulating substances also differed significantly between the cutting and grafting groups in the ZZ and SG cultivars. In the AY cultivar, significant differences were observed between the cutting and grafting groups for MDA, proline, and chlorophyll levels, suggesting that different propagation methods may affect waterlogging stress tolerance through regulation of osmotic substances. Similar trends were observed in the other two cultivars.

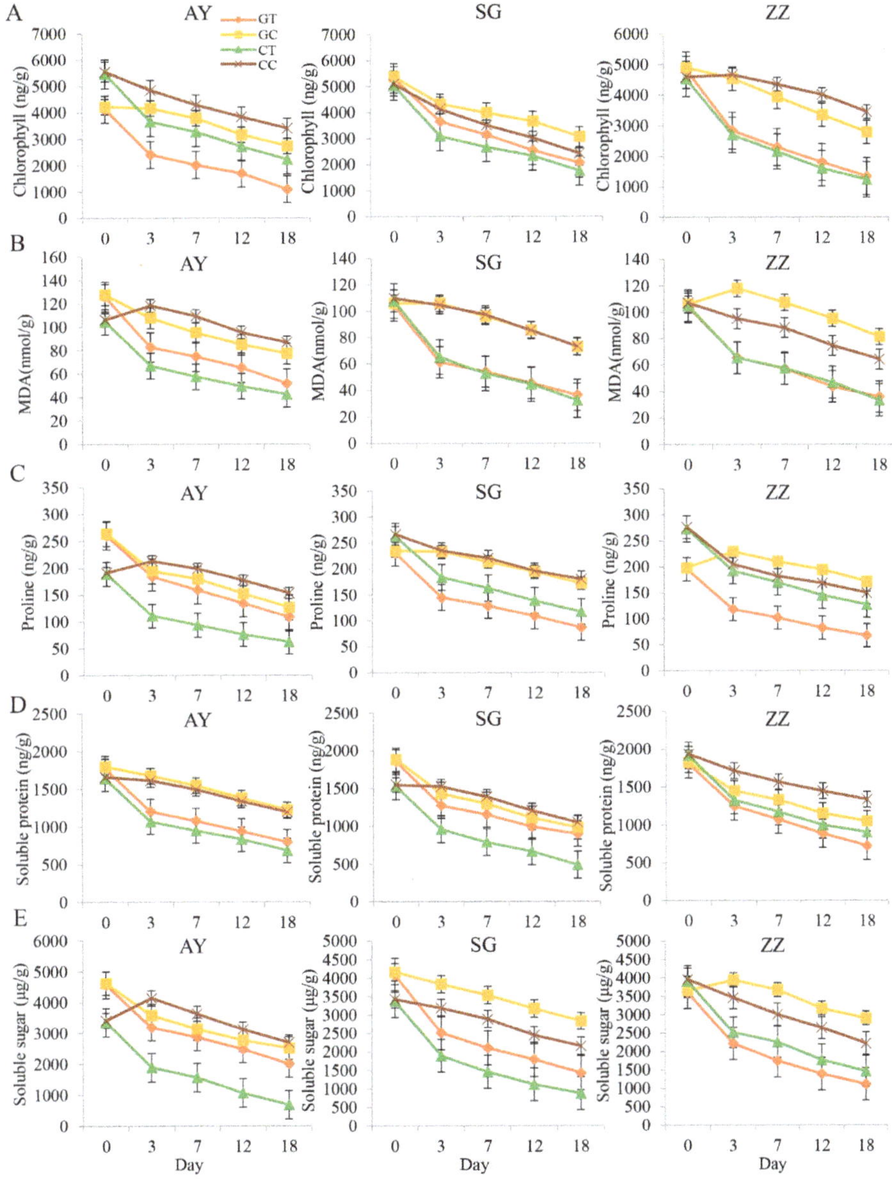

Figure 1. The contents of osmotic regulation substances in three cultivars between waterlogging and control groups, including chlorophyll (**A**), MDA (**B**), proline (**C**), soluble protein (**D**), and soluble sugar (**E**) in three mulberry cultivars (AY, SG and ZZ) between waterlogging and control groups. CC, CT, GC, and GT in the legend indicate cut mulberry under control and waterlogging treatment, and grafted mulberry under control and waterlogging treatment, respectively.

2.2. Dynamics of Enzyme Activities after Waterlogging Treatments

We found that waterlogging treatments significantly reduced APX, POD, and CAT activities (Figure 2). Interestingly, SOD activities in the waterlogged treatment groups were significantly higher than those in the control groups in all three cultivars. SOD activities also gradually increased within 20 days after waterlogging treatments. No significant difference in enzyme activities was observed between the cutting and grafting groups in any of the three mulberry cultivars. A smaller difference between propagation methods was observed for the AY cultivar compared to the other two cultivars (SG and ZZ).

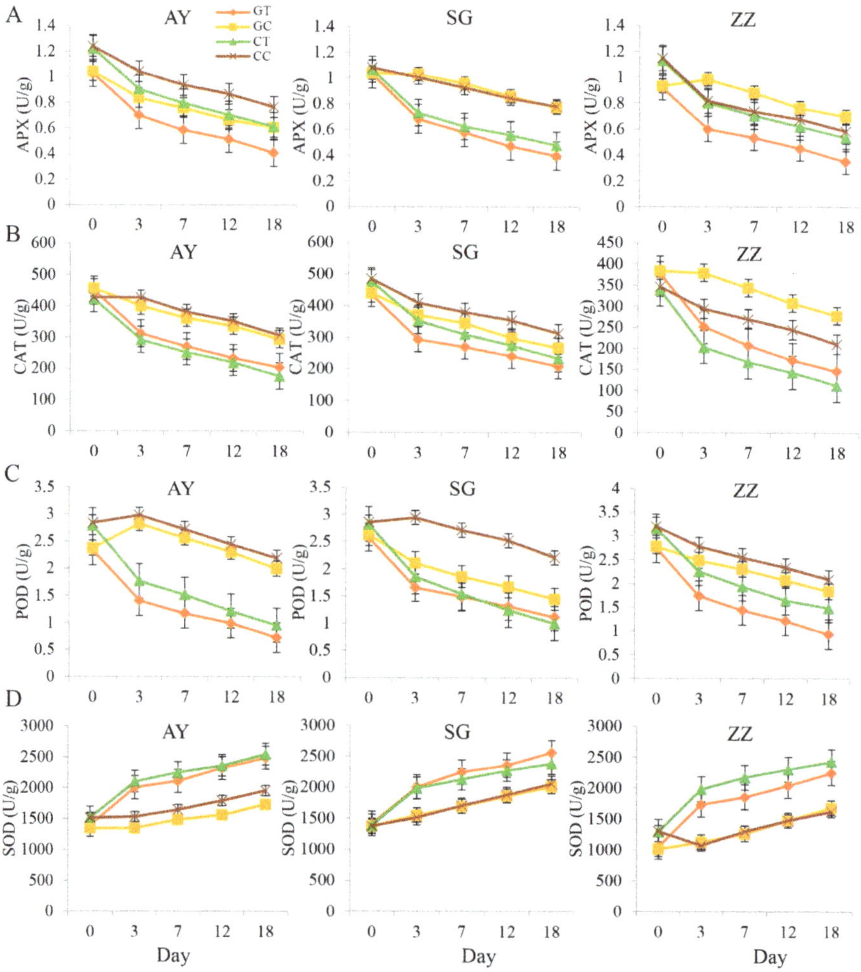

Figure 2. The changes of enzyme activities in three cultivars after waterlogging treatments, including APX (**A**), CAT (**B**), POD (**C**), and SOD (**D**) in three mulberry cultivars (AY, SG and ZZ) between waterlogging and control groups. CC, CT, GC, and GT in the legend indicate cut mulberry under control and waterlogging treatment, and grafted mulberry under control and waterlogging treatment, respectively.

2.3. Dynamic Changes of Photosynthetic Characters in Three Mulberry Cultivars after Waterlogging Treatment

The rates of photosynthesis (Pn) for the three mulberry cultivars from 6:00 a.m. to 6:00 p.m. are shown in Figure 3. In all three cultivars, the Pn values for the waterlogged treatment groups were significantly lower than those for the control groups, indicating that waterlogging stress can impair photosynthesis in mulberry (Figure 3A). In the control groups for the AY cultivar, the grafted groups showed significantly higher Pn values than the cutting groups, suggesting that the cutting method may cause physiological damage to this cultivar (Figure 3A). Exposure to waterlogging resulted in a decrease in Pn, and no apparent peak was observed. No difference was observed between the cutting and grafting methods in the waterlogged treatment groups. The dynamics of stomatal conductance (Gs) showed that waterlogging treatment affected stomatal conductivity in all three mulberry cultivars (Figure 3B). In the AY cultivar, the grafted groups showed higher Gs values than the cutting groups, but a completely different trend was observed for the ZZ cultivar (Figure 3B). Interestingly, waterlogging stress appeared to decrease intercellular CO2 concentration (Ci) in the early stages, but no significant difference was observed after 10:00 a.m. (Figure 3C). The transpiration rate (Tr) curves for all three cultivars showed that waterlogging produced a strong effect on leaf transpiration (Figure 3D). The grafting and cutting methods also affected the transpiration rate in the AY and SG cultivars under waterlogging stress, with grafted groups showing higher Tr values in these two cultivars.

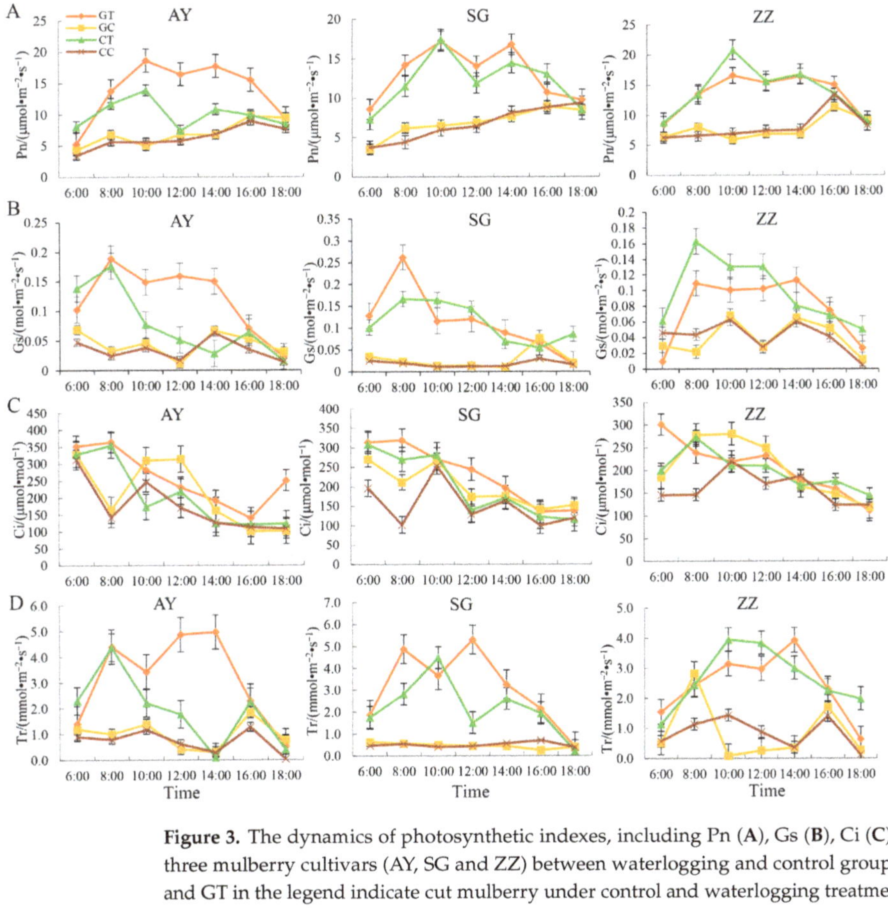

Figure 3. The dynamics of photosynthetic indexes, including Pn (**A**), Gs (**B**), Ci (**C**), and Tr (**D**) in three mulberry cultivars (AY, SG and ZZ) between waterlogging and control groups. CC, CT, GC, and GT in the legend indicate cut mulberry under control and waterlogging treatment, and grafted mulberry under control and waterlogging treatment, respectively.

To further understand the dynamics of photosynthetic properties in the three mulberry cultivars after waterlogging, we measured the indices of Pn, Gs, Ci, and Tr over 20 days following waterlogging treatments. The curves for these indices are shown in Figure S1. Overall, Pn gradually increased as the mulberry plants recovered from waterlogging stress, although there was a slight decrease during the first three days. Waterlogging treatments lowered Pn in all three cultivars, and no significant difference was observed between the grafting and cutting groups (Figure S3A). Similar results were obtained for Gs and Tr

(Figure S3B,D). Interestingly, the Ci curves showed that waterlogging stress produced only a minor effect on Ci in mulberry and on the propagation methods (Figure S3C).

2.4. The Number of DEGs in Cut and Grafted Mulberry Responding to Waterlogging Stress

To understand the gene expression patterns and molecular basis of waterlogging tolerance in mulberry propagated through cutting and grafting, we performed RNA-seq on 36 samples from both propagation methods at 0-day recovery (D0), 3-day recovery (D3), and 10-day recovery (D10) after waterlogging. After quality control of the raw reads, a total of 233.37 Gb of clean reads were mapped to the reference genome with mapping ratios ranging from 71.37% to 76.99%. The FPKM values of the genes were calculated, and the pairwise Pearson correlation coefficients between the three biological replicates were above 0.9 (Figure S2A). Principal component analysis (PCA) clearly showed clustering patterns for the three biological replicates (Figure S2B), indicating high variability between biological replicates. All samples were divided into two groups according to control and waterlogging (Figure S2B).

To identify differentially expressed genes (DEGs) between waterlogged and control groups and between cutting and grafting groups, we performed nine pairwise comparisons. The total number of up- and down-regulated genes varied between comparisons (Figure S3). For both cutting and grafting groups, the number of DEGs was over 4000 at D0, then quickly decreased at D3 before increasing again at D10. The number of DEGs between cutting and grafting groups at D0, D3, and D10 under waterlogging treatment was small and did not change significantly (Figure 4 and Figure S3). At D0, D3, and D10, 202 common DEGs were identified between grafted and cut mulberry after waterlogging stress (Figure 4A), while in the control group, 741 and 280 common DEGs were identified after waterlogging in grafted (Figure 4B) and cut (Figure 4C) mulberry, respectively. The largest number of unique DEGs from cut and grafted mulberry was observed at D0 with 1103 and 594 DEGs, respectively (Figure 4D). To understand the dynamic changes in mulberry, we performed another eight pairwise comparisons and found that the number of DEGs varied between D0 vs. D3 and D3 vs. D10 under the same conditions (Figures S3 and S4). A total of 108 and 83 common DEGs were identified in cut and grafted mulberry under waterlogging and control conditions, respectively (Figure S4A,B). Under waterlogging stress, we identified 164 and 76 DEGs in cut and grafted mulberry, respectively, while under control conditions only 42 and 31 DEGs were identified in cut and grafted mulberry, respectively (Figure S4C). In total, we identified 10,394 common DEGs in all 17 pairwise comparisons. These results indicate that waterlogging stress produces a major impact on mulberry growth and that cutting, grafting, and waterlogging all continuously affect the genetic processes in mulberry. Furthermore, cutting and grafting can enhance the waterlogging tolerance of mulberry.

2.5. Dynamics of Photosynthesis-Related Gene Expression in Mulberry with Waterlogging Stress

We performed GO enrichment analysis on the DEGs and found that several GO terms, including thylakoid, photosynthesis, chloroplast, photosystem, photosynthetic membrane, and chloroplast thylakoid, were significantly enriched in the pairwise comparisons (Table S1). Several key metabolic pathways, including photosynthesis, carbon metabolism, sesquiterpenoid and triterpenoid biosynthesis, carbon fixation in photosynthetic organisms, and porphyrin and chlorophyll metabolism were also significantly enriched for the DEGs (Table S1).

We selected DEGs identified between the waterlogged treatment and control groups to clarify mulberry's genetic response to waterlogging stress. Most DEGs were related to energy metabolism, carbohydrate metabolism, and lipid metabolism (Figure 5A). Most importantly, waterlogging stress directly or indirectly caused the downregulation of key genes related to photosynthesis, such as photosystem II genes (*PsbA*, *PsbB*, *PsbC*, etc.), photosystem I genes (*PsaA*, *PsaB*, *PsaD*, etc.), cytochrome b6/f complex genes (*PetB*, *PetD*, *PetA*, *PetC*, etc.), and photosynthetic electron transport genes (*PetE*, *PetF*, *PetJ*). These DEGs were detected at time points D0, D3, and D10 (Figure 5B). The gene expression patterns showed that most photosynthesis-related genes were downregulated after waterlogging treatment, particularly at D0 (Figure 5C). As the plants recovered from waterlogging stress at D3 and D10, some photosynthesis-related genes were expressed at levels similar to those in the control groups (Figure 5D,E).

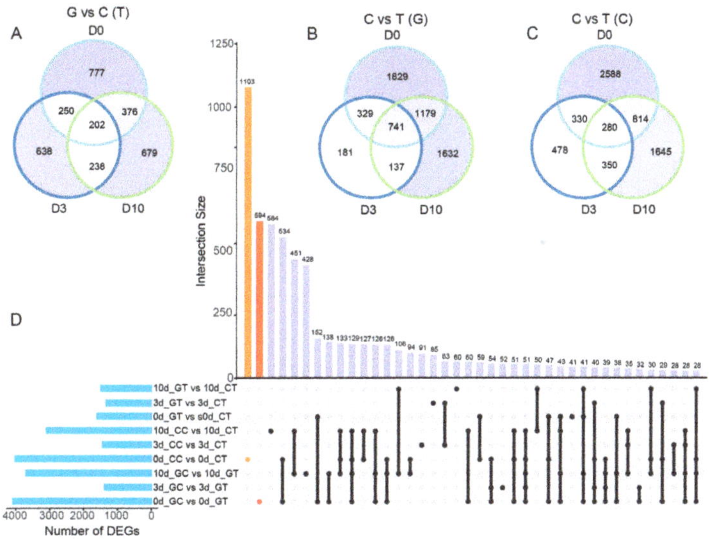

Figure 4. Venn diagram and UpSet indicated the number of differentially expressed genes (DEGs) per contrast: divided into three groups including grafting and cutting under waterlogging treatments (**A**); control and waterlogging treatments in grafting (**B**) and cutting (**C**); upset summary for nine pairwise comparisons (**D**). Numbers in intersections represent the number of DEGs shared in the intersection contrasts.

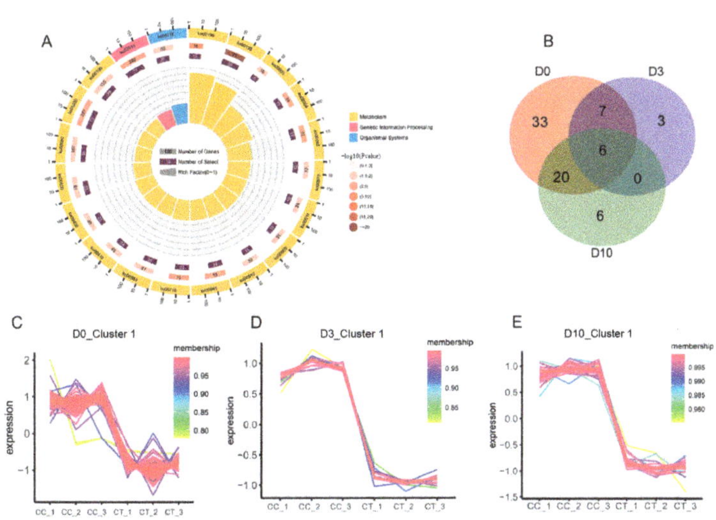

Figure 5. The KEGG enrichment analysis for DEGs between waterlogging and control group (**A**); Venn diagram of the photosynthesis-related DEGs from three comparisons (**B**); gene expression trend of the photosynthesis-related genes in the waterlogging treatments and control group in D0 (**C**), D3 (**D**), and D10 (**E**), respectively.

2.6. Propagation Methods Affect Photosynthesis-Related Gene Expression Responding to Waterlogging Stress

To focus on differences in gene regulation between grafted and cut groups, we selected DEGs identified between these groups. According to KEGG pathway enrichment analysis, several key pathways were identified at both D3 and D10 time points, including flavonoid biosynthesis, circadian rhythm, and cutin, suberin, and wax biosynthesis. For example, chalcone synthase genes (*CHS1*, *CHS2*, *CHS3*), *HD3A*, and *HD3B* showed higher expression levels in the cutting groups. Additionally, genes enriched in the MAPK pathway (*MPK3*, *MPK14*, *MKK9*, *CML45*) were upregulated in the cutting groups at the D3 time point (Figure 6A). Interestingly, the photosynthetic pathway was enriched with DEGs at the D10 time point but not at the D3 time point (Figure 6B). All photosynthesis-related genes were found to be upregulated in the cutting groups compared to the grafted groups, suggesting that the cutting groups had better photosynthetic recovery ability after waterlogging treatments. Similarly, genes associated with carbon fixation were also upregulated in the cutting groups compared to the grafted groups.

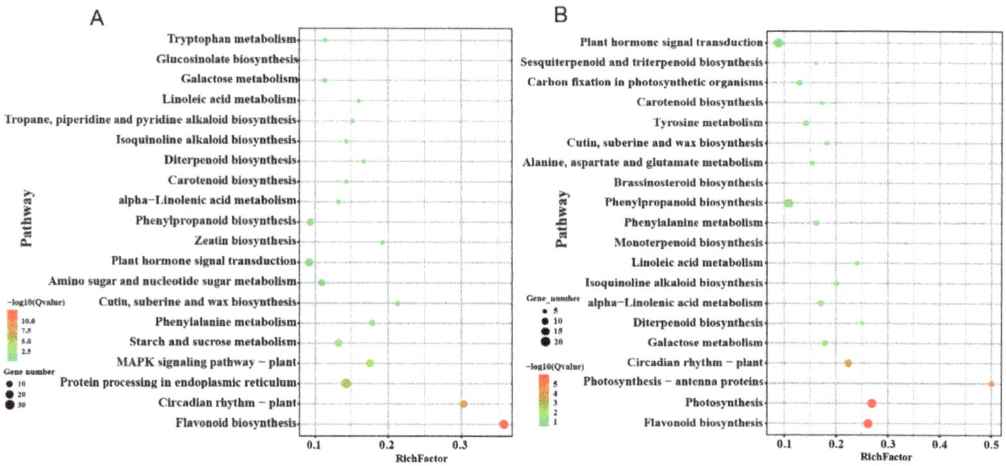

Figure 6. Top 20 enriched KEGG pathways between grafted and cutting groups under waterlogging stress at day 3 (**A**) and day 10 (**B**). The number of differentially expressed genes (DEGs) in each pathway is counted, and the enrichment factors and *p*-values are displayed.

2.7. Validation of Photosynthesis-Related Genes Expression with qPCR Method

We selected ten photosynthesis-related genes from the DEGs for qPCR analysis. Compared to the control, our result showed that all genes were down-regulated after the waterlogging stress except for *FEDA*, which was consistent with the RNA-seq data (Figure 7). *Lhca* and *Lhcb* belong to light-harvesting chlorophyll a/b binding (LHC) superfamily, which plays critical roles in photosynthesis [30]. The expression change of *Lhcb* gene family normally influenced plant growth and development under abiotic stress [31]. The down-regulation of *AtLhcb4*, *AtLhcb5*, and *AtLhcb6* expression leads to the accumulation of higher levels of superoxide and more severe oxidative stress, which further disrupts photoprotection [32,33]. Overexpression of the tomato *Lhcb2* gene in tobacco alleviated photo-oxidation of PSII and enhanced tobacco tolerance to chilling stress [34]. We hypothesize that down-regulation of these genes also may affect the photo-oxidation in photosynthesis to alter waterlogging tolerance in mulberry. *FEDA* was a ferredoxin and reduced expression of the *FEDA* gene in *Arabidopsis* leaves destroyed by redox-regulated adaptations in the photosynthetic system [35]. *FEDA* was extremely up-regulated after waterlogging recovery at D3 in our study (Figure 7). Therefore, we hypothesize that the *FEDA* gene may regulate the oxidative adaptation of the mulberry photosystem by increasing its expression in response to waterlogging stress.

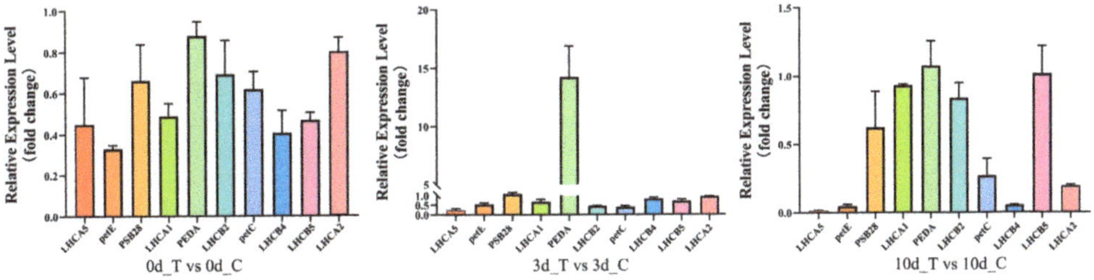

Figure 7. Validation of RNA-seq data for the expression of 10 photosynthesis-related genes using the qPCR method in cutting-propagated mulberry cultivars under waterlogging and control conditions at 0, 3, and 10 days.

3. Discussion

Waterlogging is an important water-related stress that can damage plants by rapidly reducing the rate of photosynthesis and stomatal conductivity [36]. Photosynthesis is highly sensitive to water stress, and limitations in photosynthetic carbon metabolism have been analyzed in crops [37–39]. Photosynthetic cells are highly sensitive to oxidative stress and have robust antioxidant systems to protect plants from oxidative damage [36]. At the genetic level, the negative effects of waterlogging stress on leaves may result from oxidative damage to important molecules due to an imbalance between the production of activated oxygen and its metabolism in plants [40,41]. Previous studies have documented the effects of water-related stress on various crops, including mulberry [28,42,43]. However, the mechanisms by which waterlogging stress inhibits photosynthesis in mulberry have not been extensively investigated. A better understanding of the mechanisms that allow plants to adapt to waterlogging stress and maintain growth and productivity during periods of waterlogging will ultimately aid in the selection of waterlogging-tolerant cultivars. Efficient approaches to identifying waterlogging-resistant genotypes and understanding the key periods during which plants can tolerate waterlogging has been a key goal for plant researchers.

3.1. *Dynamic Changes in Physiological Indices of Mulberry under Waterlogging Stress*

Under waterlogging conditions, plants produce antioxidants, flavonoids, and secondary metabolites that play a role in protecting the plant by detoxifying reactive oxygen species (ROS) and stabilizing proteins and amino acids. These compounds help the plant to cope with abnormal conditions caused by waterlogging [44]. Superoxide dismutase (SOD), peroxidase (POD), catalase (CAT), and ascorbate peroxidase (APX) are key enzymes involved in the detoxification of reactive oxygen species (ROS) in plants. These enzymes work together to remove toxic oxygen species and protect the plant from oxidative damage under stress conditions. SOD catalyzes the conversion of superoxide radicals into hydrogen peroxide and oxygen, while CAT, POD, and APX work to remove hydrogen peroxide by converting it into water and oxygen. This coordinated action helps to maintain the redox balance within the cell and prevent oxidative damage to cellular components such as lipids, proteins, and nucleic acids. The activity of these enzymes can be modulated by various factors, including changes in environmental conditions, developmental stages, and the presence of other stressors. Understanding the regulation of these enzymes and their role in plant stress responses can provide valuable insights into the mechanisms underlying plant adaptation to changing environments.

In this study, we found that after waterlogging treatment in mulberry, the activities of APX, CAT, and POD decreased, except for SOD. Additionally, chlorophyll content was also reduced by waterlogging (Figure 2). Chlorophyll plays a crucial role in capturing light energy and converting it into chemical energy through photosynthesis. As such, a reduction in chlorophyll content can directly lead to a decrease in the rate of photosynthesis in plants [45]. Furthermore, waterlogging resulted in significant decreases in leaf malondialdehyde, soluble protein, soluble sugars, and proline production in mulberry. Additionally, there was a reduction in leaf photosynthetic rate, stomata conductance, and transpiration rate. These results are largely consistent with previous studies [46,47] that reported a reduction in leaf protective enzymes and photosynthetic rate when plants were subjected to waterlogging stress. Notably, we found that unlike other protective enzymes, SOD activities gradually increased after waterlogging treatment. This finding suggests that SOD plays a crucial role in responding to active oxygen activities caused by waterlogging in mulberry. It is possible that the increase in SOD activity stimulates the cellular protective mechanism to mitigate damage. Meanwhile, we found that the early reduction in photosynthetic efficiency after waterlogging treatment in mulberry could last for 7 days and recover after that time.

The observed decrease of the content of these enzymes may indicate an alteration in the plant's ability to cope with oxidative stress caused by waterlogging. Previous studies have shown that waterlogging can lead to an increase in ROS production, due to changes in cellular metabolism and energy production [48]. Under normal conditions, plants have a range of antioxidant defense mechanisms to cope with ROS, including the enzymes APX, POD, and CAT [49]. However, under stress conditions such as waterlogging, these defense mechanisms may be overwhelmed, leading to oxidative damage and changes in enzyme activity [50]. In addition to changes in ROS production and antioxidant defense mechanisms, waterlogging can also affect other aspects of plant physiology. For example, waterlogging can lead to changes in nutrient uptake and transport, as well as alterations in hormone signaling and gene expression [51]. These changes can affect plant growth and development, as well as their ability to cope with stress.

In conclusion, our results suggest that waterlogging can alter the antioxidant defense mechanisms of mulberry trees, as indicated by the decrease in the content of APX, POD, and CAT. Further studies are needed to elucidate the mechanisms underlying this response and to determine its impact on plant growth and productivity.

3.2. Cutting Propagation Methods Displayed Better Recovery Capacity from Waterlogging Stress than Grafting

Grafting and cutting are widely used methods of plant propagation and play a crucial role in improving the yield and quality of crop trees and vegetables. Grafting can typically alter several physiological and biochemical reactions between the rootstock and scion, affecting the growth, development, and resilience of plants [52,53]. Previous studies have reported that grafting improves photosynthetic capacity [54] and antioxidant system activity of crops under salt stress [55].

However, in this study, we found that the mulberry grafting method showed no advantage in reducing the inhibitory effect of waterlogging stress on photosynthesis compared to the cutting method (Figure 3). Waterlogging stress occurs when the soil becomes saturated with water, depriving the roots of oxygen and making it difficult for the plant to take up nutrients. This condition can inhibit photosynthesis and reduce plant growth. It seems that in this case, the cutting method was more effective in helping mulberry trees recover from waterlogging stress.

At the genetic level, photosynthesis in the early phase after waterlogging treatment showed no difference between grafting and cutting groups. However, after 10 days of waterlogging treatment, photosynthesis-related genes were up-regulated in the cutting groups (Figure 4). Photosynthesis, carbon metabolism, sesquiterpenoid and triterpenoid biosynthesis, carbon fixation in photosynthetic organisms, and porphyrin and chlorophyll metabolism were significantly enriched for the DEGs under waterlogging stress (Table S1).

Waterlogging stress directly or indirectly caused the downregulation of key genes related to photosynthesis, such as photosystem II genes (*PsbA*, *PsbB*, *PsbC*, etc.), photosystem

I genes (*PsaA*, *PsaB*, *PsaD*, etc.), cytochrome b6/f complex genes (*PetB*, *PetD*, *PetA*, *PetC*, etc.), and photosynthetic electron transport genes (*PetE*, *PetF*, *PetJ*) (Figure 5).

Interestingly, the cutting groups showed higher expression levels of chalcone synthase genes (*CHS1*, *CHS2*, *CHS3*), *HD3A*, *HD3B* and the MAPK pathway (*MPK3*, *MPK14*, *MKK9*, *CML45*) than the grafting group (Figure 6). Chalcone synthase (CHS) is a crucial rate-limiting enzyme in the flavonoid biosynthetic pathway that plays an important role in regulating plant growth, development, and abiotic stress tolerance [56]. The MAPK pathway is also involved in stress responses in plants. It is a signaling pathway that can be activated by various stimuli, including abiotic stresses such as drought, cold, and salt stress. Activation of the MAPK pathway can lead to changes in gene expression that help the plant cope with the stress. Abiotic stress can cause damage to plants at the cellular level and disrupt their normal growth and development. In response to stress, plants activate various defense mechanisms to protect themselves. One such mechanism is the production of secondary metabolites such as flavonoids, which can help protect the plant against stress-induced damage [57,58]. The increased expression of chalcone synthase genes under stress may be part of this defense mechanism. The *HD3A* and *HD3B* genes are involved in the regulation of flowering time in plants [59]. Abiotic stress can affect the timing of flowering and disrupt the normal reproductive cycle of plants. The increased expression of *HD3A* and *HD3B* genes under stress may be part of the plant's response to ensure proper flowering and reproduction despite the stressful conditions.

In addition to the fact that the expression of some key genes in the cutting group was higher than the grafting group, we also found that cutting groups had better recovery capacity from waterlogging stress than grafted mulberry. It should be noted that the genetic trait of different rootstocks leading to varying waterlogging tolerance may result in different recovery abilities of grafted mulberry trees in response to waterlogging or other environmental stresses, and the more combinations should be investigated in further studies.

4. Materials and Methods

4.1. Plants and Sample Preparation

The experiment was conducted at the Industrial Crops Institute of the Hubei Academy of Agricultural Sciences. In this study, three representative mulberry fruit cultivars (AY, SG, and ZZ) were selected for propagation by grafting and cutting. Grafting and cutting trials were carried out in March 2019 using two-year-old healthy mulberry plants. After grafting and cutting, the mulberry seedlings were planted in freshly prepared pots ($28 \times 20 \times 20$ cm) containing well-dried, pulverized garden soil, decomposed sand, and well-rotted manure in a ratio of 5:3:2. The seedlings were cared for consistently. The experiment was carried out with 10 replicates per variety.

Waterlogging treatments were carried out in June 2019. Before the treatments, each pot was placed in a 150 cm diameter trough. Waterlogging treatments were performed by filling the outer tank with water up to 5 cm above the sand surface. The control groups were watered normally. The waterlogging treatment was carried out with 10 repetitions. Mulberry samples were collected in the morning after 0, 3, 7, 12, and 18 days.

4.2. Measurements of Osmotic Regulation Substances and Chlorophyll Content

To measure the dynamics of osmotically regulating substances in mulberry, we analyzed the levels of soluble protein, sugar, proline, and malondialdehyde (MDA) in mulberry leaves.

The soluble protein content was measured using the Coomassie brilliant blue method. The amount of 0.1 mL of sample extract was taken into a test tube and 5 mL of G250 reagent was added. After mixing well for 2 min, distilled water was used as the blank, and absorbance was measured at 595 nm. The absorbance value was recorded, and the protein concentration was obtained through the standard curve. The soluble protein content (ng/g) was further calculated.

The soluble sugar content was measured using the anthrone colorimetric method. The amount of 0.5 mL of extract was taken into a 10 mL test tube, with distilled water as the control. Subsequently, 0.5 mL of distilled water and 5 mL of anthrone reagent were added and mixed well. Absorbance was then measured at a wavelength of 620 nm. The optical density value was recorded, and the corresponding sugar content (μg/g) was obtained through the standard curve.

Proline content was measured using the sulfosalicylic acid method. Next, 2 mL of supernatant was taken, and 2 mL of ice acetic acid plus 3 mL of coloring solution were added. The mixture was boiled in a water bath for 40 min and cooled. The amount of 5 mL of toluene was then added to the solution, which was shaken well. After standing for layering, the toluene layer of absorbance was measured at 520 nm, with toluene as the zero adjustment to calculate the proline content (ng/g).

Malondialdehyde (MDA) content was measured using the thiobarbituric acid method. The absorbance of the supernatant extract at 450 nm, 532 nm, and 600 nm was measured. The MDA content in the tissue (nmol/g) was then calculated.

Chlorophyll content was measured using the acetone extraction method. Fresh mulberry leaves in the amount of 0.5 g were weighed and placed in a mortar. Some quartz sand and 80% acetone were added and ground into a homogenate. The homogenate was filtered into a volumetric flask. The mortar, glass rod, funnel, and filter paper were rinsed with 80% acetone until all the chlorophyll was rinsed into the volumetric flask and finally diluted to 10 mL. The mixture was mixed well, and the absorbance was measured at 645 nm and 663 nm to calculate the chlorophyll content (ng/g).

4.3. Measurements of Enzyme Activities

Fresh leaves with three replicates were collected from the upper shoots to measure the activities of four enzymes (SOD, CAT, POD, APX).

Superoxide dismutase (SOD) activity was measured using the nitroblue tetrazolium (NBT) method. The amount of 0.1 mL of crude enzyme extract was taken for enzyme activity measurement. Phosphate buffer solution replaced enzyme solution as the control, with one light control and one dark control. Each reaction solution was added to a 10 mL test tube to a final volume of 3.3 mL. The test tube and control test tube were placed under a 4000 lx daylight lamp at a temperature of 25–35 °C for 20–30 min. The dark control reacted for the same length of time in the dark. The reaction was immediately terminated by covering the test tube after it was over. During measurement, the solution was diluted appropriately, and the dark control was used as the blank. The absorbance value was measured with a UV-visible spectrophotometer at a wavelength of 550 nm and recorded to calculate SOD enzyme activity (U/g).

Catalase (CAT) activity was measured using the UV absorption method. Phosphate buffer in the amount of 0.1 mL of 2% H_2O_2 and 2 mL was taken into a 1 cm quartz cuvette, and 0.1 mL of crude enzyme extract was added. The mixture was mixed well at room temperature and immediately measured for changes in optical density within 5 min using a UV-visible spectrophotometer at a wavelength of 240 nm until the optical density reduction per minute reached stability. CAT activity (U/g) was then calculated.

Peroxidase (POD) activity was measured using the guaiacol method [60]. Enzyme activity was measured with 0.1 mL of crude enzyme extract. According to the reaction system, 2.9 mL of 0.05 mol/L phosphate buffer solution, 0.5 mL of 2% H_2O_2, 0.1 mL of 2% guaiacol solution, and 0.1 mL of enzyme solution were added respectively. The mixture was mixed well at room temperature, and absorbance was measured at 420 nm until the optical density reduction per minute reached stability. POD activity (U/g) was then calculated.

Ascorbic acid peroxidase (APX) activity was measured using an Elisa kit. The absorbance was measured at a wavelength of 290 nm using an enzyme-linked immunosorbent assay instrument. The plant ascorbic acid peroxidase activity (U/g) in the sample was calculated through the standard curve.

4.4. Measurements of Photosynthetic Characters

The net photosynthesis rate (Pn), stroma conductance (Gs), intercellular CO_2 concentration (Ci), and transpiration rate (Tr) of mulberry leaves were measured simultaneously from 6:00 a.m. to 6:00 p.m. using a LI-6400XT portable photosynthetic analyzer manufactured by LI-COR (Lincoln, NE, USA). The measurements were carried out with three repetitions. The diurnal fluctuations in photosynthesis were measured from 12 May to 13 May. Leaves were selected from three well-lit top shoots of mulberry trees. A leaf with normal function was selected from each shoot. Measurements were taken once every 2 h and repeated three times each time.

4.5. Statistical Analysis

Means and standard deviation of replicates were calculated using Microsoft (Redmond, WA, USA) Office Excel 2010. Statistical analysis was performed using SPSS19.0 software (SPSS, Inc., Chicago, IL, USA) with p-value < 0.05 as significant difference in this study.

4.6. RNA-Seq of Mulberry Leaves

After waterlogging treatment, mulberry leaves from two groups (cutting and grafting) were collected at 0, 3, and 10 days, with corresponding controls set for each time point. Total RNA was extracted using RNAiso (TaKaRa, Dalian, China). Three biological replicates were performed for RNA-seq. The concentration and quality were checked with NanoDrop 2000 (Life Technology, Waltham, MA, USA) and Agilent 2100 Bioanalyzer System (Agilent, Santa Clara, CA, USA). After quality control of the RNA samples, a total of 36 libraries were constructed using the NEB library kit, according to the instructions. Sequencing was performed using the BGI MGIseq 2000 system. The raw reads generated by the MGIseq platform were filtered using fastp [61] with default parameters.

The HISAT2 was used to align the clean data to the mulberry reference genome [62], then RSEM software was used to calculate gene expression levels based the alignment file. Based on Fragments Per Kilobase of exon model per million mapped fragments (FPKM) files, the R function cor and prcomp were used to perform Pairwise Pearson correlation coefficient and principal component analysis (PCA), respectively. Gene expression patterns were evaluated using the TCseq package (https://github.com/MengjunWu/TCseq, accessed on 10 December 2022). The differentially expressed genes (DEGs) were identified using the DEseq2 program [63] with the criterion of |Fold Change| > 2 and FDR < 0.05. The Venn Diagram and UpSetR R package were used to draw the Venn diagram for these DEGs. Gene ontology (GO), and Kyoto Encyclopedia of Genes and Genomes (KEGG) pathway enrichment analysis of DEGs were performed using OmicShare tools (https://www.omicshare.com, accessed on 10 December 2022).

4.7. Validation of RNA-Seq Data

First strand cDNA was synthesized using PrimeScript RT reagent kit (TaKaRa, China). RT-qPCR was performed with Light Cycler 480 (Roche, Basel, Switzerland) in 20 L with SYBR Premix Ex Taq Kit (TakaRa, China). Reaction conditions were 95 for 3 min followed by 40 cycles of 94 for 10 s, 55 for 10 s, and 72 for 30 s. Gene expression levels were calculated relative to the expression levels of -actin and GADPH using the 2-Ct method. Primers were designed using the NCBI primer design program. The sequences of the primers are shown in Table S2.

5. Conclusions

In conclusion, our study revealed that the osmotically regulating substances and enzyme activities in three cultivars showed little difference between the cutting and grafting groups, with the exception of the AY cultivar which showed a clear difference from the ZZ and SG cultivars. Our results also demonstrated that waterlogging can affect the photosynthetic rate in mulberry, with a gradual increase in photosynthetic rate observed as mulberry recovered from waterlogging stress over a 20-day period. At the genetic level, waterlogging stress was found to directly or indirectly cause the down-regulation of key genes associated with photosynthesis, such as *petC* and *petE*. Interestingly, cutting groups were found to have a better ability to recover from waterlogging stress than grafted mulberries. These findings provide valuable insights into the underlying mechanisms of dual-method mulberry propagation in responding to waterlogging stress and highlight the potential for developing waterlogging-tolerant mulberry cultivars.

Our study contributes to the understanding of the physiological and genetic responses of mulberry to waterlogging stress and provides a foundation for future research on improving waterlogging tolerance in this economically important crop. The striking finding that cutting groups can better recover from waterlogging stress than grafted mulberries offers a promising avenue for further investigation and may have important implications for mulberry cultivation in waterlogged areas.

Supplementary Materials: The following supporting information can be downloaded at: https://www.mdpi.com/article/10.3390/plants12112066/s1, Figure S1: Dynamics of the photosynthetic characters; Figure S2: The number of DEGs identified with 17 pairwise comparisons; Figure S3: Pairwise Pearson correlation coefficient (A) and PCA analysis for 36 sample with FPKM (B); Figure S4: Venn diagram and UpSet indicated the number of differentially expressed genes (DEGs) per contrast; Table S2: The primer sequences used for RT-qPCR.

Author Contributions: Y.L., C.Z. and W.D. designed the research; Y.L., J.H., C.Y., R.M., Z.Z., Z.D. and X.H. performed the experiments and analyzed data; Y.L., C.Z. and W.D. wrote and revised the manuscript. All authors have read and agreed to the published version of the manuscript.

Funding: This study was funded by the Key R&D program of Hubei Province (2022BBA0065), and the China Agriculture Research System of MOF and MARA (No: CARS-18-SYZ10).

Institutional Review Board Statement: Not applicable.

Informed Consent Statement: Not applicable.

Data Availability Statement: The details of the DEGs (Table S1) have been deposited in figshare (https://doi.org/10.6084/m9.figshare.14343932.v1).

Conflicts of Interest: The authors declare no conflict of interest.

References

1. Kaur, G.; Singh, G.; Motavalli, P.P.; Nelson, K.A.; Orlowski, J.M.; Golden, B.R. Impacts and management strategies for crop production in waterlogged or flooded soils: A review. *Agron. J.* **2020**, *112*, 1475–1501. [CrossRef]
2. Hasanuzzaman, M.; Bhuyan, M.H.M.B.; Zulfiqar, F.; Raza, A.; Mohsin, S.M.; Al Mahmud, J.; Fujita, M.; Fotopoulos, V. Reactive Oxygen Species and Antioxidant Defense in Plants under Abiotic Stress: Revisiting the Crucial Role of a Universal Defense Regulator. *Antioxidants* **2020**, *9*, 681. [CrossRef]
3. Irfan, M.; Hayat, S.; Hayat, Q.; Afroz, S.; Ahmad, A. Physiological and biochemical changes in plants under waterlogging. *Protoplasma* **2010**, *241*, 3–17. [CrossRef] [PubMed]
4. Fukao, T.; Barrera-Figueroa, B.E.; Juntawong, P.; Pena-Castro, J.M. Submergence and Waterlogging Stress in Plants: A Review Highlighting Research Opportunities and Understudied Aspects. *Front. Plant Sci.* **2019**, *10*, 340. [CrossRef] [PubMed]
5. Sharma, J.K.; Sihmar, M.; Santal, A.R.; Singh, N.P. Impact assessment of major abiotic stresses on the proteome profiling of some important crop plants: A current update. *Biotechnol. Genet. Eng.* **2019**, *35*, 126–160. [CrossRef] [PubMed]
6. Xie, L.J.; Zhou, Y.; Chen, Q.F.; Xiao, S. New insights into the role of lipids in plant hypoxia responses. *Prog. Lipid Res.* **2021**, *81*, 101072. [CrossRef] [PubMed]
7. Adak, M.K.; Saha, I.; Dolui, D.; Hasanuzzaman, M. An updated overview of the physiological and molecular responses of rice to anoxia. *Front. Biosci.-Landmrk* **2021**, *26*, 1240–1255.
8. Bhusal, N.; Kim, H.S.; Han, S.-G.; Yoon, T.-M. Photosynthetic traits and plant–water relations of two apple cultivars grown as bi-leader trees under long-term waterlogging conditions. *Environ. Exp. Bot.* **2020**, *176*, 104111. [CrossRef]
9. Jacobsen, A.L.; Agenbag, L.; Esler, K.J.; Pratt, R.B.; Ewers, F.W.; Davis, S.D. Xylem density, biomechanics and anatomical traits correlate with water stress in 17 evergreen shrub species of the Mediterranean-type climate region of South Africa. *J. Ecol.* **2007**, *95*, 171–183. [CrossRef]
10. Lee, Y.-H.; Kim, K.-S.; Jang, Y.-S.; Choi, I.-H. Nitric oxide production and scavenging in waterlogged roots of rape seedlings. *Genes Genom.* **2014**, *36*, 691–699. [CrossRef]

11. Özçubukçu, S.; Ergün, N.; Ilhan, E. Waterlogging and nitric oxide induce gene expression and increase antioxidant enzyme activity in wheat (*Triticum aestivum* L.). *Acta Biol. Hung.* **2014**, *65*, 47–60. [CrossRef] [PubMed]
12. Jia, L.; Qin, X.; Lyu, D.; Qin, S.; Zhang, P. ROS production and scavenging in three cherry rootstocks under short-term waterlogging conditions. *Sci. Hortic.* **2019**, *257*, 108647. [CrossRef]
13. Peng, Y.; Zhou, Z.; Tong, R.; Hu, X.; Du, K. Anatomy and ultrastructure adaptations to soil flooding of two full-sib poplar clones differing in flood-tolerance. *Flora* **2017**, *233*, 90–98. [CrossRef]
14. Oliveira, A.S.d.; Ferreira, C.S.; Graciano-Ribeiro, D.; Franco, A.C. Anatomical and morphological modifications in response to flooding by six Cerrado tree species. *Acta Bot. Bras.* **2015**, *29*, 478–488. [CrossRef]
15. Li, M.; López, R.; Venturas, M.; Pita, P.; Gordaliza, G.G.; Gil, L.; Rodríguez-Calcerrada, J. Greater resistance to flooding of seedlings of Ulmus laevis than Ulmus minor is related to the maintenance of a more positive carbon balance. *Trees* **2015**, *29*, 835–848. [CrossRef]
16. Olorunwa, O.J.; Adhikari, B.; Brazel, S.; Shi, A.; Popescu, S.C.; Popescu, G.V.; Barickman, T.C. Growth and Photosynthetic Responses of Cowpea Genotypes under Waterlogging at the Reproductive Stage. *Plants* **2022**, *11*, 2315. [CrossRef]
17. Barickman, T.C.; Simpson, C.R.; Sams, C.E. Waterlogging Causes Early Modification in the Physiological Performance, Carotenoids, Chlorophylls, Proline, and Soluble Sugars of Cucumber Plants. *Plants* **2019**, *8*, 160. [CrossRef]
18. Chen, C.C.; Li, M.S.; Chen, K.T.; Lin, Y.H.; Ko, S.S. Photosynthetic and Morphological Responses of Sacha Inchi (*Plukenetia volubilis* L.) to Waterlogging Stress. *Plants* **2022**, *11*, 249. [CrossRef]
19. Sharma, S.; Bhatt, U.; Sharma, J.; Darkalt, A.; Mojski, J.; Soni, V. Effect of different waterlogging periods on biochemistry, growth, and chlorophyll a fluorescence of *Arachis hypogaea* L. *Front Plant. Sci.* **2022**, *13*, 1006258. [CrossRef]
20. Phukan, U.J.; Mishra, S.; Shukla, R.K. Waterlogging and submergence stress: Affects and acclimation. *Crit. Rev. Biotechnol.* **2016**, *36*, 956–966. [CrossRef]
21. Salazar, C.; Hernández, C.; Pino, M.T. Plant water stress: Associations between ethylene and abscisic acid response. *Chil. J. Agric. Res.* **2015**, *75*, 71–79. [CrossRef]
22. Chen, W.; Yao, Q.; Patil, G.B.; Agarwal, G.; Deshmukh, R.K.; Lin, L.; Wang, B.; Wang, Y.; Prince, S.J.; Song, L. Identification and comparative analysis of differential gene expression in soybean leaf tissue under drought and flooding stress revealed by RNA-Seq. *Front. Plant Sci.* **2016**, *7*, 1044. [CrossRef] [PubMed]
23. Dossa, K.; You, J.; Wang, L.; Zhang, Y.; Li, D.; Zhou, R.; Yu, J.; Wei, X.; Zhu, X.; Jiang, S. Transcriptomic profiling of sesame during waterlogging and recovery. *Sci. Data* **2019**, *6*, 204. [CrossRef]
24. Yu, C.; Huang, S.J.; Hu, X.M.; Deng, W.; Xiong, C.; Ye, C.H.; Li, Y.; Peng, B. Changes in photosynthesis, chlorophyll fluorescence, and antioxidant enzymes of mulberry (*Morus* ssp.) in response to salinity and high-temperature stress. *Biologia* **2013**, *68*, 404–413. [CrossRef]
25. Yuan, Q.; Zhao, L. The Mulberry (*Morus alba* L.) Fruit A Review of Characteristic Components and Health Benefits. *J. Agric. Food Chem.* **2017**, *65*, 10383–10394. [CrossRef] [PubMed]
26. Hosali, R.; Murthy, C. To analyse the cost of mulberry and cocoon production in Haveri district. *Int. J. Commer. Bus. Manag.* **2015**, *8*, 58–63. [CrossRef]
27. Zenginbal, H.; Eşitken, A. Effects of the application of various substances and grafting methods on the grafting success and growth of black mulberry (*Morus nigra* L.). *Acta Sci. Pol. Hortorum Cultus* **2016**, *15*, 99–109.
28. Zhang, H.H.; Li, X.; Zhang, S.B.; Yin, Z.P.; Zhu, W.X.; Li, J.B.; Meng, L.; Zhong, H.X.; Xu, N.; Wu, Y.N.; et al. Rootstock Alleviates Salt Stress in Grafted Mulberry Seedlings: Physiological and PSII Function Responses. *Front. Plant Sci.* **2018**, *9*, 1806. [CrossRef]
29. Zahoor, R.; Dong, H.; Abid, M.; Zhao, W.; Wang, Y.; Zhou, Z. Potassium fertilizer improves drought stress alleviation potential in cotton by enhancing photosynthesis and carbohydrate metabolism. *Environ. Exp. Bot.* **2017**, *137*, 73–83. [CrossRef]
30. Lan, Y.; Song, Y.; Zhao, F.; Cao, Y.; Luo, D.; Qiao, D.; Cao, Y.; Xu, H. Phylogenetic, Structural and Functional Evolution of the LHC Gene Family in Plant Species. *Int. J. Mol. Sci.* **2022**, *24*, 488. [CrossRef]
31. Jiang, Q.; Xu, Z.S.; Wang, F.; Li, M.Y.; Ma, J.; Xiong, A.S. Effects of abiotic stresses on the expression of Lhcb1 gene and photosynthesis of Oenanthe javanica and Apium graveolens. *Biol. Plant.* **2014**, *58*, 256–264. [CrossRef]
32. de Bianchi, S.; Betterle, N.; Kouril, R.; Cazzaniga, S.; Boekema, E.; Bassi, R.; Dall'Osto, L. Arabidopsis mutants deleted in the light-harvesting protein Lhcb4 have a disrupted photosystem II macrostructure and are defective in photoprotection. *Plant Cell* **2011**, *23*, 2659–2679. [CrossRef] [PubMed]
33. Chen, Y.E.; Ma, J.; Wu, N.; Su, Y.Q.; Zhang, Z.W.; Yuan, M.; Zhang, H.Y.; Zeng, X.Y.; Yuan, S. The roles of Arabidopsis proteins of Lhcb4, Lhcb5 and Lhcb6 in oxidative stress under natural light conditions. *Plant Physiol. Biochem.* **2018**, *130*, 267–276. [CrossRef] [PubMed]
34. Deng, Y.S.; Kong, F.Y.; Zhou, B.; Zhang, S.; Yue, M.M.; Meng, Q.W. Heterology expression of the tomato LeLhcb2 gene confers elevated tolerance to chilling stress in transgenic tobacco. *Plant Physiol. Biochem.* **2014**, *80*, 318–327. [CrossRef] [PubMed]
35. Voss, I.; Koelmann, M.; Wojtera, J.; Holtgrefe, S.; Kitzmann, C.; Backhausen, J.E.; Scheibe, R. Knockout of major leaf ferredoxin reveals new redox-regulatory adaptations in Arabidopsis thaliana. *Physiol. Plant* **2008**, *133*, 584–598. [CrossRef]
36. Yamori, W.; Kusumi, K.; Iba, K.; Terashima, I. Increased stomatal conductance induces rapid changes to photosynthetic rate in response to naturally fluctuating light conditions in rice. *Plant Cell Environ.* **2020**, *43*, 1230–1240. [CrossRef]

37. Yang, X.H.; Chen, L.S.; Cheng, L.L. Leaf Photosynthesis and Carbon Metabolism Adapt to Crop Load in 'Gala' Apple Trees. *Horticulturae* **2021**, *7*, 47. [CrossRef]
38. Treves, H.; Kuken, A.; Arrivault, S.; Ishihara, H.; Hoppe, I.; Erban, A.; Hohne, M.; Moraes, T.A.; Kopka, J.; Szymanski, J.; et al. Carbon flux through photosynthesis and central carbon metabolism show distinct patterns between algae, C-3 and C-4 plants. *Nat. Plants* **2022**, *8*, 78–91. [CrossRef]
39. Rho, H.; Yu, D.J.; Kim, S.J.; Lee, H.J. Limitation Factors for Photosynthesis in 'Bluecrop' Highbush Blueberry (*Vaccinium corymbosum*) Leaves in Response to Moderate Water Stress. *J. Plant Biol.* **2012**, *55*, 450–457. [CrossRef]
40. Hazrati, S.; Tahmasebi-Sarvestani, Z.; Modarres-Sanavy, S.A.M.; Mokhtassi-Bidgoli, A.; Nicola, S. Effects of water stress and light intensity on chlorophyll fluorescence parameters and pigments of *Aloe vera* L. *Plant Physiol. Bioch.* **2016**, *106*, 141–148. [CrossRef]
41. Pinnola, A.; Bassi, R. Molecular mechanisms involved in plant photoprotection. *Biochem. Soc. T* **2018**, *46*, 467–482. [CrossRef] [PubMed]
42. Liu, Y.; Ji, D.F.; Turgeon, R.; Chen, J.; Lin, T.B.; Huang, J.; Luo, J.; Zhu, Y.; Zhang, C.K.; Lv, Z.Q. Physiological and Proteomic Responses of Mulberry Trees (*Morus alba.* L.) to Combined Salt and Drought Stress. *Int. J. Mol. Sci.* **2019**, *20*, 2486. [CrossRef] [PubMed]
43. Mohan, R.; Kaur, T.; Bhat, H.A.; Khajuria, M.; Pal, S.; Vyas, D. Paclobutrazol Induces Photochemical Efficiency in Mulberry (*Morus alba* L.) Under Water Stress and Affects Leaf Yield Without Influencing Biotic Interactions. *J. Plant Growth Regul.* **2020**, *39*, 205–215. [CrossRef]
44. Bhusal, N.; Adhikari, A.; Lee, M.; Han, A.; Han, A.R.; Kim, H.S. Evaluation of growth responses of six gymnosperm species under long-term excessive irrigation and traits determining species resistance to waterlogging. *Agric. For. Meteorol.* **2022**, *323*, 109071. [CrossRef]
45. Bhusal, N.; Han, S.-G.; Yoon, T.-M. Impact of drought stress on photosynthetic response, leaf water potential, and stem sap flow in two cultivars of bi-leader apple trees (*Malus* × domestica Borkh.). *Sci. Hortic.* **2019**, *246*, 535–543. [CrossRef]
46. Ren, B.Z.; Hu, J.; Zhang, J.W.; Dong, S.T.; Liu, P.; Zhao, B. Effects of urea mixed with nitrapyrin on leaf photosynthetic and senescence characteristics of summer maize (*Zea mays* L.) waterlogged in the field. *J. Integr. Agric.* **2020**, *19*, 1586–1595. [CrossRef]
47. Wollmer, A.C.; Pitann, B.; Muhling, K.H. Waterlogging events during stem elongation or flowering affect yield of oilseed rape (*Brassica napus* L.) but not seed quality. *J. Agron. Crop. Sci.* **2018**, *204*, 165–174. [CrossRef]
48. Blokhina, O.; Virolainen, E.; Fagerstedt, K.V. Antioxidants, oxidative damage and oxygen deprivation stress: A review. *Ann. Bot.* **2003**, *91*, 179–194. [CrossRef]
49. Alscher, R.G.; Erturk, N.; Heath, L.S. Role of superoxide dismutases (SODs) in controlling oxidative stress in plants. *J. Exp. Bot.* **2002**, *53*, 1331–1341. [CrossRef]
50. Apel, K.; Hirt, H. Reactive oxygen species: Metabolism, oxidative stress, and signal transduction. *Annu. Rev. Plant Biol.* **2004**, *55*, 373–399. [CrossRef]
51. Bailey-Serres, J.; Voesenek, L.A. Flooding stress: Acclimations and genetic diversity. *Annu. Rev. Plant Biol.* **2008**, *59*, 313–339. [CrossRef]
52. Zhang, Z.H.; Liu, Y.; Cao, B.L.; Chen, Z.J.; Xu, K. The effectiveness of grafting to improve drought tolerance in tomato. *Plant Growth Regul.* **2020**, *91*, 157–167. [CrossRef]
53. Pagliarani, C.; Vitali, M.; Ferrero, M.; Vitulo, N.; Incarbone, M.; Lovisolo, C.; Valle, G.; Schubert, A. The Accumulation of miRNAs Differentially Modulated by Drought Stress Is Affected by Grafting in Grapevine. *Plant Physiol.* **2017**, *173*, 2180–2195. [CrossRef] [PubMed]
54. Nawaz, M.A.; Chen, C.; Shireen, F.; Zheng, Z.H.; Jiao, Y.Y.; Sohail, H.; Afzal, M.; Imtiaz, M.; Ali, M.A.; Huang, Y.; et al. Improving vanadium stress tolerance of watermelon by grafting onto bottle gourd and pumpkin rootstock. *Plant Growth Regul.* **2018**, *85*, 41–56. [CrossRef]
55. Jia, C.S.; Cao, D.D.; Ji, S.P.; Lin, W.T.; Zhang, X.M.; Muhoza, B. Whey protein isolate conjugated with xylo-oligosaccharides via maillard reaction: Characterization, antioxidant capacity, and application for lycopene microencapsulation. *LWT-Food Sci. Technol.* **2020**, *118*, 108837. [CrossRef]
56. Dao, T.T.; Linthorst, H.J.; Verpoorte, R. Chalcone synthase and its functions in plant resistance. *Phytochem. Rev.* **2011**, *10*, 397–412. [CrossRef]
57. Sinha, A.K.; Jaggi, M.; Raghuram, B.; Tuteja, N. Mitogen-activated protein kinase signaling in plants under abiotic stress. *Plant Signal Behav.* **2011**, *6*, 196–203. [CrossRef]
58. Zhang, H.; Zhu, J.; Gong, Z.; Zhu, J.K. Abiotic stress responses in plants. *Nat. Rev. Genet* **2022**, *23*, 104–119. [CrossRef]
59. Hori, K.; Matsubara, K.; Yano, M. Genetic control of flowering time in rice: Integration of Mendelian genetics and genomics. *Theor. Appl. Genet* **2016**, *129*, 2241–2252. [CrossRef]
60. Zhao, Y.; Li, Y.S.; Gao, X.F. A New Method for Accurate Determination of Peroxidase Activity Based on Fluorescence Decrease of Guaiacol. *Chinese J. Anal. Chem.* **2015**, *43*, 1040–1046.
61. Chen, S.F.; Zhou, Y.Q.; Chen, Y.R.; Gu, J. fastp: An ultra-fast all-in-one FASTQ preprocessor. *Bioinformatics* **2018**, *34*, 884–890. [CrossRef] [PubMed]

62. He, N.; Zhang, C.; Qi, X.; Zhao, S.; Tao, Y.; Yang, G.; Lee, T.H.; Wang, X.; Cai, Q.; Li, D.; et al. Draft genome sequence of the mulberry tree *Morus notabilis*. *Nat. Commun.* **2013**, *4*, 2445. [CrossRef] [PubMed]
63. Love, M.I.; Huber, W.; Anders, S. Moderated estimation of fold change and dispersion for RNA-seq data with DESeq2. *Genome Biol.* **2014**, *15*, 550. [CrossRef] [PubMed]

plants

MDPI

Article

Transcriptional Memory in *Taraxacum mongolicum* in Response to Long-Term Different Grazing Intensities

Yalin Wang [1,2], Wenyan Zhu [3], Fei Ren [4], Na Zhao [2], Shixiao Xu [2,*] and Ping Sun [1,4,*]

1 College of Animal Science and Technology, Henan University of Science and Technology, Luoyang 471003, China
2 Northwest Institute of Plateau Biology, Chinese Academy of Sciences, Xining 810008, China
3 College of Horticulture and Plant Protection, Henan University of Science and Technology, Luoyang 471003, China
4 State Key Laboratory of Plateau Ecology and Agriculture, Qinghai University, Xining 810016, China
* Correspondence: sxxu@nwipb.cas.cn (S.X.); pingsunny@msn.com (P.S.); Tel.: +86-13997163501 (S.X.); +86-13525415882 (P.S.)

Abstract: Grazing, as an important land use method in grassland, has a significant impact on the morphological and physiological traits of plants. However, little is known about how the molecular mechanism of plant responds to different grazing intensities. Here, we investigated the response of *Taraxacum mongolicum* to light grazing and heavy grazing intensities in comparison with a non-grazing control. Using de novo transcriptome assembly, *T. mongolicum* leaves were compared for the expression of the different genes under different grazing intensities in natural grassland. In total, 194,253 transcripts were de novo assembled and comprised in nine leaf tissues. Among them, 11,134 and 9058 genes were differentially expressed in light grazing and heavy grazing grassland separately, with 5867 genes that were identified as co-expression genes in two grazing treatments. The Nr, SwissProt, String, GO, KEGG, and COG analyses by BLASTx searches were performed to determine and further understand the biological functions of those differentially expressed genes (DEGs). Analysis of the expression patterns of 10 DEGs by quantitative real-time RT-PCR (qRT-PCR) confirmed the accuracy of the RNA-Seq results. Based on a comparative transcriptome analysis, the most significant transcriptomic changes that were observed under grazing intensity were related to plant hormone and signal transduction pathways, carbohydrate and secondary metabolism, and photosynthesis. In addition, heavy grazing resulted in a stronger transcriptomic response compared with light grazing through increasing the of the secondary metabolism- and photosynthesis-related genes. These changes in key pathways and related genes suggest that they may synergistically respond to grazing to increase the resilience and stress tolerance of *T. mongolicum*. Our findings provide important clues for improving grassland use and protection and understanding the molecular mechanisms of plant response to grazing.

Keywords: transcriptional memory; grazing; *T. mongolicum*

Citation: Wang, Y.; Zhu, W.; Ren, F.; Zhao, N.; Xu, S.; Sun, P. Transcriptional Memory in *Taraxacum mongolicum* in Response to Long-Term Different Grazing Intensities. *Plants* **2022**, *11*, 2251. https://doi.org/10.3390/plants11172251

Academic Editor: Małgorzata Nykiel

Received: 30 July 2022
Accepted: 23 August 2022
Published: 30 August 2022

1. Introduction

Grassland is a major part of the terrestrial ecosystem, covering 40% of the global land area [1]. It plays an irreplaceable role in human living by regulating the climate, conserving soil and water resources, maintaining biodiversity, and providing biological products [2,3]. The quality of herbage decides the safe production and efficiency of animal husbandry through growth rate, nutritional value, and yield [4,5]. However, herbage has to tolerate several damages in nature, including adverse natural factors, inappropriate use, and human activities, leading to the imbalance of the grassland ecosystem. In particular, herbage needs to tolerate several biotic and abiotic stresses, such as salt [6], drought [7], wounding [8], and foraging [9], because of the geographical distribution and growth characteristics. These factors seriously threaten the survival, growth, production, and value of the grassland vegetation.

In recent years, studies on the response of herbage to external stress (biotic or abiotic stress) and extreme environments have gradually increased, and the point involves many aspects, such as morphological, physiological, and molecular levels [10–12]. The research with the emergence of genomic tools and high-throughput phenotyping technologies focuses on investigating the genetic and molecular mechanisms of phenotypic plasticity after different kinds or degrees of stress [13,14]. Phenotype is an important reflection of plant response to plant structure, function, and interactions with the environment, which is influenced by individual traits and the diversity of environmental conditions [15]. Under stress, plants usually exhibit a series of transcriptional responses. For example, a study indicated that cold-stress was related to plant hormone and signal transduction pathways, primary and secondary metabolism, photosynthesis, and members of transcription factors in *Magnolia wufengensis* [16]. Meanwhile, another research identified the metabolic pathways, such as carbohydrate metabolism, photosynthesis, and lipid metabolism, for distinct genotypes in chickpea response to drought and salinity [11]. Plants can acquire the ability to adaptive environmental changes through learning processes, which can be defined as stress memory [17]. Stress memory is influenced by stress times, stress degrees, and different phenological periods of plants [18,19]. A common theme about plant response to a range of biotic and abiotic stresses is the phenomenon of priming whereby previous exposure makes a plant more resistant to future exposure [20]. After priming, plants display a faster or stronger defense response through rapid change in the gene expression if the stress recurs [21]. It also would be reserved for a period of time by the information storage and signal expression in case the harming happens again [22]. The memory span may often be in the range of days to weeks in the acclimated plants, and sometimes may extend to the offspring [21]. The study that concentrated on clonal *Leymus Chinensis* response to long-term overgrazing-induced memory showed that livestock grazing induces a transgenerational effect on growth inhibition through altered gene expression of defense and immune responses, pathogenic resistance, and cell development [23].

Grazing, which is a crucial environmental stress factor, influences plants by repeatedly altering the essentially availability of resources for plant growth [24,25]. The growth conditions, including soil nutrients [2], water [26], and light, are frequently changed by the interaction of foraging, trampling, and fecal accumulation, which further affects herbage growth time [9], productivity [5], palatability [27], and many other aspects. After grazing, leaves were fed to animals with trampling, thus restricting normal growth, which also decreases photosynthesis [28], but increases antioxidant capacity [29] and secondary metabolites synthesis [30]. Long-time grazing plants also perform different resistance strategies to avoid or reduce damage through morphological or biochemical changes, or rapid regrowth/reproduction [31,32]. Meanwhile, herbage usually exhibits dwarfing [23], adjusting the proportion of aboveground and underground [33], and some herbage have low palatability after grazing [27]. The research about transcriptome-wide gene expression plasticity conducted long-term different grazing intensities resulted in gene expression plasticity in the recovery stage of *Stipa grandis*, affecting diverse biological processes and metabolic pathways that were involved in the Calvin–Benson cycle and photorespiration metabolic pathways [13]. Wang et al. researched alfalfa with different grazing tolerance and showed that the DEGs were related to the ribosome and translation-related activities, cell wall processes, and oxygen levels [30]. Although these studies show the transcription process of herbage after grazing, the complex transcription changes that were decided in species were unclear in stress memory during the recovery period of herbage. Therefore, to better understand the transcription changes with different grazing intensities in nature, we focus on the DEGs and metabolism pathways at grazing recovery periods under different grazing conditions.

T. mongolicum, one of the main associated species in the alpine meadow, is an edible medicinal herb that is primarily used as an anti-inflammatory, antibacterial, anti-allergy, and antioxidant owing to its bioactive metabolites such as phenolic compounds, polysaccharides, and flavonoids [34]. As a medicinal plant, it has been regarded as a potential substitute for traditional antibiotics in the raising dairy diet in recent years. Li et al. found the supplementation impacts, the kind of ruminal microorganisms, and metabolites that rumen fermentation was enhanced in cows [35]. Therefore, we chose it as the objective to study how transcriptomes respond in natural grazing grassland. We set up three treatments, including light-grazing, heavy-grazing, and non-grazing, and sequenced, assembled, and compared the transcriptomes of *T. mongolicum* to identify gene expression dynamics and DEGs in different grazing intensities. The present study will facilitate further functional genomics studies in *T. mongolicum* and aid in the understanding of the molecular mechanisms that are behind the long-term grazing response in plants and the associated morphological changes.

2. Materials and Methods

2.1. Plant materials and Treatments

The grazing experiment was performed at the Haibei Alpine Meadow Ecosystem Research Station that is managed by the Northwest Institute of Plateau Biology, Chinese Academy of Sciences. The station is located in the northeastern portion of the Qinghai–Tibet Plateau, China (N37°29′, E 101°12′; 3220 m). This region has a typical continental climate for plateaus, with a short, cool summer and a long, cold winter, and its annual average temperature is −1.7 °C (the maximum and minimum temperatures are 9.9 °C in July and −15.2 °C in January, respectively). The annual mean precipitation is 500 mm, and more than 80% of precipitation occurs during the growing seasons (from May to September). The vegetation in this region mainly consists of *Kobresia humilis Sergiev.*, *Elymus nutans Griseb.*, *Poa pratensis* L., *Carex scabrirostris Kukenth.*, *Gentiana straminea Maxim.*, and *Potentilla nivea* L. The soil has been identified as alpine meadow soil [36].

There were two grazing treatments and fenced treatments that were launched in 2009. The grazing treatments, light-grazing (utilization 30%) and heavy-grazing (utilization 60%), were grazed by sheep for 2 days per month during the vegetation growing season (June-September) every year. The three plots are adjacent to each other, and the terrain and soil conditions are the same, and other influencing factors can be ignored. The vegetation type of the plot is the same as that of natural grassland. In order to identify gene expression responses of *T. mongolicum* in the recovery growth stage after grazing, we chose mid-August for grazing, which was the period of maximum plant biomass in herbage. The new leaves of *T. mongolicum* were randomly collected from three individual plants from 9:00 a.m. to 11:00 a.m. after two weeks of grazing at the beginning of September, and fresh leaves were immediately frozen in liquid nitrogen and then stored at −80 °C. Meanwhile, we collected three biological replicates in the fenced plot as the non-grazing sample.

2.2. RNA-Seq Library Preparation and ILLUMNA sequencing

A total of 0.1 g leaves were taken from each plant for total RNA extraction using Trizol reagent (Invitrogen, Carlsbad, CA, USA). The quality and quantity were evaluated using Nanodrop 2000 (NanoDrop, Wilmington, DE, USA), Agilent 2100 Bioanalyzer (Agilent Technologies, Inc, Santa Clara, CA, USA). The RNA samples with a ratio of 260/280 nm greater than 1.8 and RNA integrity number (RIN) > 7 were selected for subsequent processing. The poly(A) mRNA was isolated using Oligo (dT) Beads. The mRNA fragmentation and subsequent RNA-Seq library conversion were carried out using a TruseqTM RNA sample prep Kit (Illumina, San Diego, CA, USA) in accordance with the manufacturer's instructions. There were nine cDNA libraries that were constructed and each cDNA library was sequenced using the Illumina HiSeq 2000 platform (Shanghai Origingene Bio-pharm Technology Co., Ltd., Shanghai, China).

2.3. Quality and De Novo Assembled

The total RNA was extracted from the fresh leaves from 3 treated groups with 3 biological replicates, designated non-grazing (NG1, NG2, NG3), light grazing (LG1, LG2, LG3), and heavy grazing (HG1, HG2 and HG3). Cutadapt (Version 1.16) was used to filter the raw reads, including removing the reads containing adaptor sequences, "N" percentages that were greater than 5%, and low-quality reads (<20% low-quality nucleotides). FastQC (Version 0.11.4) analysis was further performed to estimate the quality of raw reads (www.bioinformatics.babraham.ac.uk/bugzilla/; accessed on 4 December 2019). Then, clean and high-quality reads were do novo assembled into a transcriptome using Trinity software (Version 2.6.6) with three steps (Inchworm, Chrysalis, and Butterfly) [37]. Finally, the results were evaluated by different quality metrics including N50 length of the transcriptome assemblies, the total number of bps in the assembly, and percent of GC.

2.4. Functional Annotation and Identification of DEGs

Functional annotation of all the unigenes was conducted by several approaches. First, using the open reading frames (ORFs) process to predict all the assembled transcripts. Then, functional annotation of the transcripts was performed using BLASTX between the unigenes and various protein databases (E-value < 10^{-5}) [38], such as the non-redundant protein (NR) database (http://www.ncbi.nlm.nih.gov; accessed on 4 December 2019), the Swiss-Prot protein database (http://www.expasy.ch/sprot; accessed on 4 December 2019), the Kyoto Encyclopedia of Genes and Genomes (KEGG) pathway database (http://www.genome.jp/kegg/; accessed on 4 December 2019), and the Cluster of Orthologous Groups (COG) database (http://www.ncbi.nlm.nih.gov/COG/; accessed on 4 December 2019). The GO distribution for all the unigenes whose expression was significantly altered in the *T. mongolicum* transcriptome were classified using the Blast2GO program [39].

The FPKM (fragments per kilo base of exon per million) method was used to determine the expression levels of the unigenes. The DEGs between grazing and non-grazing groups were identified using the edgeR package (FDR < 0.05 and $|\log_2 FC| \geq 1$) [40]. Correlation analysis was carried out to assess the correlation between the replicates and treatments. For the functional and pathway enrichment analysis, the DEGs were then mapped into GO terms (p-value ≤ 0.05) and the KEGG database (p-value ≤ 0.05).

2.5. RNA-Seq Result Validation by qRT-PCR

A total 10 DEGs were selected to verify our RNA sequencing expression data by qRT-PCR. The primer sequences are listed in Table S1. The Glyceraldehyde 3-phosphate dehydrogenase (GAPDH) gene was used as an internal control to normalize the measured gene expression levels. The total RNA was reverse transcribed into cDNA using the HiScript 1st Strand cDNA Synthesis Kit (Takara, Nanjing, China). Real-time PCR was performed in a real-time PCR platform (CFX96, Bio-Rad, USA) using the AceQ qPCR SYBR Green Master Mix (Takara, Shanghai, China). The cycling conditions were as initial denaturation at 95 °C for 5 min followed by 40 cycles of 95 °C for 10 s and 60 °C for 30 s. For each qPCR analysis, three technical replicates were performed. The $2^{-\Delta\Delta CT}$ method was used to verify the relative quantity of 10 DEGs in the transcription profile.

2.6. Statistical Analysis

The tables in this study were performed using Excel 2020, and the graphs were drawn by GraphPad Prism 8.

3. Results

3.1. Illumina Sequencing and Reads Assembly

The total raw reads of 377,827,065 from the grazing and non-grazing samples with three biological replicates were obtained by pair-end RNA-Seq sequencing (Table 1). After filtering, 99.8% of the raw reads were retained as clean reads for subsequent transcriptome analysis. All the samples performed high quality results as reflected by all Q20 that were larger than 96.0%. The GC content was approximately 46% in the nine samples. High correlation coefficients were obtained for each treatment, indicating that the data were reliable for further analysis (Figure 1).

Table 1. Sequencing the *T. mongolicum* transcriptome in nine samples from plants that were light grazing treated (LG−1, LG−2, LG−3), heavy grazing treated (HG−1, HG−2, HG−3), and non−grazing−treated (NG−1, NG−2, NG−3).

Sample	Raw Reads	Clean Reads	Clean Assembled Bases	Q20 (%)	GC (%)
LG−1	34,877,558	34,826,890	5,179,996,953	96.94	46.51
LG−2	37,462,782	37,404,946	5,562,407,266	96.81	46.48
LG−3	42,165,774	42,080,700	6,247,968,369	96.94	46.02
HG−1	50,461,110	50,385,728	7,483,079,145	97.13	48.55
HG−2	41,660,330	41,560,876	6,177,856,152	96.56	47.49
HG−3	44,534,628	44,395,988	6,593,904,050	96.9	48.35
NG−1	38,969,368	38,908,870	5,795,654,929	97.06	48.16
NG−2	39,309,852	39,241,864	5,835,824,125	96.91	46.58
NG−3	48,385,668	48,308,606	7,164,002,702	97.24	46.76
All−unigene	377,827,065	377,114,472	56,040,693,687	96.94	47.21

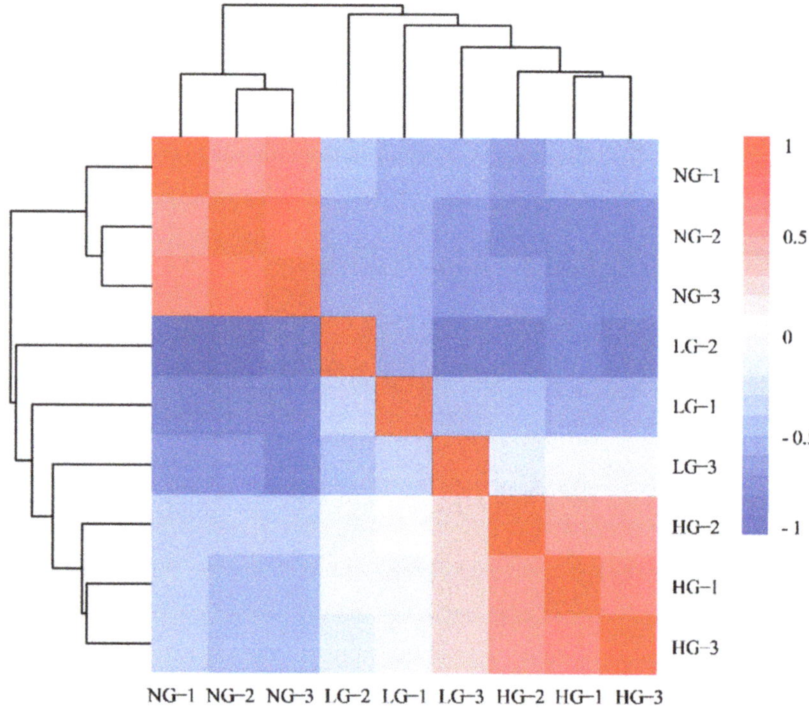

Figure 1. Pearson correlation coefficients of transcript levels in nine samples. NG: non−grazing; LG: light grazing; HG: heavy grazing.

A total of 194,253 transcripts were de novo assembled from all paired clean reads in Trinity, with a contig N50 of 1341 bp and average length of 869.48 bp (Table 2). The size distribution of the assembled transcripts showed that 91.52% (177,762) of sequence length ranged from 201 to 2000 bp and the percentage of sequences that were >2000 bp was 8.48% (16,476) (Figure 2). The total number of unigenes of 88,598 was generated. The N50 of the unigenes was 1118 bp, and the average length was 659.36 bp (Table 2). There were 83,558 unigenes with 94.30% in total were <2000 bp in length (Figure 2).

Table 2. Statistics of sequencing and assembly results.

Type	Unigene	Transcript
Total sequence number	88,598	194,253
Total sequence base	58,417,702	168,900,058
Largest length (bp)	10,562	10,562
Smallest length (bp)	201	186
Average length (bp)	659.36	869.48
N50 length (bp)	1118	1341
N90 length (bp)	259	371

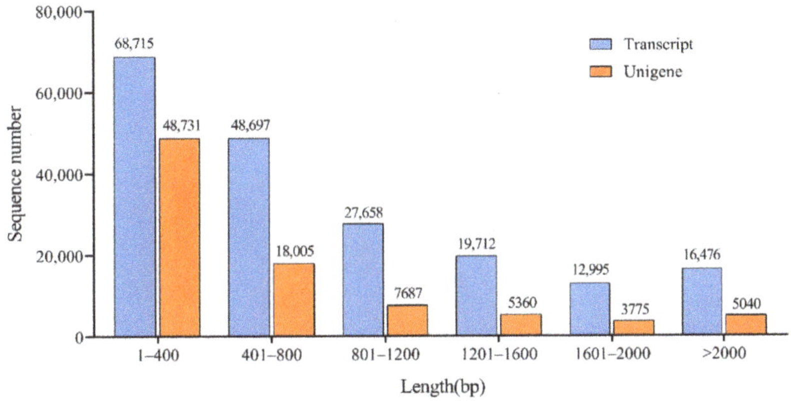

Figure 2. Transcript and unigene length distribution.

3.2. Annotation and Classification of *T. mongolicum* Unigenes

To identify the putative function of the *T. mongolicum* unigenes, six complementary methods were utilized. The assembled unigenes were searched against the NCBI non-redundant (Nr), SwissProt, String, GO, KEGG, and Pfam databases using the BLASTX algorithm, with an E-value < 10^{-5}. Among these unigenes, 24,825 exhibited significant hits in the Nr database and 16,215 in the SwissProt database. In the other four databases (String GO, KEGG and Pfam), 14,402, 16,039, 10,437, and 10,761 unigenes, respectively, were successfully aligned to known proteins in the nine databases (Table 3). According to the Nr database, the unigene sequences exhibited the most similar BLASTx matches to gene sequences from *Helianthus annuus* (8297), followed by *Cynara cardunculus var. scolymus* (6880) and *Oryza sativa Japonica* Group (1094) (Figure 3).

Table 3. Annotation statistics of *T. mongolicum* unigenes.

Type	Unigene	Transcript
Nr	24,825	69,093
SwissProt	16,215	43,153
String	14,402	35,730
GO	16,039	42,629
KEGG	10,437	28,358
Pfam	10,761	33,095

The functions of the *T. mongolicum* unigenes were classified via GO analysis. In total, 16,039 unigenes were successfully categorized into 67 functional groups, and these groups were classified into the following three major GO categories using BLAST2GO (Figure 4, only show the unigenes >1000): 'biological processes', 'cell component', and 'molecular function'. In the dominant subcategories of the 'biological processes' category, 'cellular process' (14,083), 'metabolic process' (12,536), 'response to stimulus' (9991), 'biological regulation' (9255), 'regulation of biological process' (8546) were the top five groups. 'Cell' (14,991), 'cell part' (14,984), 'organelle' (13,261), 'membrane' (9399), 'organelle part' (9303) were categorized in the 'cell component' category; 'binding' (11,972), 'catalytic activity' (9553), 'transporter activity' (1573), 'transcription regulator activity' (1324), and 'structural molecule activity' (1170) in the 'molecular function' category.

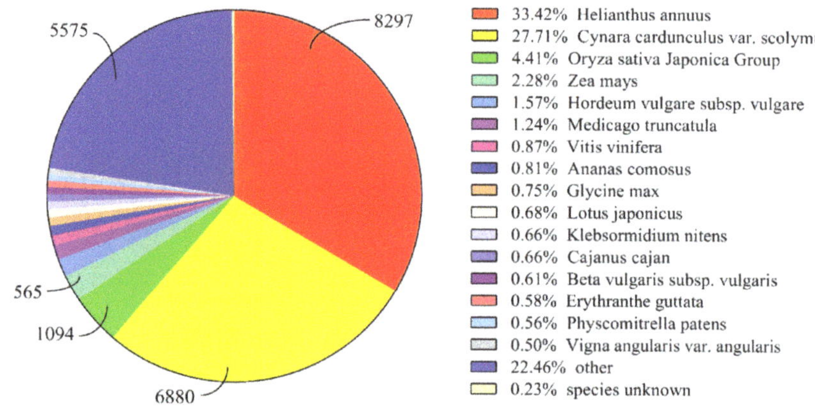

Figure 3. Species distribution of top BLAST hits for each unigene.

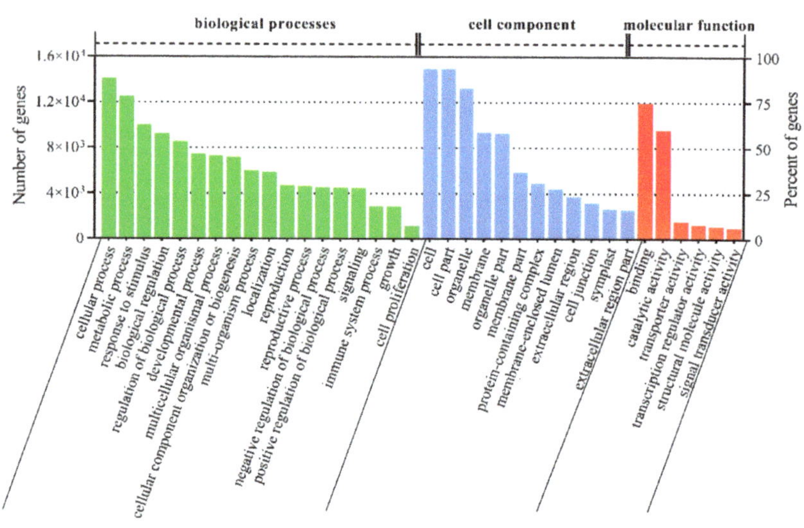

Figure 4. Functional annotation of unigenes based on Gene Ontology (GO) categorization. Green histograms represent "biological process", blue histograms represent "cell component", red histograms represent "molecular function".

The functions of *T. mongolicum* unigenes were predicted and classified by searching the COG database (Figure 5). Assuming that each protein in the COG database independently evolved from an ancestral protein, we classified the 8168 unigenes based on String annotation into 25 groups of COG classifications. Among these classifications, 'General function prediction only' (873, 16.86%) accounted for the largest proportion, followed by 'Translation, ribosomal structure and biogenesis' (833, 10.20%), 'Signal transduction mechanisms' (734, 8.99%), 'Posttranslational modification, protein turnover, chaperones' (721, 8.83%), and

'Carbohydrate transport and metabolism' (666, 8.15%). However, few genes were clustered as 'Nuclear structure' (10, 0.12%) or 'Extracellular structures' (1, 0.01%).

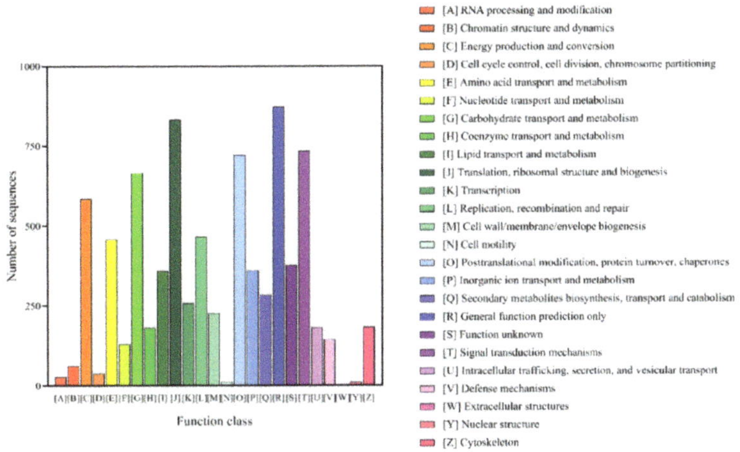

Figure 5. Clusters of Orthologous Group (GOC) classification.

To identify the active biological pathways in *T. mongolicum*, pathway annotations of the unigenes were performed using the KEGG pathway tool. The KEGG annotated unigenes (12,180) were distributed into five groups with 36 sub-categories (Figure 6). The main groups that were found were 'Metabolism' (A, 7010), followed by 'Genetic Information Processing' (B, 2501), 'Environmental Information Processing' (C, 1655), 'Cellular Processes' (D, 1498), and 'Organismal Systems' (E, 3246). In these five groups, the most abundant was 'metabolism' with 13 sub-groups. 'Global and overview maps' (2554) was the highest in these sub-groups, followed by 'Carbohydrate metabolism' (987), 'Energy metabolism' (653), 'Amino acid metabolism' (596), and 'Lipid metabolism' (512). 'Signal transduction' (1576) and 'Translation' (1262) were the largest class in 'Environmental Information Processing' and 'Genetic Information Processing'.

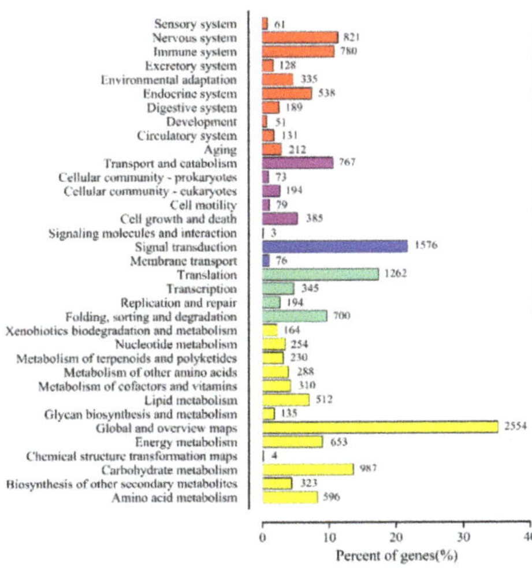

Figure 6. The KEGG metabolic pathway genes that were involved are divided into five branches: A: Cellular Process, B: Environmental Information Processing, C: Genetic Information Processing, D: Metabolism, and E: Organismal Systems.

3.3. DEGs in Different Grazing Intensity

The DEGs were analyzed relative to non-grazing treatments. A total of 11,134 and 9058 DEGs (p-value $\leq$ 0.05 and | log2 (fold change) | > 1) were identified and analyzed for the LG and HG treatments in our results (Figure 7). In the two treatments, 5867 unigenes were identified that were commonly with induced/repressed, 5266 and 3190 unigenes were induced/repressed exclusively at light grazing intensity and at heavy grazing intensity, respectively.

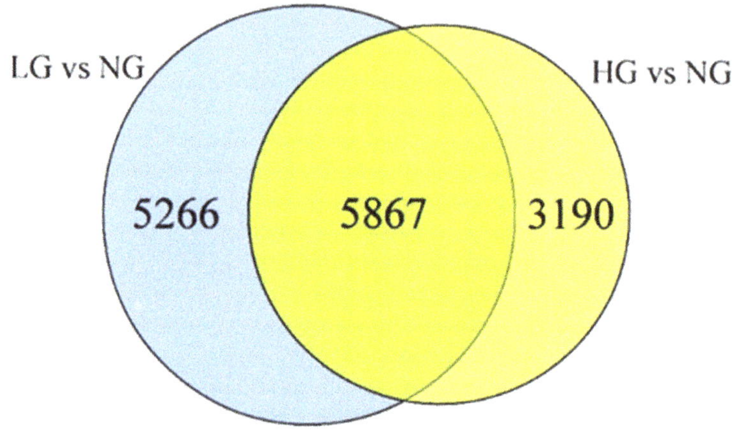

Figure 7. Venn diagram of the functional annotation where each number represents the number of genes in different databases. Blue shows DEGs in light grazing vs non−grazing, light yellow shows DEGs in heavy grazing vs non−grazing, deep yellow indicates co-expression DEGs in two treats.

3.4. Pathways Enrichment Analysis of DEGs

Using the KEGG database, pathways displaying significant changes (p-value $\leq$ 0.05) in response to grazing treatment were identified in the two treatments (Table 4). Compared with non-grazing samples, 9 and 15 KEGG pathways were significantly enriched in LG and HG treatments, respectively. The pathways, including 'Plant–pathogen interaction', 'Glycolysis/Gluconeogenesis', 'Pentose phosphate pathway', 'MAPK signaling pathway-plant', 'Starch and sucrose metabolism', and 'Circadian rhythm-plant', were significantly enriched in both grazing treatments, suggesting that plant carbohydrate metabolism and signaling transduction play significant roles in resistance to grazing stress in *T. mongolicum.* The 'Vasopressin-regulated water reabsorption' was only enriched in LG treatment; 'Photosynthesis-antenna proteins', 'Calcium signaling pathway', 'Biosynthesis of secondary metabolites', 'Carbon fixation in photosynthetic organisms', and 'Plant hormone signal transduction' were only significantly enriched in the HG treatment, meaning that the process of resistance to grazing in *T. mongolicum* was complex. The pathway of 'Biosynthesis of secondary metabolites' exhibited the most DEGs, suggesting that the HG treatment resulted in more secondary metabolites compared with LG treatment in *T. mongolicum.*

Table 4. Significantly enriched gene pathways involving differentially expressed genes (DEGs) following the grazing treatment.

ID	Pathways	Q-Value	No. of Genes
LG vs. NG			
ko04626	Plant–pathogen interaction	0.001	20
ko04962	Vasopressin-regulated water reabsorption	0.003	6
ko00010	Glycolysis/Gluconeogenesis	0.008	19
ko00030	Pentose phosphate pathway	0.010	11

Table 4. *Cont.*

ID	Pathways	Q-Value	No. of Genes
ko04016	MAPK signaling pathway-plant	0.019	23
ko03018	RNA degradation	0.022	21
ko00511	Other glycan degradation	0.033	6
ko00500	Starch and sucrose metabolism	0.034	16
ko04712	Circadian rhythm-plant	0.050	11
HG vs. NG			
ko00196	Photosynthesis-antenna proteins	0.000	10
ko00500	Starch and sucrose metabolism	0.003	17
ko00010	Glycolysis/Gluconeogenesis	0.004	18
ko04016	MAPK signaling pathway-plant	0.008	22
ko04020	Calcium signaling pathway	0.017	7
ko04712	Circadian rhythm-plant	0.019	11
ko04626	Plant–pathogen interaction	0.021	16
ko00680	Methane metabolism	0.022	10
ko04922	Glucagon signaling pathway	0.022	10
ko01110	Biosynthesis of secondary metabolites	0.030	128
ko04391	Hippo signaling pathway-fly	0.041	5
ko04915	Estrogen signaling pathway	0.041	5
ko00030	Pentose phosphate pathway	0.042	9
ko00710	Carbon fixation in photosynthetic organisms	0.047	13
ko04075	Plant hormone signal transduction	0.051	18

3.5. Validation of Gene Expression Profiles by qRT-PCR

To confirm the accuracy and reproducibility of this RNA-Seq data, 10 DEGs were chosen randomly for qRT-PCR with two grazing treatments, including seven up-regulated genes and three down-regulated genes in the unigene dataset (Figure 8). All but one of the 10 genes in the two treatments had similar trends in expression patterns in the qRT-PCR assays as in the RNA-Seq data, confirmed the reliability of our RNA-Seq data.

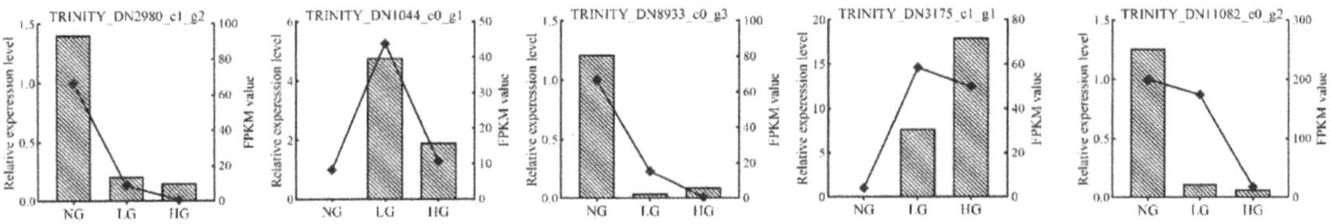

Figure 8. qRT-PCR validation of the RNA-Seq results of *T. mongolicum* in response to grazing.

3.6. Identification of Signal Transduction-Related Unigenes

According to KEGG enrichment analyses of the DEGs, signal transduction in plants was evidently influenced by grazing stress. In the pathway of 'MAPK signaling pathway-plant', we found four mitogen-activated protein kinase (MAPKs) genes were up-regulated in two treatments, which play important roles in signal transduction that is induced by abiotic stress. Meanwhile, we found more than 240 genes were predicted to encode protein kinases with varying expression levels in plants under grazing-treatment conditions

(Table S1). There are 110 genes that were common in the two treatments compared with NG. A total of 32 genes and 28 genes were concluded in the group of serine/threonine-protein kinases and receptor-like protein kinase (RLKs) with most being up-regulated, and 11 genes belonging to RLKs in the group of serine/threonine-protein kinases. We also found two wall-associated receptor kinases and two cysteine-rich receptor-like protein kinases in the RLKs group. There are further three Ca^{2+}-related protein kinases, seven ATP binding-related kinases, and three pyruvate kinases in all the protein kinases.

Calcium, a metal ion, is a secondary messenger in plant cells and plays essential roles in many signaling pathways. The calcium binding proteins of the EF-hand super-family are associated with the regulation of all aspects of cell function. In our data, in total 42 genes were predicted to encode Ca^{2+}-related proteins with most of the genes being up-regulated compared with NG in different treatments (Table S1). We identified five calcium-binding EF-hand proteins (CBL), five calcium-dependent protein kinases (CDPKs) and their related proteins, and two sodium/calcium exchanger-related proteins in two treatments. Meanwhile, we found three CBL-interacting protein kinases (CIPKs) were unique in the LG treatments.

The DEGs that were related to reactive oxygen species (ROS) metabolism were found in 39 genes in two treatments. It included peroxidase (POD), catalase (CAT), glutathione S-transferase (GST), glutaredoxin, and thioredoxin, which play important roles in ROS scavenging. A total of 16 co-expressing DEGs were found in two treatments with 13 up-regulated, while 16 DEGs were only in the LG treatment with 12 up-regulated and 7 DEGs were only in the HG treatment with 4 down-regulated, which means that the LG treatment had more activated scavenging (Table S1).

3.7. Identification of Plant Hormone-Related Unigene

Plant hormones play a major role in plant growth and development and regulate various metabolic processes. EREBP, the most downstream element in the ethylene signal transduction, was from AP2/ERF transcription factor superfamily. We found 14 genes about EREBP in two treatments with five genes down-regulated and nine genes that were up-regulated, while interestingly four genes only in LG were all up-regulated and five genes were down-regulated with five out of six in HG, exhibiting ethylene may express in different ways in the two treatments (Table S2). Auxin closely interacts with ethylene, thus jointly regulating many biological processes in plants. We found 30 genes in two treatments. There were 13 genes that were discovered in two treatments with four genes that were down-regulated and nine genes that were up-regulated. There were five SAUR and one IAA and one auxin response factor (ARF) that were included in the 13 genes, which were important proteins in the regulatory process of auxin. In addition, LG and HG treatments also included several genes that were involved in auxin-related genes, and most of the genes were up-regulated, meaning that the expression of auxin had increasing trends after grazing.

The genes, which are associated with the biosynthesis of the defense-type hormone jasmonic acid (JA), were found in two treatments with 13 genes, including one lipoxygenase (LOX), one 12-oxophytodienoate reductase, two allene oxide cyclase (AOC), eight Acyl-CoA-related proteins, and one fatty acid desaturase (FAD), and generally interacted in JA biosynthesis (Table S2). In these genes, the majority of these genes were up-regulated, except the 12-oxophytodienoate reductase and three Acyl-CoA-related genes. We also found three LOX and one allene oxide synthase (AOS) with high expression exclusively in HG, which may mean that high intensity grazing had a significant effect on plant growth and defense.

3.8. DEGs Involved in Metabolism and Biosynthesis

Based on the KEGG pathway enrichment analyses, many DEGs are associated with metabolism and biosynthesis. We found that most of the enriched pathways were included in carbohydrate metabolism. Compared with NG treatments, the crucial enriched pathways in LG and HG were "Starch and sucrose metabolism", "Pentose phosphate pathway", and

"Glycolysis/Gluconeogenesis" (Table S3). In the "Starch and sucrose metabolism" pathway, we found a total of 16 co-expressing DEGs in LG and HG treatments with 13 up-regulated genes, exhibiting a trend that converting to sucrose and other forms of sugar. We found a starch synthase gene (TRINITY_DN17902_c0_g1) was decreased by 6.9-fold, while an alpha-amylase gene (TRINITY_DN4625_c0_g1) was up-regulated compared with NG by 6.6-fold. Meanwhile, genes that were involved in the monosaccharide conversion and hydrolysis process were obviously up-regulated, meaning that glucose metabolism was activated after two weeks of grazing. In the "Glycolysis/Gluconeogenesis" pathway, we found a total of 20 co-expressing DEGs with 18 up-regulated genes. One pyruvate dehydrogenase, 3 pyruvate kinase, and 4 alcohol dehydrogenase had higher expression compared with NG, which means the "Glycolysis/Gluconeogenesis" process in our results mainly carried out the glycolysis process after grazing. Meanwhile, three phosphoglycerate-related proteins were found, which also confirmed the occurrence of glycolysis in *T. mongolicum*. "Pentose phosphate pathway" was found in 11 co-expressing genes that were up-regulated. One transketolase gene, which plays an important catalytic role in pentose biosynthesis, had a high expression compared with NG by 10.1-fold. One carbohydrate kinase and one glucose-6-phosphate 1-dehydrogenase were also noted in this pathway which was up-regulated by 7.4-fold and 6.1-fold, respectively. There were four fructose phosphate-related proteins with up-regulated expression that were common in the "Pentose phosphate pathway" and "Glycolysis/Gluconeogenesis" pathway. One phosphoglucomutase and one glucose-6-phosphate isomerase, the principal enzymes of glucose metabolism were present in all three pathways and were up-regulated 9.9-fold and 4.4-fold, respectively.

The "Biosynthesis of secondary metabolites" pathways were significantly enriched in the HG treatment without LG treatment. High intensity grazing resulted in secondary-related genes rapidly rising and total 188 genes that were found to be involved in a number of pathways, which means that high intensity grazing may have an important effect on plant resistance against outside stress (Table S3). The main concerned pathways of secondary metabolism were phenylpropanoid biosynthesis. The phenylpropanoid-related protein, including three phenylalanine ammonia-lyase, one 4-coumarate—CoA ligase, and three polyphenol oxidase, which are key enzymes of phenylpropanoid pathways, had high expression compared with NG, were down-regulated. We also found several genes that were involved in shikimate pathways, which are the precursor to synthesize phenylalanine. The numbers of secondary metabolism-related proteins, including peroxidase, pyruvate kinase, alcohol dehydrogenase, isopentenyl diphosphate isomerase, glyoxylate/hydroxypyruvate reductase, and linoleate 13S-lipoxygenase, were also up-regulated with high expression. The protein family of cytochrome P450 was found in a total of 42 genes with the most of up-regulation, which not only acted on secondary metabolism but also played an important role in multiple metabolic processes of the plant.

3.9. Identification of Photosynthesis Involved in the Response to Grazing Stress

Grazing caused a serious loss of plant leaves, which was bound to affect photosynthesis. Depending on our results, we identified many DEGs that were related to photosynthesis and chloroplast. KEGG analysis also showed the pathways of photosynthesis were enriched in the HG treatments, including "Photosynthesis-antenna proteins" and "Carbon fixation in photosynthetic organisms". The "Photosynthesis-antenna proteins" pathway was the top enriched pathway in HG, which means the photosynthesis-related process had an intense change in the HG treatments of *T. mongolicum*. We found a total of 24 DEGs that were related to the photosynthesis pathway, including photosystem I (PSI), photosystem II (PSII), ferredoxin, oxygen-evolving enhancer protein, and cytochrome b6-f protein. There were also 10 genes jointly in LG and HG that were all up-regulated with high expression (Table S4). Interestingly, five out of six genes that were related to PSI and PSII were down-regulated in the LG treatment, while the related genes in HG treatment had an opposite trend and were up-regulated. A total of 22 genes that were involved in the "Photosynthesis-antenna proteins" pathway were found in our result with 4 and 12 genes that were only classified in LG and HG, respectively (Figure 9). The main protein of this pathway, chlorophyll A-B binding protein, had a similar tendency to be down-regulated in LG compared with NG, while nearly all the proteins were up-regulated with high expression in the HG treatment, which may mean the different responses of plant resistance to different grazing intensities. Most photosynthesis-related proteins and photosynthetic pigment genes were up-regulated compared with NG, which may mean that *T. mongolicum* significantly increased the capacity of photosynthesis to balance plant growth after two weeks of grazing.

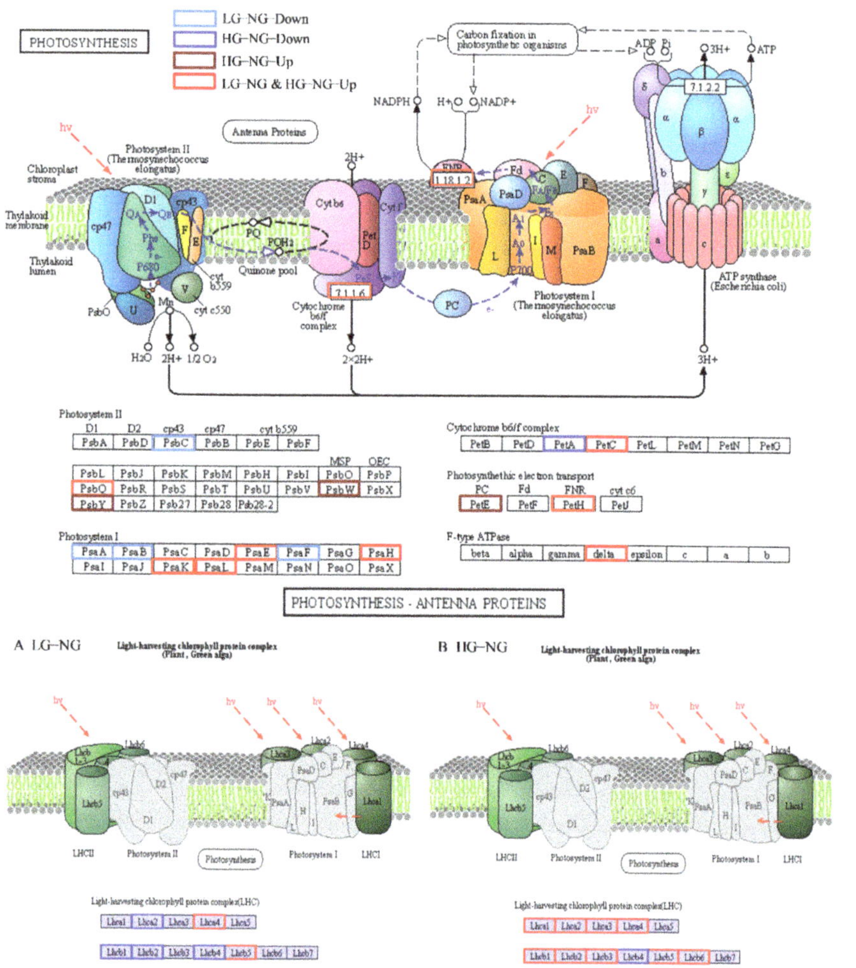

Figure 9. The DEGs about photosynthesis pathway and photosynthesis-antenna proteins in two grazing treatments. Blue blocks mean down-regulated genes in light grazing treatment vs. non-grazing treatment, dark blue blocks represent down-regulated genes in heavy grazing treatment vs. non-grazing treatment, dark red blocks indicate up-regulated genes in heavy grazing treatment vs non-grazing treatment, red blocks mean up-regulated co-expression genes in light grazing treatment vs. non-grazing treatment and heavy grazing treatment vs non-grazing treatment.

4. Discussion

The Qinghai-Tibet Plateau was famous for its special geographical location and highly vulnerable climate. Over the last decades, most of the areas have been grazed by Tibetan sheep and yaks at different intensities. Grazing is the major factors affecting the ecosystem through long-term changes including vegetation loss. Many studies were focused on the influence of vegetation through different grazing intensities [41,42], but, as the transcriptome RNA-Seq analysis developed, some had concern about the RNA-level effects on vegetation by grazing [23,43]. While, prior to this study, grazing was a complex stress for plants with biotic and abiotic effects resulting in the mechanisms of grazing that are not clear yet, most scholars have paid attention to the effect of stress with design and treatment in the lab but not in herbage/plants in natural habitat. The research on wild conditions about how grazing influences plant transcriptomes was lacking, so we chose *T. mongolicum*, a typical species that is found in the grassland of the Tibet Plateau, as our research object, and using transcriptomic RNA-Seq analysis, to probe the molecular mechanisms that were influenced by different grazing intensity. This report obtained new insights into the molecular mechanisms of the grazing stress response in *T. mongolicum*.

For all the samples from the treatment plants, more than 37 million high-quality reads were obtained, which were de novo assembled into 88,598 unigenes with an N50 of 1118 bp and the average length of the assembled unigenes was 659 bp, indicating that the assembly was of high quality. Among the functional annotations, 24,825, 16,039, and 10,437 genes were annotated with the NR GO, and KEGG databases, respectively. We found a total of more than 14 thousand DEGs in LG and HG traits, but only 5867 genes existed in the two treatments. Most of the genes were especially distributed in LG or HG traits, which indicated the molecular mechanisms of *T. mongolicum* to respond to stress by grazing were complex and discrepant.

4.1. Signal Mediate Responses under Different Grazing Intensities

Under stress, plants can trigger the expression of genes that are involved in multiple signal transduction pathways and further activate downstream regulatory pathways that are associated with physiological adaptation. Ca^{2+}, as an important second messenger in plants, which is changed by nearly all signals about developmental, hormonal, and stresses [44,45], was activated in *T. mongolicum* in response to the leaf loss due to grazing. In our results, we found several genes relating to Ca^{2+}-related proteins. The expression of the main proteins, such as calcium exchanger-related protein, calcium-binding EF-hand proteins (CBLs), calcium-dependent protein kinases (CDPKs), and CBL-interacting protein kinases (CIPKs), were all up-regulated after grazing. Both CBLs and CIPKs, as the key step of Ca^{2+} signal transduction, belonged to the group of serine/threonine protein kinases, that interact and regulate plant responses to various environmental stress [46,47]. After being stimulated by biotic and abiotic stress, plants form specific Ca^{2+} signals in the cells that directly binds to the EF-hand domain to change the conformation of CDPKs with kinase active sites that are exposed and kinase activity is activated, which play a key role in decoding Ca^{2+} signatures and transducing signals [44]. CDPKs also can interact with different kinds of pathways to control plant growth and development, hormone signal transduction, and adaptation to stress [45]. A previous study also found that CDPKs and MARKs could interact to mediate signal transduction in wound-induced ethylene production [48]. In our results, we found four MAPKs in two treatments that were enriched to the "MAPK signaling pathway". The MAPK signaling pathway is considered to act on a variety of biological processes in plants, including the transcriptional activation of defense genes, the synthesis of plant resistance-related hormones, and the outbreak of ROS that is induced by pathogens and the thickening of cell walls [49,50].

Besides, we also found other kinds of protein kinases, more than 250, which may play an important role in the perception of grazing in *T. mongolicum*. RLKs, one kind of important plant protein kinase that is located in the cell membrane, could receive stimulus signal and participated in intracellular signal transduction processes as the receptor of signaling molecules [51]. In our result, we mainly found LRR-RLKs, wall-associated receptor kinase, and cysteine-rich receptor-like protein kinase to participate in the signal transduction to defend the effect of grazing. We also found several ATP binding-related kinases, carbohydrate kinases, and pyruvate kinases, which may act on the energy transfer and secondary metabolism process of *T. mongolicum* during adaption of grazing.

ROS are not only normal byproducts of plant cellular oxidative metabolism, but also well-known secondary messengers in various cellular processes in plants. Studies have found that ROS had a dual function in plants, and the transformation depends on its concentration: it could participate in the regulation of plant growth and development and response to adversity stress, as the important signal molecule, in low concentrations; it could become a cell killer when the concentration exceeds the limit that the cell can withstand [52]. Genes encoding POD, GST, GPX, and PPO were identified with up-regulated expression in our results, which means *T. mongolicum* activated the ROS system to defend against damage after grazing.

4.2. Phytohormone Signals under Different Grazing Intensities

Plant hormones play key roles during stress through the interaction of multiple hormones and coordination with various signal transduction pathways [53]. We found that grazing mainly regulated three hormones, auxin, JAs, and ethylene, to adapt to the stress and sustain growth in *T. mongolicum*. Ethylene plays a major role in plant growth and development and response to biotic or abiotic stresses [54]. In our results, we found several genes relating to EREBP and ERF. As one of the plant-specific transcription factors, the ERF subfamily transcription factors are key regulators downstream of the ethylene-mediated stress response signal pathway [8]. But interestingly, most of the genes that were only involved ERF in HG were up-regulated, while in LG they were down-regulated, which may mean different grazing intensities changed the ethylene expression level to adapt to the environment. Ethylene could also interact with auxin to regulate plant growth and development. Auxin affects plant development, coordinates leaf senescence, and participates in multiple signaling pathways with other hormones [55]. The ARF is considered to be a key protein that directly affects auxin downstream response genes. Studies have shown that ARF19 is a positive auxin signal regulating factor, and its expression increases with leaf senescence [56]. We found one ARF19 gene (TRINITY_DN12575_c0_g1) in two treatments had up-regulated expression by 4.86-fold, which may mean that after grazing *T. mongolicum* decreased senescence and increased leaf growth. We also found two IAA proteins and five SAUR proteins with up-regulated transcripts, which jointly participated in the regulation of synthesis and transport to affect the plant growth and signal transduction.

JA signal pathways were the other essential pathways in our results to defend against stress. The response of jasmonic acid to plant defense is very rapid after being injured or eaten by herbivores. It immediately activates the biosynthesis of jasmonic acid and initiates downstream genes through signal transduction, such as inhibiting the digestion process of the eaten leaves by producing defensive molecules [57,58]. We found several genes in Acyl-CoA protein, FAD, AOC, and LOX, which were key enzymes in JA signal pathways. Meanwhile, JAs also could regulate the secondary metabolic process and synthesize the secondary metabolites to resist the harm that is caused by external pressure [59]. However, more and more studies have shown that activating the JA pathway in plants can reduce the activity of plant cell cycle proteins and block the cell cycle, thereby weakening the cell division and elongation of plants, and slowing plant vegetative growth [60]. Some studies have also shown that the resources that are available to the plant and its ecological environment are limited, and growth and defense will consume these limited resources by competing and promoting each other [61]. While the resources that are accumulated by plant growth can be used for defense, and the resources that are recovered by the defense can be used for plant growth [61]. Therefore, under natural conditions, in order to better adjust to the external environment, plants must "balance" between growth and defense to keep the two in a balanced state [28]. As the most typical hormone of growth and defense, ethylene, auxin, and JAs showed different expression levels in our results after different grazing treatments, which adjusted the *T. mongolicum* growth and defense to adapt to external stress.

4.3. Metabolism and Biosynthesis under Different Grazing Intensities

In the process of plant growth and metabolism, carbohydrates are not simply a product of plant photosynthesis, but also a substrate for respiration, providing carbon framework and energy for plant growth and development, and enhancing plant resistance to stress [62]. Carbohydrate metabolism is the center of the entire biological metabolism, connecting the metabolism of proteins, lipids, nucleic acids, and secondary substances [63]. Especially, when plants lose lots of leaves after grazing, which both source and sink of carbohydrates, carbohydrate metabolism became a crucial process to maintain biological functions and adapt to environmental stress. In our result, we found several pathways of carbohydrates metabolism, including "Starch and sucrose metabolism" "Pentose phosphate pathway", and "Glycolysis/Gluconeogenesis", were all enriched in LG and HG based on KEGG analysis. In the "Starch and sucrose metabolism" process, we found carbohydrates tended to convert to sucrose and other monosaccharides, increasing the content of soluble sugars in *T. mongolicum*. Studies have shown that soluble sugars could increase plant stress resistance and regeneration rate to regulate growth [64]. In the "Glycolysis/Gluconeogenesis" pathways, we found several genes that were involved in glycolysis with up-regulated expression, such as glyceraldehyde-3-phosphate dehydrogenase, pyruvate kinase, phosphoglycerate kinase, enolase, aldolase, and hexokinase, participating in the important process of glycolysis. The pyruvate kinase and hexokinase were two of the three irreversible enzymes in the glycolysis process that significantly affect the direction of sugar metabolism [65]. We also found several alcohol dehydrogenases were enriched in this pathway which were up-regulated, which was the key enzymes of anaerobic respiration [66]. This means that *T. mongolicum* had an anaerobic process in sugar metabolism. Meanwhile, alcohol dehydrogenase is also regulated by mechanical injury and disease resistance, and it has an obvious connection with secondary metabolism in plant resistance [67,68]. In "Pentose phosphate pathway", all genes that are enriched in these pathways were up-regulated, and few of genes were the same as "Glycolysis/Gluconeogenesis" and "Starch and sucrose metabolism". There were three pathways that interacted and regulated the carbohydrate metabolism of *T. mongolicum* to adapt to external environment change.

The secondary metabolism plays a significant role in defense against external adversity and increases defense ability after plant injury [69]. Facing feeding or injury of animals and humans, plants increase the secondary metabolites by changing the physiological and biochemical metabolic pathways, forming a large number of compounds such as terpenes, phenols, and alkaloids, increasing their own chemical defense capabilities [70]. Meanwhile, damage is also caused phenylalanine ammonia lyase (PAL), peroxidase, glucanase, ubiquitin protease system, and transcription factor response [59]. In our results, we found several genes in the phenylalanine metabolic pathway with significantly up-regulated expression. The increased activity of PAL is conducive to the synthesis of lignin, flavonoids, and other secondary metabolites, and it improves plant resistance in terms of disease resistance and stress resistance [71]. Glucosinolates that were produced in the metabolic pathway of phenylalanine are also markedly induced in mechanical damage and can resist the destruction of herbivores [72]. Early studies also have found that the total phenol content in the leaves of damaged plants has increased dramatically [71]. In our results, we found a number of the genes of polyphenol oxidase in HG treatment, which may mean that heavy grazing will result in high defense ability. Previous research has suggested that after the wound-induced damage, plants could form a sterile wound skin to protect the underlying tissue from drying out and resist the invasion of pathogens, and the cell wall matrix at the wound site needs polyphenols and fatty compounds for nitrosation to make the wound heal [73]. In total, *T. mongolicum* was increasing defense ability through phenylalanine metabolism and increasing healing speed through polyphenols synthesis.

4.4. Photosynthesis under Different Grazing Intensities

Photosynthesis is the basic physiological process underlying plant growth and development, but grazing can significantly affect this process. Previous research showed that after grazing, photosynthesis was inhibited for a few days, subsequently, it would increase by compensatory photosynthesis [74]. Genes of porphyrin and chlorophyll metabolism, participating in chlorophyll synthesis, were up-regulated in two treatments, implying that the photosynthesis increased after two weeks of grazing. In our research, however, we found the gene expression level that was related to photosynthesis was different. For instance, the HG treatment had more significantly increased expression than the LG treatment with almost all single genes being up-regulated in HG while down-regulated in LG. This mainly includes the expression of genes that are involved in PSI, PSII, and light-harvesting chlorophyll proteins. The efficiency of photosynthesis is closely connected with the activities of PSI and PSII, converting light energy to chemical energy via electron transport [75]. Light-harvesting chlorophyll protein is the main photoreceptor in the plant photosynthesis process [76]. A total of 10 co-expressing genes in two treatments were up-regulated, while eight genes in HG were up-regulated and six genes in LG were down-regulated, which means that the photosystem functions were increased after two weeks of grazing and the compensatory photosynthesis process was more activated in the HG treatment. Furthermore, the photosynthesis process could adjust the generation of ROS, and thus affect the efficiency of photosynthesis [77]. In our research, the scavenging oxidase was up-regulated, which means that after two weeks of grazing, the plant photosynthesis system recovered gradually.

5. Conclusions

After grazing, various metabolic and regulatory processes are altered to adapt to grazing injuries, and these processes often do not disappear immediately but are accompanied by a long stage of plant growth. In this study, we provide a comprehensive description of the transcriptomic responses of plants under natural grazing conditions, and found the genes that were related to the changes in signal and hormone transduction mechanism, carbohydrate metabolism, and photosynthesis in plants after grazing for 2 weeks at maximum biomass. Based on the transcriptomic data, grazing may first stimulate induced signal transduction, including Ca^{2+}, phytohormone, and ROS signaling. All signaling pathways act on protein kinases that can switch on numerous proteins. Then, several metabolism pathways were activated to reduce the damage from leaf loss after grazing through accelerating carbohydrate metabolism and increasing the synthesis of secondary metabolites. The photosynthesis process was also significantly improved to accumulate photosynthetic products and scavenge excess ROS. Meanwhile, the *T. mongolicum* had a stronger response in the photosynthesis process and secondary metabolism, which may mean a compensatory growth in high intensity grazing than light intensity grazing. The proposed model may facilitate future studies on the molecular mechanisms underlying grazing stress responses in plants.

Supplementary Materials: The following supporting information can be downloaded at: https://www.mdpi.com/article/10.3390/plants11172251/s1, Table S1: The primer sequences in qRT-PCR; Table S2: Ca^{2+}- and ROS-related DEGs and protein kinases under different grazing intensities; Table S3: DEGs that were associated with plant hormone under different grazing intensities. Table S4: DEGs that were related to metabolism and biosynthesis under different grazing intensities. Table S5: DEGs that were associated with photosynthesis under different grazing intensities.

Author Contributions: Conceptualization, P.S. and W.Z.; methodology, P.S. and W.Z.; software, Y.W.; validation, P.S., W.Z. and Y.W.; formal analysis, Y.W.; investigation, Y.W.; resources, P.S., S.X. and N.Z.; data curation, Y.W.; writing—original draft preparation, Y.W.; writing—review and editing, F.R.; visualization, Y.W.; supervision, P.S.; project administration, P.S. and S.X.; funding acquisition, P.S. and S.X. All authors have read and agreed to the published version of the manuscript.

Funding: This research was funded by the Qinghai Research and Development and Transformation Project (219-NK-173), the National Natural Science Foundation of China (31901171), and the Open Project of State Key Laboratory of Plateau Ecology and Agriculture, Qinghai University (2022-KF-10).

Institutional Review Board Statement: Not applicable.

Informed Consent Statement: Not applicable.

Data Availability Statement: Not applicable.

Acknowledgments: We sincerely appreciate the technical support provided by OriginGene company in transcription sequencing.

Conflicts of Interest: The authors declare no conflict of interest.

References

1. Petermann, J.S.; Buzhdygan, O.Y. Grassland biodiversity. *Curr. Biol.* **2021**, *31*, R1195–R1201. [CrossRef]
2. Shirazi, S.; Ahmadi, A.; Abdi, N.; Toranj, H.; Khaleghi, M.R. Long-term grazing exclosure: Implications on water erosion and soil physicochemical properties (case study: Bozdaghin Rangelands, North Khorasan, Iran). *Environ. Monit. Assess.* **2021**, *193*, 51. [CrossRef]
3. Li, W.; Li, X.; Zhao, Y.; Zheng, S.; Bai, Q. Ecosystem structure, functioning and stability under climate change and grazing in grasslands: Current status and future prospects. *Curr. Opin. Environ. Sust.* **2018**, *33*, 124–135. [CrossRef]
4. Pavlu, V.; Hejcman, M.; Pavlu, L.; Gaisler, J.; Nezerkova, P. Effect of continuous grazing on forage quality, quantity and animal performance. *Agric. Ecosyst. Environ.* **2016**, *113*, 349–355. [CrossRef]
5. Li, C.; Hao, X.; Willms, W.D.; Zhao, M.; Han, G. Seasonal response of herbage production and its nutrient and mineral contents to long-term cattle grazing on a rough fescue grassland. *Agric. Ecosyst. Environ.* **2009**, *132*, 32–38. [CrossRef]
6. Lei, L.; Zheng, H.; Bi, Y.; Yang, L.; Zou, D. Identification of a major QTL and candidate gene analysis of salt tolerance at the bud burst stage in rice (*Oryza sativa* L.) using QTL-seq and RNA-seq. *Rice* **2020**, *13*, 55. [CrossRef]
7. Yates, S.A.; Swain, M.T.; Hegarty, M.J.; Chernukin, I.; Lowe, M.; Allison, G.G.; Ruttink, T.; Abberton, M.T.; Jenkins, G.; Skot, L. De Novo assembly of red clover transcriptome based on RNA-seq data provides insight into drought response, gene discovery and marker identification. *BMC Genom.* **2014**, *15*, 453. [CrossRef]
8. Cheong, Y.H. Transcriptional profiling reveals novel interactions between wounding, pathogen, abiotic stress, and hormonal responses in *Arabidopsis*. *Plant Physiol.* **2001**, *129*, 661–677. [CrossRef] [PubMed]
9. Zhang, Z.; Gong, J.; Wang, B.; Li, X.; Ding, Y.; Yang, B.; Zhu, C.; Liu, M.; Zhang, W. Regrowth strategies of *Leymus chinensis* in response to different grazing intensities. *Ecol. Appl.* **2020**, *30*, e02113. [CrossRef] [PubMed]
10. Atkinson, N.J.; Urwin, P.E. The interaction of plant biotic and abiotic stresses: From genes to the field. *J. Exp. Bot.* **2012**, *63*, 3523–3543. [CrossRef] [PubMed]
11. Garg, R.; Shankar, R.; Thakkar, B.; Kudapa, H.; Krishnamurthy, L.; Mantri, N.; Varshney, R.K.; Bhatia, S.; Jain, M. Transcriptome analyses reveal genotype- and developmental stage-specific molecular responses to drought and salinity stresses in chickpea. *Sci. Rep.* **2016**, *6*, 19228. [CrossRef] [PubMed]
12. Ding, N.; Huertas, R.; Torres-Jerez, I.; Liu, W.; Watson, B.; Scheible, W.R.; Udvardi, M. Transcriptional, metabolic, physiological and developmental responses of switchgrass to phosphorus limitation. *Plant Cell Environ.* **2021**, *44*, 186–202. [CrossRef]
13. Dang, Z.; Jia, Y.; Tian, Y.; Li, J.; Zhang, Y.; Huang, L.; Liang, C.; Lockhart, P.J.; Matthew, C.; Li, F. Transcriptome-wide gene expression plasticity in *Stipa grandis* in response to grazing intensity differences. *Int. J. Mol. Sci.* **2021**, *22*, 11882. [CrossRef]
14. Kenkel, C.D.; Matz, M.V. Gene expression plasticity as a mechanism of coral adaptation to a variable environment. *Nat. Ecol. Evol.* **2016**, *1*, 14. [CrossRef]
15. Watt, M.; Fiorani, F.; Usadel, B.; Rascher, U.; Muller, O.; Schurr, U. Phenotyping: New windows into the plant for breeders. *Annu. Rev. Plant Biol.* **2020**, *71*, 689–712. [CrossRef]
16. Deng, S.; Ma, J.; Zhang, L.; Chen, F.; Sang, Z.; Jia, Z.; Ma, L. De Novo transcriptome sequencing and gene expression profiling of *Magnolia wufengensis* in response to cold stress. *BMC Plant Biol.* **2019**, *19*, 321. [CrossRef]
17. Nikiforou, C.; Manetas, Y. Ecological stress memory: Evidence in two out of seven species through the examination of the relationship between leaf fluctuating asymmetry and photosynthesis. *Ecol. Indic.* **2017**, *74*, 530–534. [CrossRef]
18. Amaral, M.N.; Arge, L.W.P.; Auler, P.A.; Rossatto, T.; Milech, C.; Magalhaes, A.M., Jr.; Braga, E.J.B. Long-term transcriptional memory in rice plants submitted to salt shock. *Planta* **2020**, *251*, 111. [CrossRef]
19. Gonzalez, A.P.; Chrtek, J.; Dobrev, P.I.; Dumalasova, V.; Fehrer, J.; Mraz, P.; Latzel, V. Stress-induced memory alters growth of clonal offspring of white clover (*Trifolium repens*). *Am. J. Bot.* **2016**, *103*, 1567–1574. [CrossRef]
20. Hilker, M.; Schmulling, T. Stress priming, memory, and signaling in plants. *Plant Cell Environ.* **2019**, *42*, 753–761. [CrossRef]
21. Wang, X.; Zhang, X.; Chen, J.; Wang, X.; Cai, J.; Zhou, Q.; Dai, T.; Cao, W.; Jiang, D. Parental drought-priming enhances tolerance to post-anthesis drought in offspring of wheat. *Front. Plant Sci.* **2018**, *9*, 261. [CrossRef]
22. Baurle, I. Can't remember to forget you: Chromatin-based priming of somatic stress responses. *Semin. Cell Dev. Biol.* **2018**, *83*, 133–139. [CrossRef] [PubMed]
23. Ren, W.; Hou, X.; Wu, Z.; Kong, L.; Guo, H.; Hu, N.; Wan, D.; Zhang, J. De Novo transcriptomic profiling of the clonal *Leymus chinensis* response to long-term overgrazing-induced memory. *Sci. Rep.* **2018**, *8*, 17912. [CrossRef]
24. Ritchie, M.E. Grazing management, forage production and soil carbon dynamics. *Resources* **2020**, *9*, 49. [CrossRef]

25. Sitters, J.; Wubs, E.R.J.; Bakker, E.S.; Crowther, T.W.; Adler, P.B.; Bagchi, S.; Bakker, J.D.; Biederman, L.; Borer, E.T.; Cleland, E.E.; et al. Nutrient availability controls the impact of mammalian herbivores on soil carbon and nitrogen pools in grasslands. *Glob. Chang. Biol.* **2020**, *26*, 2060–2071. [CrossRef] [PubMed]
26. Döbert, T.F.; Bork, E.W.; Apfelbaum, S.; Carlyle, C.N.; Chang, S.X.; Khatri-Chhetri, U.; Silva Sobrinho, L.; Thompson, R.; Boyce, M.S. Adaptive multi-paddock grazing improves water infiltration in canadian grassland soils. *Geoderma* **2021**, *401*, 115314. [CrossRef]
27. Nakano, T.; Bat-Oyun, T.; Shinoda, M. Responses of palatable plants to climate and grazing in semi-arid grasslands of Mongolia. *Glob. Ecol. Conser.* **2020**, *24*, e01231. [CrossRef]
28. Liu, M.; Gong, J.; Li, Y.; Li, X.; Yang, B.; Zhang, Z.; Yang, L.; Hou, X. Growth-defense trade-off regulated by hormones in grass plants growing under different grazing intensities. *Physiol. Plant.* **2019**, *166*, 553–569. [CrossRef] [PubMed]
29. Kerchev, P.I.; Fenton, B.; Foyer, C.H.; Hancock, R.D. Plant responses to insect herbivory: Interactions between photosynthesis, reactive oxygen species and hormonal signaling pathways. *Plant. Cell Environ.* **2012**, *35*, 441–453. [CrossRef]
30. Wang, J.; Zhao, Y.; Ray, I.; Song, M. Transcriptome responses in Alfalfa associated with tolerance to intensive animal grazing. *Sci. Rep.* **2016**, *6*, 19438. [CrossRef]
31. Zhang, Z.; Gong, J.; Shi, J.; Li, X.; Song, L.; Zhang, W.; Li, Y.; Zhang, S.; Dong, J.; Liu, Y. Multiple herbivory pressures lead to different carbon assimilation and allocation strategies: Evidence from a perennial grass in a typical steppe in northern China. *Agric. Ecosyst. Environ.* **2022**, *326*, 107776. [CrossRef]
32. Agrawal, A.A. Overcompensation of plants in response to herbivory and the by-product benefits of mutualism. *Trends Plant Sci.* **2003**, *5*, 309–313. [CrossRef]
33. Sun, J.; Ma, B.; Lu, X. Grazing enhances soil nutrient effects: Trade-offs between Aboveground and belowground biomass in alpine grasslands of the Tibetan plateau. *Land Degrad. Dev.* **2018**, *29*, 337–348. [CrossRef]
34. Li, W.; Lee, C.; Kim, Y.H.; Ma, J.Y.; Shim, S.H. Chemical constituents of the aerial part of *Taraxacum mongolicum* and their chemotaxonomic significance. *Nat. Prod. Res.* **2017**, *31*, 2303–2307. [CrossRef]
35. Li, Y.; Lv, M.; Wang, J.; Tian, Z.; Yu, B.; Wang, B.; Liu, J.; Liu, H. Dandelion (*Taraxacum mongolicum Hand.-Mazz.*) supplementation-enhanced rumen fermentation through the interaction between ruminal microbiome and metabolome. *Microorganisms* **2020**, *9*, 83. [CrossRef]
36. Zhou, H.; Zhou, L.; Zhao, X.; Liu, W.; Li, Y.; Gu, S.; Zhou, X. Stability of alpine meadow ecosystem on the Qinghai-Tibetan plateau. *Sci. Bull.* **2006**, *51*, 320–327. [CrossRef]
37. Grabherr, M.G.; Haas, B.J.; Yassour, M.; Levin, J.Z.; Thompson, D.A.; Amit, I.; Adiconis, X.; Fan, L.; Raychowdhury, R.; Zeng, Q.; et al. Full-length transcriptome assembly from RNA-seq data without a reference genome. *Nat. Biotechnol.* **2011**, *29*, 644–652. [CrossRef]
38. Camacho, C.; Coulouris, G.; Avagyan, V.; Ma, N.; Papadopoulos, J.; Bealer, K.; Madden, T.L. BLAST+: Architecture and applications. *BMC Bioinform.* **2009**, *10*, 421. [CrossRef]
39. Conesa, A.; Gotz, S.; Garcia-Gomez, J.M.; Terol, J.; Talon, M.; Robles, M. Blast2GO: A universal tool for annotation, visualization and analysis in functional genomics research. *Bioinformatics* **2005**, *21*, 3674–3676. [CrossRef]
40. Robinson, M.D.; McCarthy, D.J.; Smyth, G.K. EdgeR: A bioconductor package for differential expression analysis of digital gene expression data. *Bioinformatics* **2010**, *26*, 139–140. [CrossRef]
41. Koncz, P.; Vadász-Besnyői, V.; Csathó, A.I.; Nagy, J.; Szerdahelyi, T.; Tóth, Z.; Pintér, K.; Fóti, S.; Papp, M.; Balogh, J. Carbon uptake changed but vegetation composition remained stable during transition from grazing to mowing grassland management. *Agric. Ecosyst. Environ.* **2020**, *304*, 107161. [CrossRef]
42. Mendes, C.; Dias, E.; Rochefort, L.; Azevedo, J. Regenerative succession of Azorean peatlands after grazing: Vegetation path to self-recovery. *Wetl. Ecol. Manag.* **2020**, *28*, 177–190. [CrossRef]
43. Chen, S.; Cai, Y.; Zhang, L.; Yan, X.; Cheng, L.; Qi, D.; Zhou, Q.; Li, X.; Liu, G. Transcriptome analysis reveals common and distinct mechanisms for Sheepgrass (*Leymus Chinensis*) responses to defoliation compared to mechanical wounding. *PLoS ONE* **2014**, *9*, e89495. [CrossRef] [PubMed]
44. Billker, O.; Lourido, S.; Sibley, L.D. Calcium-dependent signaling and kinases in Apicomplexan parasites. *Cell Host Microbe.* **2009**, *5*, 612–622. [CrossRef] [PubMed]
45. Valmonte, G.R.; Arthur, K.; Higgins, C.M.; MacDiarmid, R.M. Calcium-dependent protein kinases in plants: Evolution, expression and function. *Plant Cell Physiol.* **2014**, *55*, 551–569. [CrossRef] [PubMed]
46. Wan, D.; Li, R.; Zou, B.; Zhang, X.; Cong, J.; Wang, R.; Xia, Y.; Li, G. Calmodulin-binding protein CBP60g is a positive regulator of both disease resistance and drought tolerance in *Arabidopsis*. *Plant Cell Rep.* **2012**, *31*, 1269–1281. [CrossRef] [PubMed]
47. Grabarek, Z. Structural basis for diversity of the EF-hand calcium-binding proteins. *J. Mol. Biol.* **2006**, *359*, 509–525. [CrossRef]
48. Li, S.; Han, X.; Yang, L.; Deng, X.; Wu, H.; Zhang, M.; Liu, Y.; Zhang, S.; Xu, J. Mitogen-activated protein kinases and calcium-dependent protein kinases are involved in wounding-induced ethylene biosynthesis in *Arabidopsis Thaliana*. *Plant Cell Environ.* **2018**, *41*, 134–147. [CrossRef]
49. Xu, J.; Zhang, S. Mitogen-activated protein kinase cascades in signaling plant growth and development. *Trends Plant Sci.* **2015**, *20*, 56–64. [CrossRef]
50. Zhang, M.; Zhang, S. Mitogen-activated protein kinase cascades in plant signaling. *J. Integr. Plant Biol.* **2022**, *64*, 301–341. [CrossRef]
51. Ye, Y.; Ding, Y.; Jiang, Q.; Wang, F.; Sun, J.; Zhu, C. The role of receptor-like protein kinases (RLKs) in abiotic stress response in plants. *Plant Cell Rep.* **2017**, *36*, 235–242. [CrossRef] [PubMed]

52. Gilroy, S.; Suzuki, N.; Miller, G.; Choi, W.-G.; Toyota, M.; Devireddy, A.R.; Mittler, R. A tidal wave of signals: Calcium and ROS at the forefront of rapid systemic signaling. *Trends Plant Sci.* **2014**, *19*, 623–630. [CrossRef]
53. Nambara, E.; Van Wees, S.C.M. Plant hormone functions and interactions in biological systems. *Plant J.* **2021**, *105*, 287–289. [CrossRef]
54. Wang, K.L.-C.; Li, H.; Ecker, J.R. Ethylene biosynthesis and signaling networks. *Plant Cell* **2002**, *14*, S131–S151. [CrossRef]
55. Chen, H.; Bullock, D.A.; Alonso, J.M.; Stepanova, A.N. To fight or to grow: The Balancing role of ethylene in plant abiotic stress responses. *Plants* **2021**, *11*, 33. [CrossRef] [PubMed]
56. Huang, K.-L.; Ma, G.-J.; Zhang, M.-L.; Xiong, H.; Wu, H.; Zhao, C.-Z.; Liu, C.-S.; Jia, H.-X.; Chen, L.; Kjorven, J.O.; et al. The ARF7 and ARF19 transcription factors positively regulate PHOSPHATE STARVATION RESPONSE1 in *Arabidopsis* Roots. *Plant Physiol.* **2018**, *178*, 413–427. [CrossRef]
57. Liu, N.; Avramova, Z. Molecular mechanism of the priming by jasmonic acid of specific dehydration stress response genes in *Arabidopsis*. *Epigenet. Chromatin.* **2016**, *9*, 8. [CrossRef]
58. Lyons, R.; Manners, J.M.; Kazan, K. Jasmonate biosynthesis and signaling in monocots: A comparative overview. *Plant Cell Rep.* **2013**, *32*, 815–827. [CrossRef]
59. Afrin, S.; Huang, J.; Luo, Z. JA-mediated transcriptional regulation of secondary metabolism in medicinal plants. *Sci. Bull.* **2015**, *60*, 1062–1072. [CrossRef]
60. Campos, M.L.; Yoshida, Y.; Major, I.T.; Ferreira, D.D.; Weraduwage, S.M.; Froehlich, J.E.; Johnson, B.F.; Kramer, D.M.; Jander, G.; Sharkey, T.D.; et al. Rewiring of Jasmonate and Phytochrome B signalling uncouples plant growth-defense tradeoffs. *Nat. Commun.* **2016**, *7*, 12570. [CrossRef] [PubMed]
61. Barto, E.K.; Cipollini, D. Testing the optimal defense theory and the growth-differentiation balance hypothesis in *Arabidopsis Thaliana*. *Oecologia* **2005**, *146*, 169–178. [CrossRef] [PubMed]
62. Kou, S.-M.; Jin, R.; Wu, Y.-Y.; Huang, J.-W.; Zhang, Q.-Y.; Sun, N.-J.; Yang, Y.; Guan, C.-F.; Wang, W.-Q.; Zhu, C.-Q.; et al. Transcriptome analysis revealed the roles of carbohydrate metabolism on differential Acetaldehyde production capacity in persimmon fruit in response to high-CO_2 treatment. *J. Agric. Food Chem.* **2021**, *69*, 836–845. [CrossRef] [PubMed]
63. McLoughlin, F.; Marshall, R.S.; Ding, X.; Chatt, E.C.; Kirkpatrick, L.D.; Augustine, R.C.; Li, F.; Otegui, M.S.; Vierstra, R.D. Autophagy plays prominent roles in amino acid, nucleotide, and carbohydrate metabolism during fixed-carbon starvation in maize. *Plant Cell* **2020**, *32*, 2699–2724. [CrossRef] [PubMed]
64. Couée, I.; Sulmon, C.; Gouesbet, G.; El Amrani, A. Involvement of soluble sugars in reactive oxygen species balance and responses to oxidative stress in plants. *J. Exp. Bot.* **2006**, *57*, 449–459. [CrossRef]
65. Plaxton, W.C. The organization and regulation of plant glycolysis. *Annu. Rev. Plant Physiol. Plant Mol. Biol.* **1996**, *47*, 185–214. [CrossRef]
66. Ventura, I.; Brunello, L.; Iacopino, S.; Valeri, M.C.; Novi, G.; Dornbusch, T.; Perata, P.; Loreti, E. *Arabidopsis* phenotyping reveals the importance of Alcohol Dehydrogenase and Pyruvate Decarboxylase for aerobic plant growth. *Sci. Rep.* **2020**, *10*, 16669. [CrossRef]
67. Kato-Noguchi, H. Wounding stress induces Alcohol Dehydrogenase in maize and lettuce seedlings. *Plant Growth Regul.* **2001**, *35*, 285–288. [CrossRef]
68. Tesniere, C.; Torregrosa, L.; Pradal, M.; Souquet, J.-M.; Gilles, C.; Dos Santos, K.; Chatelet, P.; Gunata, Z. Effects of genetic manipulation of Alcohol Dehydrogenase levels on the response to stress and the synthesis of secondary metabolites in grapevine leaves. *J. Exp. Bot.* **2006**, *57*, 91–99. [CrossRef]
69. Alsamadany, H. De Novo leaf transcriptome assembly of *Bougainvillea spectabilis* for the identification of genes involves in the secondary metabolite pathways. *Gene* **2020**, *746*, 144660. [CrossRef]
70. Obata, T. Metabolons in plant primary and secondary metabolism. *Phytochem. Rev.* **2019**, *18*, 1483–1507. [CrossRef]
71. Yoshikawa, M.; Luo, W.; Tanaka, G.; Konishi, Y.; Matsuura, H.; Takahashi, K. Wounding stress induces Phenylalanine Ammonia Lyases, leading to the accumulation of Phenylpropanoids in the model Liverwort *Marchantia Polymorpha*. *Phytochemistry* **2018**, *155*, 30–36. [CrossRef]
72. Guan, Y.; Hu, W.; Xu, Y.; Yang, X.; Ji, Y.; Feng, K.; Sarengaowa. Metabolomics and physiological analyses validates previous findings on the mechanism of response to wounding stress of different intensities in broccoli. *Food Res. Int.* **2021**, *140*, 110058. [CrossRef]
73. Zhang, J.; Sun, X. Recent advances in Polyphenol Oxidase-mediated plant stress responses. *Phytochemistry* **2021**, *181*, 112588. [CrossRef]
74. Liu, M.; Gong, J.; Yang, B.; Ding, Y.; Zhang, Z.; Wang, B.; Zhu, C.; Hou, X. Differences in the photosynthetic and physiological responses of *Leymus Chinensis* to different levels of grazing intensity. *BMC Plant Biol.* **2019**, *19*, 558. [CrossRef]
75. Bag, P.; Chukhutsina, V.; Zhang, Z.; Paul, S.; Ivanov, A.G.; Shutova, T.; Croce, R.; Holzwarth, A.R.; Jansson, S. Direct energy transfer from photosystem II to photosystem I confers winter sustainability in scots pine. *Nat. Commun.* **2020**, *11*, 6388. [CrossRef]
76. Ren, W.; Hu, N.; Hou, X.; Zhang, J.; Guo, H.; Liu, Z.; Kong, L.; Wu, Z.; Wang, H.; Li, X. Long-term overgrazing-induced memory decreases photosynthesis of clonal offspring in a perennial grassland plant. *Front. Plant Sci.* **2017**, *8*, 419. [CrossRef]
77. Khorobrykh, S.; Havurinne, V.; Mattila, H.; Tyystjärvi, E. Oxygen and ROS in photosynthesis. *Plants* **2020**, *9*, 91. [CrossRef]

plants

MDPI

Article

Metabolic Changes in Seed Embryos of Hypoxia-Tolerant Rice and Hypoxia-Sensitive Barley at the Onset of Germination

Jayamini Jayawardhane [1,2,*], M. K. Pabasari S. Wijesinghe [1,3], Natalia V. Bykova [3] and Abir U. Igamberdiev [1,*]

1 Department of Biology, Memorial University of Newfoundland, St. John's, NL A1C 5S7, Canada; mkpswijesing@mun.ca
2 Department of Botany, Faculty of Science, University of British Columbia, Vancouver, BC V6T 1Z4, Canada
3 Morden Research and Development Centre, Agriculture and Agri-Food Canada, Morden, MB R6M 1Y5, Canada; natalia.bykova@agr.gc.ca
* Correspondence: jjayawardhan@mun.ca (J.J.); igamberdiev@mun.ca (A.U.I.)

Abstract: Rice (*Oryza sativa* L.) and barley (*Hordeum vulgare* L.) are the cereal species differing in tolerance to oxygen deficiency. To understand metabolic differences determining the sensitivity to low oxygen, we germinated rice and barley seeds and studied changes in the levels of reactive oxygen species (ROS) and reactive nitrogen species (RNS), activities of the enzymes involved in their scavenging, and measured cell damage parameters. The results show that alcohol dehydrogenase activity was higher in rice than in barley embryos providing efficient anaerobic fermentation. Nitric oxide (NO) levels were also higher in rice embryos indicating higher NO turnover. Both fermentation and NO turnover can explain higher ATP/ADP ratio values in rice embryos as compared to barley. Rice embryos were characterized by higher activity of *S*-nitrosoglutathione reductase than in barley and a higher level of free thiols in proteins. The activities of antioxidant enzymes (superoxide dismutase, ascorbate peroxidase, monodehydroascorbate reductase, dehydroascorbate reductase) in imbibed embryos were higher in rice than in barley, which corresponded to the reduced levels of ROS, malonic dialdehyde and electrolyte leakage. The observed differences in metabolic changes in embryos of the two cereal species differing in tolerance to hypoxia can partly explain the adaptation of rice to low oxygen environments.

Keywords: *Oryza sativa*; *Hordeum vulgare*; hypoxia tolerance; nitric oxide; imbibition; reactive oxygen species; ATP/ADP ratio

Citation: Jayawardhane, J.; Wijesinghe, M.K.P.S.; Bykova, N.V.; Igamberdiev, A.U. Metabolic Changes in Seed Embryos of Hypoxia-Tolerant Rice and Hypoxia-Sensitive Barley at the Onset of Germination. *Plants* **2021**, *10*, 2456. https://doi.org/10.3390/plants10112456

Academic Editors: Beata Prabucka, Mateusz Labudda, Marta Gietler, Justyna Fidler and Małgorzata Nykiel

Received: 30 September 2021
Accepted: 11 November 2021
Published: 14 November 2021

Publisher's Note: MDPI stays neutral with regard to jurisdictional claims in published maps and institutional affiliations.

1. Introduction

Rice (*Oryza sativa* L.) and barley (*Hordeum vulgare* L.) are economically important cereal crops. Apart from being the staple foods in many countries, rice and barley are also ideal plant models for studying monocot seed germination because of the availability of genomic resources with annotated reference genomes and well-studied physiological, morphological and metabolic traits [1,2]. Rice is an anoxia/hypoxia tolerant species, while barley is hypoxia/anoxia intolerant [3], which makes these species important for elucidating the genetic, physiological and biochemical background of hypoxia tolerance. Since the germinating seeds of all cereals and many other plants are highly hypoxic upon imbibition and before radicle protrusion [4], we used rice and barley as contrasting plant species to study the differences in their metabolism in order to explain the metabolic basis of coping with hypoxia tolerance at the early stages of germination.

Seed germination is a vital stage in the plant life cycle, and it begins with seed rehydration and imbibition [5]. In general, the germination process can be distinguished by three major phases, which include rapid water uptake by a dry seed upon imbibition (phase I), reactivation of metabolism (phase II), and radicle protrusion (phase III) [4]. The second phase is the most critical stage where important physiological and biochemical processes that initiate the germination process reactivate [6]. Due to imbibition, the cell wall enlarges, the seed coat becomes softened [7], and the water availability directs the enzymatic hydrolysis of proteins, lipids and carbohydrates, and the transportation of metabolites [8].

Seed germination and dormancy are under the control of both genetic and biochemical processes [9]. Weakening of the endosperm during germination via by α-xylosidase activity, biosynthesis of xyloglucan in the endosperm, arrangement of cutin coat in the endosperm-testa interface play critical roles in determining dormancy and germination [9]. Some structures of the seed, such as seed coat (or testa), act as a physical barrier for gas exchange [10]. Due to the resumption of respiratory activity following imbibition, the oxygen content in the seed tissue is rapidly diminished (reviewed in [11]). Therefore, the supply of the oxygen through the seed coat to the embryo becomes limited, which generates the hypoxic environment in the seed [12]. Consequently, aerobic respiration in the seed is suppressed and anaerobic respiration is developed to maintain the energy status in cells [13]. Initiation of the fermentation process under hypoxic condition is considered as an adaptive mechanism for ATP synthesis [14,15]. The increased ethanol fermentation in seeds is linked with oxygen deprivation and is catalyzed primarily with the participation of alcohol dehydrogenase (ADH) [14,16]. Consequently, the ATP level and energy charge inside seeds remain high during early germination [17,18].

Moreover, the balance between reactive nitrogen species (RNS) that include nitric oxide (NO) and reactive oxygen species (ROS), acting as the important signaling substances under stress, plays a crucial role in breaking dormancy of seeds and induction of germination [19–23]. Imbibition induces the formation of ROS and RNS [21,24] and leads to the changes in thiol redox-sensitive seed proteome [25,26]. Hypoxic environment within the seed triggers the production of NO under the seed coat [27]. ROS are formed during the restarting of metabolism by the increased oxidative processes leading to the activation of electron transport, in particular at the levels of mitochondrial electron transport chain, of the plasma membrane NADPH oxidase, xanthine oxidase and peroxidases as part of an oxidative burst during rehydration [28]. Antioxidant systems and proteins that scavenge reactive radicals and NO reduce oxidative stress damage in seeds to prevent loss of germination capacity [29,30] and these biochemical events are activated immediately upon rehydration [31–35].

In this study, we evaluated metabolic differences that control the sensitivity to oxygen deficiency in germinating rice and barley seeds. Changes in the levels of ROS and RNS, in activities of the enzymes involved in their generation and scavenging were determined, and parameters measuring cell damage such as malonic dialdehyde and electrolyte leakage were assessed. The observed differences in metabolic responses in embryos of the cultivars of two cereal species differing in tolerance to hypoxia are discussed in relation to the development of mechanisms of metabolic adjustments to low oxygen conditions occurring in germinating seeds.

2. Results

2.1. Germination Rates of Rice and Barley Seeds

Barley seeds started radicle protrusion at 15 h while in rice the protrusion of radicle was delayed until 48 h of post-imbibition. Both species showed a similar germination percentage at the end of five days after imbibition (barley 97%, rice 93%). However, the rate of germination in rice at the initial stage was lower than in barley. After three days, a similar germination percentage was reached in both species, and was comparable until the end of five days of germination assay (Figure 1).

2.2. Alcohol Dehydrogenase Activity and the Level of Adenylates during Germination

Embryonic ATP content exhibited the initial fluctuation in both species (Figure 2A). In barley, the peak of ATP level was observed at 9 h from the start of imbibition while in rice at 15 h. The following decline was sharp in barley and smoother in rice. Further increase of ATP content by 48 h was pronounced in barley but was not observed in rice (Figure 2A).

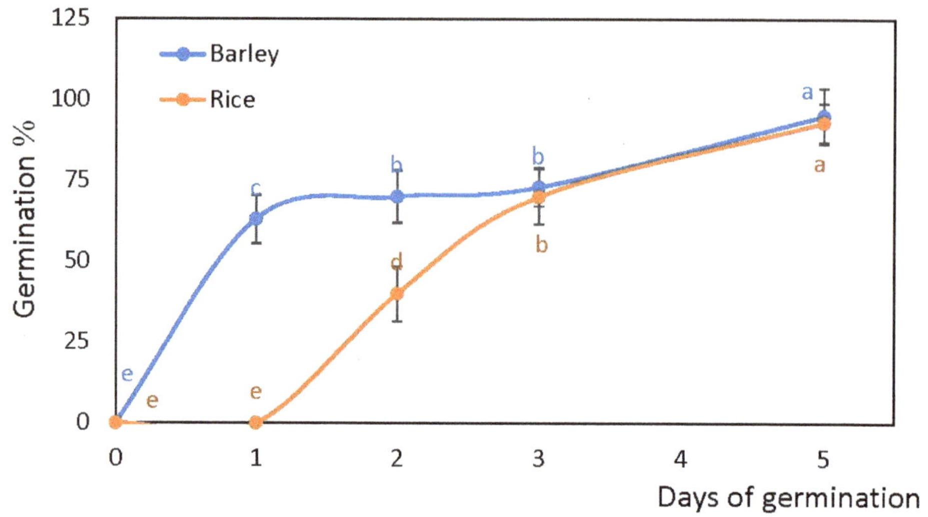

Figure 1. Germination of rice and barley seeds. Germination rates of two species within five days of imbibition. Vertical bars represent standard deviations ($n = 3$). Different letters indicate significant differences between the two species and the time points at $p < 0.05$, (one-way ANOVA test, Tukey comparison).

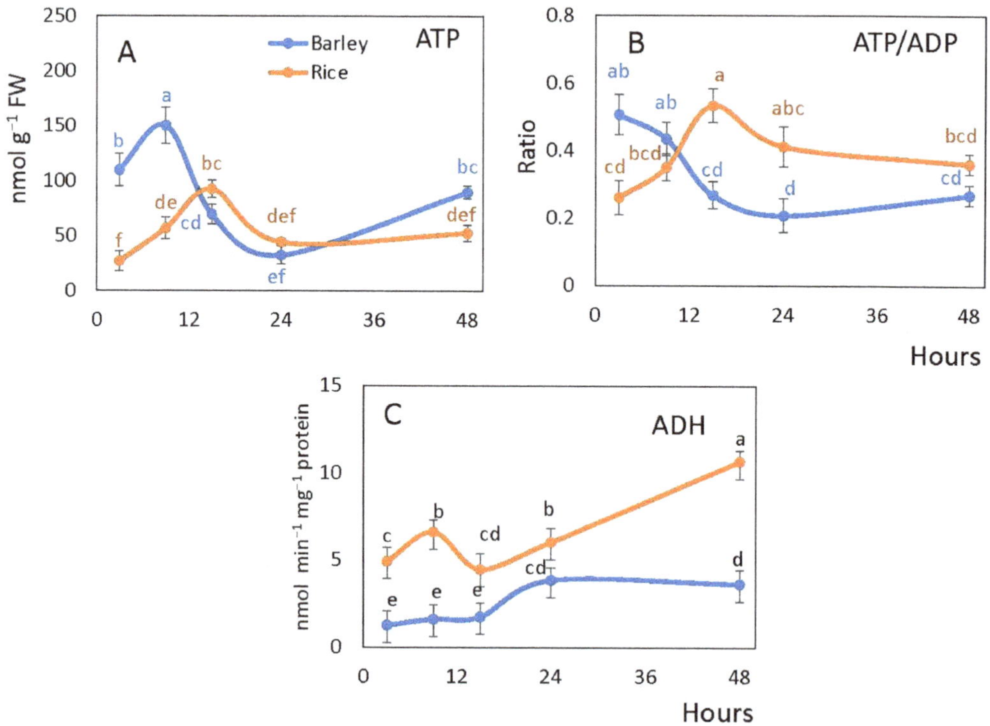

Figure 2. Total ATP (**A**), ATP/ADP ratio (**B**) and ADH activity (**C**) in barley and rice seeds following imbibition. Vertical bars represent standard deviations ($n = 3$). Different letters indicate significant differences between the two species and the time points at $p < 0.05$, (one-way ANOVA test, Tukey comparison).

The ATP/ADP ratio generally followed a similar pattern to the ATP content in rice, but in barley seed embryos it gradually declined upon imbibition and then stabilized. In rice embryos, the ATP/ADP ratio increased from the values of ~0.2 lower than in barley at 3 h peaking at ~0.25 higher than in barley at 15 h, and then slightly decreased (Figure 2B).

ADH activity in the embryos was significantly higher in rice than barley. In rice, it markedly increased at 9 h after imbibition followed by a drop at 15 h, after which it continued to increase significantly. On the contrary, ADH activity in barley remained constant for 15 h and then increased by 24 h (Figure 2C).

2.3. Nitric Oxide, Free Thiols, S-Nitrosoglutathione Reductase and S-Nitrosylation

NO content in embryos during germination remained higher in rice than in barley throughout the whole time of observation, the difference was not significant only at 15 h. It decreased until 15 h of imbibition in rice and until 9 h in barley, and then continuously increased (Figure 3A).

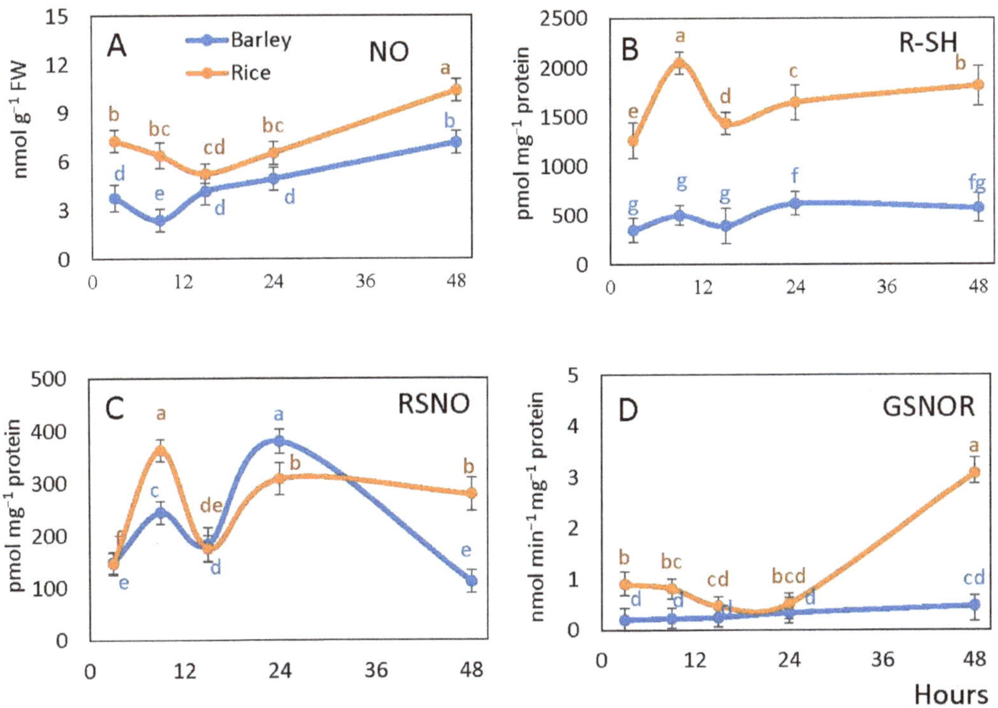

Figure 3. Levels of NO (**A**), free thiols (R-SH) (**B**), *S*-nitrosylation (RSNO) (**C**) and *S*-nitrosoglutathione (GSNOR) activity (**D**) in barley and rice embryos following imbibition. Vertical bars represent standard deviations ($n = 3$). Different letters indicate significant differences between the two species and the time points at $p < 0.05$, (one-way ANOVA test, Tukey comparison).

Free thiol (RSH) content of embryos was essentially higher in rice embryos than in barley embryos. In rice, the content of thiols exhibited an increase at 9 h after imbibition followed by a decrease at 15 h and further increase to 24 h. In barley, only a slight increase was observed from 15 to 24 h. Between 24 to 48 h of imbibition, the measured R-SH content remained stable in barley and slightly increased in rice (Figure 3B).

The content of *S*-nitrosylated (RSNO) groups exhibited significant fluctuations over time points in both species. It increased sharply in rice and more moderately in barley at 9 h, and then decreased by 15 h of imbibition. A further increase was observed at one day of imbibition, after which RSNO level remained stable in rice and decreased in barley (Figure 3C).

S-nitrosoglutathione reductase (GSNOR) activity of seed embryo was higher in rice than in barley in the first 9 h and by 48 h after imbibition. The activity level was stable in barley during the entire period of observation, while in rice it decreased by 15 h and then increased sharply from 24 to 48 h (Figure 3D).

2.4. Reactive Oxygen Species, Lipid Peroxidation and Electrolyte Leakage in Rice and Barley Seeds

The level of hydrogen peroxide (H_2O_2) in embryo was significantly higher in barley than in rice during all periods of observation, except the 15 h point. In barley, apart from the steep decrease from 9 to 15 h corresponding to the time before radicle protrusion, the level remained constant and exceeded H_2O_2 level in rice by 2–4 times. Rice embryo showed a generally stable H_2O_2 level until 24 h, followed by the increase to 48 h (Figure 4A).

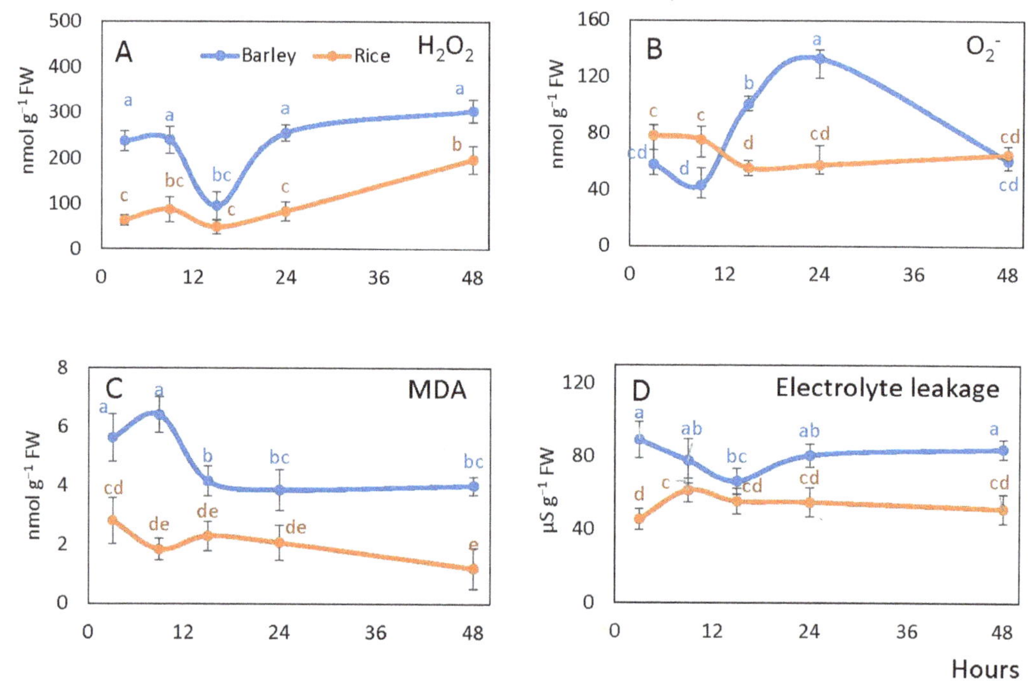

Figure 4. Barley and rice seed embryonic hydrogen peroxide (H_2O_2) (**A**), superoxide anion (**B**), lipid peroxidation in terms of MDA (**C**) and seed electrolyte leakage (**D**) following imbibition. Vertical bars represent standard deviations (n = 3). Different letters indicate significant differences between the two species and the time points at $p < 0.05$, (one-way ANOVA test, Tukey comparison).

Embryonic superoxide anion content in rice was slightly higher than in barley at 9 h after imbibition, then decreased by 15 h and did not change significantly in the subsequent hours. In barley, superoxide levels increased by 15 h peaking at 24 h at the levels threefold higher than at the beginning of imbibition, followed by further decline to 48 h (Figure 4B).

Malondialdehyde (MDA) levels in the embryos remained significantly (2–3 times) higher in barley than in rice after imbibition being the highest at 3 and 9 h of imbibition (before radicle protrusion). MDA levels in rice did not exhibit significant changes over the whole period (Figure 4C).

Electrolyte leakage of germinating barley seeds was also significantly higher in barley than in rice during almost all period of observation (except the 15 h point). The levels slightly declined in barley until 15 h and then slightly increased to 48 h. In rice seeds, the changes in the level of electrolyte leakage were statistically insignificant over the germination period tested except the initial slight increase (Figure 4D).

2.5. Antioxidant Enzyme Activities in Rice and Barley Seeds

All studied antioxidant enzymes except catalase exhibited higher activity in rice than in barley (Figure 5). SOD activity increased sharply in rice embryo during germination reaching 5–10 times higher values than in barley where it changed only slightly with the two-fold increase by 24 h (after radicle protrusion) (Figure 5A). Catalase activity strongly increased in barley embryos at the onset of germination, while in rice embryos it was always lower and increased only at 24 h of imbibition (Figure 5B).

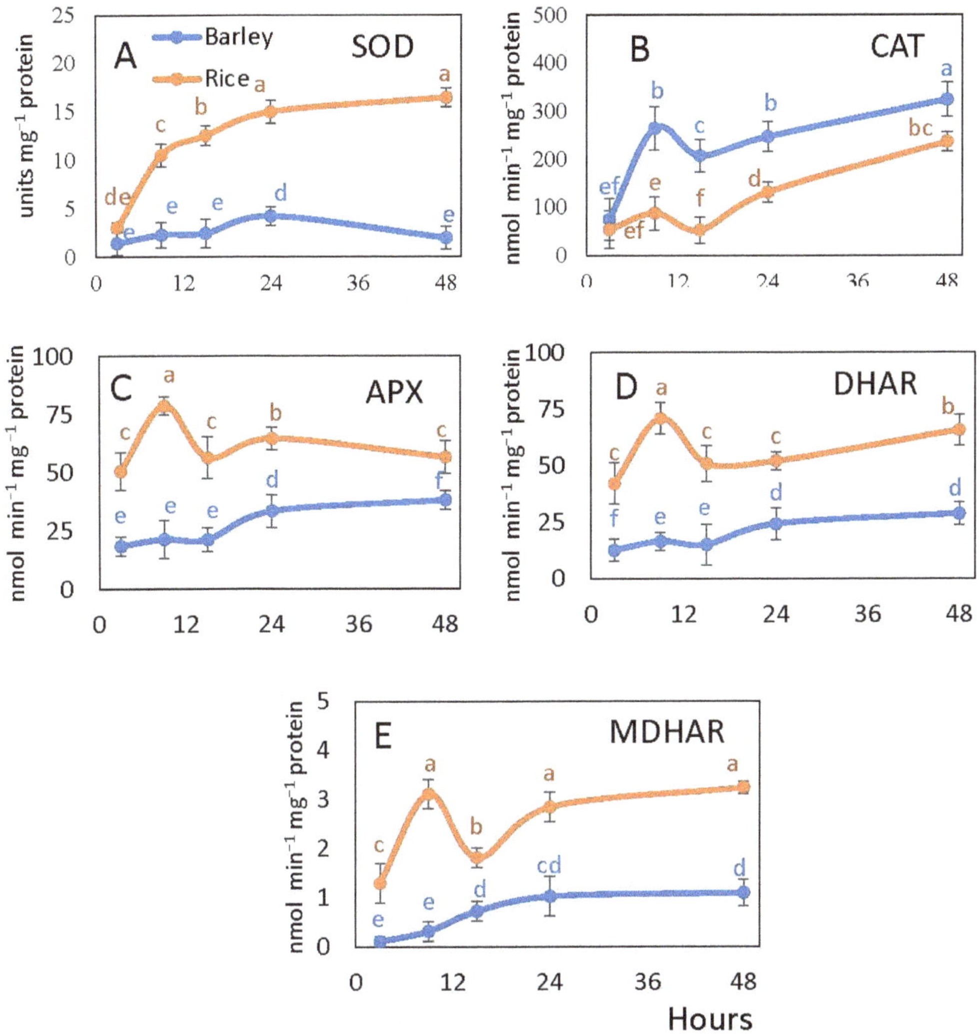

Figure 5. Embryonic antioxidant enzyme activities of barley and rice seeds following imbibition: superoxide dismutase (SOD) (**A**), catalase (CAT) (**B**), ascorbate peroxidase (APX) (**C**), dehydroascorbate reductase (DHAR) (**D**), monodehydroascorbate reductase (MDHAR) (**E**). Vertical bars represent standard deviations (n = 3). Different letters indicate significant differences between the two species and the time points at $p < 0.05$, (one-way ANOVA test, Tukey comparison).

APX, MDHAR and DHAR activities were consistently higher in rice than in barley exhibiting an increase at 9 h and decrease at 15 h. In barley embryos, they were lower with a smooth increase of APX and DHAR by 24 h (upon radicle protrusion) and the earlier continuous increase of MDHAR (Figure 5C–E).

3. Discussion

3.1. Seed Germination Phases and Hypoxic Conditions under Seed Coat

Seed imbibition triggers many biochemical and cellular processes associated with germination [6,36]. As water is taken up by the dry seed, it triggers the transition of seed to germination which completes with the emergence of embryonic radicle tissues from the seed coat [8]. Germination of cereals is controlled by complex signalling networks including both internal and external cues [23]. Phytohormones such as ABA and GA act as the hubs that connect internal and external signals, controlling germination antagonistically, whereas other phytohormones, carbohydrates, ROS, NO, microRNAs, light, and temperature also affect germination at the transcriptional, translational, and post-translational levels [23]. Bewley [4] observed radicle protrusion in rice at 50 h of imbibition and Ma et al. [35] observed radicle protrusion in barley at 12–15 h of imbibition. Our results demonstrate that barley seeds started radicle protrusion at 15 h of imbibition, while in rice it only happened between 24 and 48 h after onset of imbibition (Figure 1).

According to Gruwel et al. [37], barley seeds hydrate rapidly during very early imbibition and the absorbed water is mainly confined to the embryo tissue. A triphasic model was introduced to reflect the increase of total water content during germination [4]. During the first 20 h of imbibition (Phase I), rice seeds increased in weight rapidly without significant morphological changes [4]. Phase II showed a stable plateau until 50 h where the coleoptiles elongated [4]. Another rapid water uptake stage and radicle protrusion were shown to take place in Phase III [38].

While Phase I is characterized by the dramatic increase in respiration rates in seeds, the internal oxygen concentration due to the low permeability of seed coat is rapidly depleted [4,11]. The seed becomes hypoxic at Phase II for a prolonged period of time in rice, while in barley this phase is significantly shorter. A longer duration of the hypoxic phase in rice means the persistence of efficient anaerobic metabolism in this plant, while in barley, lower resistance to oxygen deficiency requires a fast completion of this phase and accelerated transition to Phase III initiated by radicle protrusion. From this point of view, the comparison of metabolism in rice and barley during seed germination can reveal essential metabolic differences at this stage related to hypoxia sensitivity in plants. Below we discuss these differences in relation to fermentation, nitric oxide metabolism, production and scavenging of ROS and RNS in the overall context of energy metabolism of the germinating seeds.

3.2. Respiration and Energy Availability in the Embryos of Imbibed Seeds

Due to a rapid depletion of oxygen in seed tissues upon imbibition, oxidative respiration becomes limited, resulting in the depleted energy status of early germinating seeds [39–41]. According to He et al. [42], aerobic respiration in rice seeds is quite low during the first 48 h of imbibition due to the lack of functional mitochondria. However, seed ATP levels and energy charge remain elevated because of the glycolytic fermentation and/or other adaptive and alternative mechanisms for energy generation [17,18,43]. Anaerobic respiration pathways, such as fermentation, might be the main source of energy at the early stage of germination of rice seeds [38,44]. Fermentation in terms of ADH activity was higher in rice than in barley and it increased continuously in the first nine hours and then from 15 h towards 48 h when all rice seeds become germinated (Figure 2C). This indicates more efficient anaerobic metabolism in rice than in barley and explains the observed fact that the germination of rice seeds can take more time, during which oxygen becomes almost fully depleted. During this time (starting from 15 h), rice seeds can maintain higher ATP/ADP ratio than barley seeds despite of the delayed radicle protrusion. At 24 h of imbibition, when most barley seeds are germinated and rice seeds are not, the ATP/ADP ratio remains higher in rice. In addition to the glycolytic fermentation, alternative pathways such as the phytoglobin-NO cycle, can contribute to the maintenance of high energy state under the conditions of hypoxia [45]. Further, phytoglobin (*Pgb1*) gene expression is essential to maintain redox and energy balance before radicle protrusion, when seeds experience low internal oxygen concentration [46].

3.3. Nitric Oxide and ROS Scavengers in Embryo of Imbibed Seeds

Seeds produce ROS such as H_2O_2, O_2^-, and hydroxyl radicals, and RNS, such as NO, during imbibition [21,24,47–49]. ATP production under the conditions of oxygen deficiency can be associated not only with the glycolytic fermentation but also with the phytoglobin-NO cycle [50]. NO is scavenged by class 1 phytoglobin that is expressed within 2 h after imbibition of seeds [17,45,51–53]. Our results revealed that the NO level was higher in rice than in barley embryo (Figure 3A), suggesting that the phytoglobin-NO cycle could be more active in rice, and a higher ATP/ADP ratio in rice than in barley at 15–24 h might be related to the activity of this cycle. For rice, it has been shown that the phytoglobin-NO cycle and the mitochondrial alternative oxidase play a vital role in anaerobic germination and growth of deep-water rice [54]. A sharper decrease of RSNO (which mostly refers to *S*-nitrosoglutathione) in rice than in barley (Figure 3C) after 9 h may be associated with the direction of NO towards nitrate formation in the phytoglobin-NO cycle instead of *S*-nitrosylation of glutathione and proteins. *S*-nitrosoglutathione is scavenged by GSNOR, which is the class III alcohol dehydrogenase having also the activity of formaldehyde dehydrogenase [55]. From the obtained values of RSNO levels, it can be concluded that a part of the produced NO goes to *S*-nitrosylation, and this process decreases in barley after radicle protrusion, while in rice seeds that protrude radicles later, *S*-nitrosylation remains stable at 24–48 h due to the increased GSNOR activity (Figure 3D). More efficient NO scavenging in rice may explain the observation that free SH groups are present at a higher level than in barley (Figure 3B).

3.4. Reactive Oxygen Species and Antioxidant System in Seed Embryo during Germination

Imbibition processes that induce ROS formation (mostly H_2O_2) facilitate dormancy decay and promote germination [56,57]. The balance between ROS-producing and ROS-scavenging systems plays a key role in seed germination and dormancy alleviation [21]. The ability of seeds to germinate is linked with the accumulation of a critical level of H_2O_2 [26]. Our results show higher H_2O_2 levels in barley seed embryo than in rice (equaling only at 15 h) and higher superoxide levels at 15–24 h, which correlates to the earlier onset of germination in barley seeds (Figure 4A,B). A lower NO level in barley embryo may be associated with the involvement of ROS in NO scavenging [22]. Higher ROS levels in barley embryos can also explain the increased values of cell damages in terms of lipid peroxidation and electrolyte leakages (Figure 4C,D). ROS are scavenged by the efficient antioxidant systems [29,30,33], which are activated immediately upon rehydration [31,32]. Lower ROS levels in rice embryos can be explained by higher activities of most antioxidant enzymes (except catalase) (Figure 5). Our results demonstrate that in rice embryos, APX, having higher affinity to H_2O_2 [58], is more involved in its scavenging than catalase as compared to barley. While in our study generally lower ROS levels correspond to a slower germination in rice than in barley, this correlation is comparable to the postulation of "oxidative window" for dry seeds resulting in seed dormancy decay and aging [20,59]. Germinating rice seeds are characterized by a lower level of cell damage, higher fermentation and NO turnover rates and higher ATP/ADP ratios. This results in the germination process in rice being less susceptible to stress factors as compared to barley, including the ability to germinate anaerobically.

4. Materials and Methods

4.1. Seed Germination and Isolation of Embryos

Barley (*Hordeum vulgare* L. cv. Harrington) and rice (*Oryza sativa* L. ssp. *indica*, cv. FR13A) seeds were surface sterilized with 10% NaOCl and washed three times with autoclaved distilled water. Seeds were soaked in sterile deionized water on filter papers in Petri dishes at 25 °C in darkness. The germination rate (total seeds germinated at end of trial/number of initial seeds ×100%) was calculated according to Al-Mudaris [60].

We isolated fresh embryos of imbibed seeds at different hours following imbibition (3, 9, 15, 24, 48 h), froze them in liquid nitrogen and stored at −80 °C for further analysis to investigate the biochemical changes during germination within an extensive time course, from dry seeds to radicle protrusion.

All chemicals, unless indicated otherwise, were obtained from Sigma–Aldrich, St. Louis, MO, USA.

4.2. Alcohol Dehydrogenase and Adenylate Ratios

Alcohol dehydrogenase (ADH; EC 1.1.1.1) was measured according to Blandino et al. [61] in the direction of ethanol to acetaldehyde in 50 mM Tris-HCl buffer, pH 8.0, 150 mM ethanol and 2 mM NAD^+ at 340 nm (ε = 6.22 mM^{-1} cm^{-1}).

ATP and ADP were extracted according to Joshi et al. [62] and Yuroff et al. [63] with minor modifications. The tissue powder was lysed on ice in 2.4 M perchloric acid for 60 min and centrifuged for 5 min at 20,000× *g* at 4 °C. The supernatant was neutralized with 4 M KOH, and the ratio ATP/ADP in the neutralized solution was determined according to the manufacturer's instructions using the EnzyLight^{TM} ADP/ATP Ratio Bioluminescent Assay Kit (ELDT-100) (BioAssay Systems, Hayward, CA, USA). The content of ATP and ADP was determined using ATP and ADP standards.

4.3. Nitric Oxide, Free Thiols, S-Nitrosylation and S-Nitrosoglutathione Reductase

NO levels in seed embryos were measured by the hemoglobin (Hb) method at 415 nm (ε = 131 mM^{-1} cm^{-1}), as described by Murphy and Noack [64] and Ma et al. [35]. R-SH and RSNO were measured spectrophotometrically by reducing RSNO to R-SH in the presence of ascorbate and then assaying free thiol groups using 5,5′-dithio-*bis* (2-nitrobenzoic acid) (DTNB) at 412 nm [35,65]. *S*-nitrosoglutatione reductase (GSNOR) activity was measured at 340 nm according to Sakamoto et al. [66]. Total soluble protein content was determined according to Bradford [67] using Bradford reagent (Sigma–Aldrich, St. Louis, MO, USA) and Bovine Serum Albumin (BSA) standard at 595 nm.

4.4. Hydrogen Peroxide, Electrolyte Leakage and Lipid Peroxidation

H_2O_2 content was estimated according to Velikova et al. [68] with modifications. Fresh plant biomass (600 mg) was homogenized in 3 mL 0.1% (*w/v*) TCA and centrifuged at 12,000× *g* for 20 min at 4 °C. Supernatant (300 μL) was mixed with 500 μL of 2 M KI and 200 μL of 10 mM potassium phosphate buffer (pH 7.0). The reaction mixture was incubated for 1 h in the dark at room temperature and the absorbance was recorded at 390 nm. A standard curve was prepared to quantify the H_2O_2 content.

Electrolyte leakage from seeds at different hours of post-imbibition was measured according to Tammela et al. [69] with few modifications as described by Agarie et al. [70]. The seed coat was separated and 0.2 g seeds at different hours of post-imbibition was placed in 10 mL of deionized water. The electrical conductivity of the liquid phase was measured by using a handheld portable EC and TDS meter (E-1 TDS and EC meter, Aquasana Inc., Austin, TX, USA) after keeping the tubes with seeds for 24 h in dark conditions [70].

Lipid peroxidation was measured by tracing malondialdehyde (MDA, ε = 156 mM^{-1} cm^{-1}) content using thiobarbituric acid (TBA) method as described by Heath and Parker [71].

4.5. Superoxide Dismutase, Catalase and Ascorbate-Glutathione Cycle Enzymes

Superoxide dismutase (SOD; EC 1.15.1.1) activity was assayed according to Beauchamp and Fridovich [72] by the inhibition of the photochemical reduction of nitroblue tetrazolium (NBT) at 560 nm. The amount of enzyme causing 50% inhibition of NBT reduction was used to calculate SOD activity. Catalase (EC 1.11.1.6) activity was measured according to the Guilbault [73] using the molar extinction coefficient (ε) for H_2O_2 of 43.1 M^{-1} cm^{-1}.

The enzymes of ascorbate-glutathione cycle were extracted and assayed as described by Ma et al. [35]. Ascorbate peroxidase (APX; EC 1.11.1.11) was estimated in 50 mM potassium phosphate buffer (pH 7.0) containing 0.5 mM sodium ascorbate and sample extract. The reaction was started by adding H_2O_2 (final concentration 1 mM) at 290 nm (ε = 2.8 mM^{-1} cm^{-1}). Monodehydroascorbate reductase (MDHAR; EC 1.6.5.4) activity was measured in 50 mM HEPES-KOH buffer (pH 7.6) containing 2.5 mM ascorbate, 0.25 mM NADH, and the extract. The assay was initiated by adding 0.4 U mL−1 of ascorbate oxidase and the reaction was monitored at 340 nm for 3 min (ε = 6.22 mM^{-1} cm^{-1}). Dehydroascorbate reductase (DHAR; EC 1.8.5.1) activity was measured in 50 mM HEPES-KOH buffer (pH 7.0) containing 0.1 mM EDTA, 2.5 mM GSH, and the extract. The reaction was initiated by adding freshly prepared dehydroascorbate (final concentration 0.8 mM) (ε = 14 mM^{-1} cm^{-1}).

4.6. Statistical Analysis

All data were subjected to one-way ANOVA and Tukey's multiple comparison test by using the Minitab statistical software package (Penn State University Park, State College, PA, USA, 2017). The data in the figures represent the means of three biological repeats and three technical replicates ±SD. Different letters indicate significant differences between the two species and the time points at $p < 0.05$.

5. Conclusions

Rice and barley belong to the same plant family but significantly differ in their metabolic processes activity, driving the subsequent germination steps. The embryos of rice seeds possess higher alcohol dehydrogenase activity, indicating more efficient anaerobic fermentation and elevated NO levels corresponding to higher NO turnover rates via the phytoglobin-NO cycle. Both fermentation and NO turnover result in a higher ATP/ADP ratio value in rice embryos prior to radicle protrusion, as compared to barley. The observed changes in seed metabolism following imbibition are due, in particular, to the differences in tolerance to the oxygen deficiency occurring under the seed coat. Radicle protrusion governs the subsequent transition to aerobic metabolism in the embryonic tissue. The balance and the crosstalk between NO, ROS and their scavengers determine the whole germination process under the oxygen depleted conditions of the imbibed seeds and upon aeration after radicle protrusion at the stage of seedling development.

Author Contributions: Conceptualization, A.U.I., N.V.B. and J.J.; methodology, A.U.I., N.V.B., J.J. and M.K.P.S.W.; software, J.J.; validation, A.U.I. and M.K.P.S.W.; formal analysis, J.J.; investigation, A.U.I. and J.J.; resources, A.U.I.; writing—original draft preparation, J.J. and M.K.P.S.W.; writing—review and editing, A.U.I., N.V.B., J.J. and M.K.P.S.W.; supervision, A.U.I. and N.V.B.; funding acquisition, A.U.I. and N.V.B. All authors have read and agreed to the published version of the manuscript.

Funding: This research was funded by the Natural Sciences and Engineering Research Council of Canada (NSERC), grant number RGPIN-2021-02945 (to A.U.I.), and by Agriculture and Agri-Food Canada, AAFC Project J-002600 (to N.V.B.).

Institutional Review Board Statement: Not applicable.

Informed Consent Statement: Not applicable.

Data Availability Statement: The datasets generated for this study are available upon request from the corresponding author.

Acknowledgments: The authors express their gratitude to Mihiran M. Upeksha, Somaieh Zafari and Juran C. Goyali for their support of this study.

Conflicts of Interest: The authors declare no conflict of interest regarding the publication of this paper.

References

1. Hockett, E.A.; Nilan, R.A. Genetics. In *Barley*; Rasmusson, D.C., Ed.; American Society of Agronomy: Madison, WI, USA, 1985; pp. 187–230.
2. He, D.; Wang, Q.; Wang, K.; Yang, P. Genome-Wide Dissection of the MicroRNA Expression Profile in Rice Embryo during Early Stages of Seed Germination. *PLoS ONE* **2015**, *10*, e0145424. [CrossRef]
3. Perata, P.; Guglielminetti, L.; Alpi, A. Mobilization of endosperm reserves in cereal under anoxia. *Ann. Bot.* **1997**, *79*, 49–56. [CrossRef]
4. Bewley, J.D. Seed germination and dormancy. *Plant Cell* **1997**, *9*, 1055–1066. [CrossRef] [PubMed]
5. Hasanuzzaman, M.; Kamrun, N.; Masayuki, F. Plant response to Salt Stress and Role of Exogenous Protectants to Mitigate Salt-Induced Damages. In *Ecophysiology and Responses of Plant under Salt Stress*; Ahmad, P., Azooz, M.M., Prasad, M.N.V., Eds.; Springer: New York, NY, USA, 2013; pp. 25–87. [CrossRef]
6. Bewley, J.D.; Bradford, K.J.; Hilhorst, H.W.M.; Nonogaki, H. *Seeds: Physiology of Development, Germination and Dormancy*; Springer: New York, NY, USA, 2013. [CrossRef]
7. Mwami, B.M.; Nguluu, S.N.; Kimiti, J.M.; Kimatu, J.N. Effects of water imbibition of selected bean varieties on germination. *Int. J. Innov. Sci. Res. Technol.* **2017**, *3*, 57–62.
8. Bewley, J.D.; Black, M. *Physiology of Development and Germination*, 2nd ed.; Plenum Press: New York, NY, USA, 1994; 445 p.
9. Nonogaki, H. Seed germination and dormancy: The classic story, new puzzles, and evolution. *J. Integr. Plant Biol.* **2019**, *61*, 541–563. [CrossRef] [PubMed]
10. Nutbeam, A.R.; Duffus, C.M. Oxygen exchange in the pericarp green layer of immature cereal grains. *Plant Physiol.* **1978**, *62*, 360–362. [CrossRef] [PubMed]
11. Borisjuk, L.; Rolletschek, H. The oxygen status of the developing seed. *New Phytol.* **2009**, *182*, 17–30. [CrossRef]
12. Larson, L.A. The effect of soaking pea seeds with or without seedcoats has on seedling growth. *Plant Physiol.* **1968**, *43*, 255–259. [CrossRef] [PubMed]
13. Bykova, N.V.; Hoehn, B.; Rampitsch, C.; Hu, J.; Stebbing, J.A.; Knox, R. Thiol redox-sensitive seed proteome in dormant and non-dormant hybrid genotypes of wheat. *Phytochemistry* **2011**, *72*, 1162–1172. [CrossRef]
14. Davies, D.D. Anaerobic metabolism and the production of organic acids. In *The Biochemistry of Plants*; Davies, D.D., Ed.; Academic Press: London, UK, 1980; Volume 2, pp. 581–611.
15. Kennedy, R.A.; Rumpho, M.E.; Fox, T.C. Anaerobic metabolism in plants. *Plant Physiol.* **1992**, *100*, 1–6. [CrossRef]
16. Gibbs, J.; Greenway, H. Mechanisms of anoxia tolerance in plants. I. Growth, survival and anaerobic catabolism. *Funct. Plant Biol.* **2003**, *30*, 999–1036. [CrossRef]
17. Duff, S.M.G.; Guy, P.A.; Nie, X.; Durnin, D.C.; Hill, R.D. Hemoglobin expression in germinating barley. *Seed Sci. Res.* **1998**, *8*, 431–436. [CrossRef]
18. Logan, D.C.; Millar, A.H.; Sweetlove, L.J.; Hill, S.A.; Leaver, C.J. Mitochondrial biogenesis during germination in maize embryos. *Plant Physiol.* **2001**, *125*, 662–672. [CrossRef] [PubMed]
19. Bethke, P.C.; Gubler, F.; Jacobsen, J.V.; Jones, R.L. Dormancy of *Arabidopsis* seeds barley grains can be broken by nitric oxide. *Planta* **2004**, *219*, 847–855. [CrossRef] [PubMed]
20. Bailly, C.; El-Maarouf-Bouteau, H.; Corbineau, F. From intracellular signaling networks to cell death: The dual role of reactive oxygen species in seed physiology. *C. R. Biol.* **2008**, *331*, 806–814. [CrossRef] [PubMed]
21. Oracz, K.; El-Maarouf-Bouteau, H.; Kranner, I.; Bogatek, R.; Corbineau, F.; Bailly, C. The mechanisms involved in seed dormancy alleviation by hydrogen cyanide unravel the role of reactive oxygen species as key factors of cellular signaling during germination. *Plant Physiol.* **2009**, *150*, 494–505. [CrossRef] [PubMed]
22. Wulff, A.; Oliveira, H.C.; Saviani, E.E.; Salgado, I. Nitrite reduction and superoxide-dependent nitricoxidedegradation by Arabidopsis mitochondria: Influence of external NAD(P)H dehydrogenases and alternative oxidase in the control of nitric oxide levels. *Nitric Oxide* **2009**, *21*, 132–139. [CrossRef]
23. Tai, L.; Wang, H.J.; Xu, X.J.; Sun, W.H.; Ju, L.; Liu, W.T.; Li, W.Q.; Sun, J.; Chen, K.M. Cereal pre-harvest sprouting: A global agricultural disaster regulated by complex genetic and biochemical mechanisms. *J. Exp. Bot.* **2021**, *72*, 2857–2876. [CrossRef]
24. Kranner, I.; Minibayeva, F.V.; Beckett, R.P.; Seal, C.E. What is stress? Concepts, definitions and applications in seed science. *New Phytol.* **2010**, *188*, 655–673. [CrossRef]
25. Bykova, N.V.; Hoehn, B.; Rampitsch, C.; Bykova, N.V.; Hoehn, B.; Rampitsch, C.; Banks, T.; Stebbing, J.-A.; Fan, T.; alKnox, R. Redox-sensitive proteome and antioxidant strategies in wheat seed dormancy control. *Proteomics* **2011**, *11*, 865–882. [CrossRef]
26. Bykova, N.V.; Hu, J.; Ma, Z.; Igamberdiev, A.U. The Role of Reactive Oxygen and Nitrogen Species in Bioenergetics, Metabolism and Signaling during Seed Germination. In *Reactive Oxygen and Reactive Nitrogen Species Signaling and Communication in Plants*; Gupta, K.J., Igamberdiev, A.U., Eds.; Springer: Berlin, Germany, 2015; pp. 177–195. [CrossRef]

27. Debska, K.; Krasuska, U.; Budnicka, K.; Bogatek, R.; Gniazdowska, A. Dormancy removal of apple seeds by cold stratification is Associated with fluctuation in H_2O_2, NO production and protein carbonylation level. *J. Plant Physiol.* **2013**, *170*, 480–488. [CrossRef] [PubMed]
28. Colville, L.; Kranner, I. Desiccation tolerant plants as model systems to study redox regulation of protein thiols. *Plant Growth Regul.* **2010**, *62*, 241–255. [CrossRef]
29. Davies, M.J.; Fu, S.; Wang, H.; Dean, R.T. Stable markers of oxidant damage to proteins and their application in study of human disease. *Free Radic. Biol. Med.* **1999**, *27*, 1151–1161. [CrossRef]
30. Møller, I.M.; Jensen, P.E.; Hansson, A. Oxidative modifications to cellular components in plants. *Annu Rev Plant Biol.* **2007**, *58*, 459–481. [CrossRef] [PubMed]
31. De Gara, L.; de Pinto, M.C.; Arrigoni, O. Ascorbate synthesis and ascorbate peroxidase activity during the early stage of wheat germination. *Physiol Plant.* **1997**, *100*, 894–900. [CrossRef]
32. Tommasi, F.; Paciolla, C.; de Pinto, M.C.; De Gara, L. A comparative study of glutathione and ascorbate metabolism during germination of *Pinus Pinea* L. seeds. *J. Exp Bot.* **2001**, *52*, 1647–1654. [CrossRef]
33. Müller, K.; Job, C.; Belghazi, M.; Job, D.; Leubner-Metzger, G. Proteomics reveal tissue-specific features of the cress (*Lepidium sativum* L.) endosperm cap proteome and its hormone-induced changes during seed germination. *Proteomics* **2010**, *10*, 406–416. [CrossRef]
34. Ma, Z.; Marsolais, F.; Bernards, M.A.; Sumarah, M.W.; Bykova, N.V.; Igamberdiev, A.U. Glyoxylate cycle and metabolism of organic acids in the scutellum of barley seeds during germination. *Plant Sci.* **2016**, *248*, 37–44. [CrossRef] [PubMed]
35. Ma, Z.; Marsolais, F.; Bykova, N.V.; Igamberdiev, A.U. Nitric Oxide and Reactive Oxygen Species Mediate Metabolic Changes in Barley Seed Embryo during Germination. *Front. Plant Sci.* **2016**, *7*, 138. [CrossRef]
36. Nonogaki, H.; Bassel, G.W.; Bewley, J.D. Germination–stillamystery. *Plant Science* **2010**, *179*, 574–581. [CrossRef]
37. Gruwel, M.L.H.; Chatson, B.; Yin, X.S.; Abrams, S.R. A magnetic resonance study of water uptake of whole barley kernels. *Int. J. Food Sci. Technol.* **2001**, *36*, 161–168. [CrossRef]
38. Yang, P.; Li, X.; Wang, X.; Chen, H.; Chen, F.; Shen, S. Proteomic analysis of rice (*Oryza sativa*) seeds during germination. *Proteomics* **2007**, *7*, 3358–3368. [CrossRef]
39. Geigenberger, P. Response of plant metabolism to too little oxygen. *Curr. Opin. Plant Biol.* **2013**, *6*, 247–256. [CrossRef]
40. Rolletschek, H.; Weschke, W.; Weber, H.; Wobus, U.; Borisjuk, L. Energy state and its control on seed development: Starch accumulation is associated with high ATP and steep oxygen gradients within barley grains. *J. Exp. Bot.* **2004**, *55*, 1351–1359. [CrossRef] [PubMed]
41. Weber, H.; Borisjuk, L.; Wobus, U. Molecular physiology of legume seed development. *Annu. Rev. Plant Biol.* **2005**, *56*, 253–279. [CrossRef]
42. He, D.; Yang, P. Proteomics of rice seed germination. *Front. Plant Sci.* **2013**, *4*, 246. [CrossRef] [PubMed]
43. Benamar, A.; Tallon, C.; Macherel, D. Membrane integrity and oxidative properties of mitochondria isolated from imbibing pea seeds after priming or accelerated ageing. *Seed Sci. Res.* **2003**, *13*, 35–45. [CrossRef]
44. He, D.; Han, C.; Yang, P. Gene expression profile changes in germinating rice. *J. Integr. Plant Biol.* **2011**, *53*, 835–844. [CrossRef]
45. Igamberdiev, A.U.; Hill, R.D. Nitrate, NO and haemoglobin in plant adaptation to hypoxia: An alternative to classic fermentation pathways. *J. Exp. Bot.* **2004**, *55*, 2473–2482. [CrossRef]
46. Zafari, S.; Hebelstrup, K.H.; Igamberdiev, A.U. Transcriptional and metabolic changes associated with phytoglobin expression during germination of barley seeds. *Int. J. Mol. Sci.* **2020**, *21*, 2796. [CrossRef]
47. Cakmak, I.; Strbac, D.; Marschner, H. Activities of hydrogen peroxide scavenging enzymes in germinating wheat seeds. *J. Exp. Bot.* **1993**, *44*, 127–132. [CrossRef]
48. Caliskan, M.; Cuming, A.C. Spatial specificity of H_2O_2-generating oxalate oxidase gene expression during wheat embryo germination. *Plant J.* **1998**, *15*, 165–171. [CrossRef] [PubMed]
49. Wojtyla, Ł.; Garnczarska, M.; Zalewski, T.; Bednarski, W.; Ratajczak, L.; Jurga, S. A comparative study of water distribution, free radical production and activation of antioxidative metabolism in germinating pea seeds. *J. Plant Physiol.* **2006**, *163*, 1207–1220. [CrossRef] [PubMed]
50. Gupta, K.J.; Igamberdiev, A.U. The anoxic plant mitochondrion as a nitrite: NO reductase. *Mitochondrion* **2011**, *11*, 537–543. [CrossRef]
51. Nie, X.Z.; Hill, R.D. Mitochondrial respiration and hemoglobin gene expression in barley aleurone tissue. *Plant Physiol.* **1997**, *114*, 835–840. [CrossRef]
52. Hunt, P.W.; Watts, R.A.; Trevaskis, B.; Llewelyn, D.J.; Burnell, J.; Dennis, E.S.; Peacock, W.J. Expression and evolution of functionally distinct hemoglobin genes in plants. *Plant Mol. Biol.* **2001**, *47*, 677–692. [CrossRef]
53. Ross, E.J.H.; Shearman, L.; Mathiesen, M.; Zhou, Y.J.; Arredondo-Peter, R.; Sarath, G.; Klucas, R.V. Nonsymbiotic hemoglobins in rice are synthesized during germination and in differentiating cell types. *Protoplasma* **2001**, *218*, 125–133. [CrossRef]
54. Kumari, A.; Singh, P.; Kaladhar, V.C.; Bhatoee, M.; Paul, D.; Pathak, P.K.; Gupta, K.J. Phytoglobin-NO cycle and AOX pathway play a role in anaerobic germination and growth of deepwater rice. *Plant Cell Environ.* **2021**. [CrossRef]
55. Corpas, F.J.; Alché, J.D.; Barroso, J.B. Current overview of S-nitrosoglutathione (GSNO) in higher plants. *Front. Plant Sci.* **2013**, *4*, 126. [CrossRef]

56. El-Maarouf-Bouteau, H.; Bailly, C. Oxidative signaling in seed germination and dormancy. *Plant Signal. Behav.* **2008**, *3*, 175–182. [CrossRef] [PubMed]
57. Kubala, S.; Wojtyla, Ł.; Quinet, M.; Lechowska, K.; Lutts, S.; Garnczarska, M. Enhanced expression of the proline synthesis gene P5CSA in relations to seed osmopriming improvement of *Brassica napus* germination under salinity stress. *J. Plant Physiol.* **2015**, *183*, 1–12. [CrossRef] [PubMed]
58. Anjum, N.A.; Sharma, P.; Gill, S.; Hasanuzzaman, M.; Khan, E.A.; Kachhap, K.; Mohamed, A.; Thangavel, P.; Devi, G.D.; Vasudhevan, P.; et al. Catalase and ascorbate peroxidase-representative H_2O_2-detoxifying heme enzymes in plants. *Environ. Sci. Pollut. Res. Int.* **2016**, *23*, 19002–19029. [CrossRef] [PubMed]
59. Guha, T.; Das, H.; Mukherjee, A.; Kundu, R. Elucidating ROS signaling networks and physiological changes involved in nanoscale zero valent iron primed rice seed germination sensu stricto. *Free Radic. Biol. Med.* **2021**, *171*, 11–25. [CrossRef]
60. Al-Mudaris, M. Notes on various parameters recording the speed of seed germination. *Der Trop.* **1998**, *99*, 147–154.
61. Blandino, A.; Caro, I.; Cantero, D. Comparative study of alcohol dehydrogenase activity in flor yeast extracts. *Biotechnol. Lett.* **1997**, *19*, 651–654. [CrossRef]
62. Joshi, A.K.; Ahmed, S.; Ames, G.F.L. Energy coupling in bacterial periplasmic transport systems. *J. Biol. Chem.* **1989**, *264*, 2126–2133. [CrossRef]
63. Yuroff, A.S.; Sabat, G.; Hickey, W.J. Transporter-mediated uptake of 2-chloro- and 2-hydroxybenzoate by Pseudomonas huttiensis strain D1. *Appl. Environ. Microbiol.* **2003**, *69*, 7401–7408. [CrossRef]
64. Murphy, M.M.; Noack, E. Nitric oxide assay using hemoglobin method. *Methods Enzymol.* **1994**, *233*, 240–250. [CrossRef]
65. Cochrane, D.W.; Shah, J.K.; Hebelstrup, K.H.; Igamberdiev, A.U. Expression of phytoglobin affects nitric oxide metabolism and energy state of barley plants exposed to anoxia. *Plant Sci.* **2017**, *265*, 124–130. [CrossRef]
66. Sakamoto, M.; Ueda, H.; Morikawa, H. Arabidopsis glutathione-dependent formaldehyde dehydrogenase is an S-nitrosoglutathione reductase. *FEBS Lett.* **2002**, *515*, 20–24. [CrossRef]
67. Bradford, M.M. A rapid and sensitive method for the quantification of microgram quantities of protein utilizing the principle of protein-dye binding. *Anal. Biochem.* **1976**, *72*, 248–254. [CrossRef]
68. Velikova, V.; Yordanov, I.; Edreva, A. Oxidative stress and some antioxidant systems in acid rain-treated bean plants. *Plant Sci.* **2000**, *151*, 59–66. [CrossRef]
69. Tammela, P.; Vaananen, P.S.; Lakso, I.; Hopia, A.; Vourela, H.; Nygrem, M. Tocopherols, tocotrienols and fatty acids as indicators of natural ageing in *Pinus sylvestris* seeds. *Scand. J. Forest Res.* **2005**, *20*, 378–384. [CrossRef]
70. Agarie, S.; Hanaoka, N.; Kubota, F.; Agata, W.; Kaufman, B. Measurement of cell membrane stability evaluated by electrolyte leakage as a drought and heat tolerance test in rice (*Oryza sativa* L.). *J. Fac. Agr. Kyushu Univ.* **1995**, *40*, 233–240. [CrossRef]
71. Heath, R.L.; Parker, L. Photoperoxidation in isolated chloroplasts. I. Kinetics and stoichiometry of fatty acid peroxidation. *Arch Biochem. Biophys.* **1968**, *125*, 189–198. [CrossRef]
72. Beauchamp, C.; Fridovich, I. Superoxide dismutase: Improved assays and assay applicable to acrylamide gels. *Anal. Biochem.* **1971**, *44*, 276–278. [CrossRef]
73. Guilbault, G.G. Glucose Oxidase. In *Handbook of Enzymatic Methods of Analysis*; Schwartz, M.K., Ed.; Marcel Dekker: New York, NY, USA, 1976; pp. 106–109.

MDPI
St. Alban-Anlage 66
4052 Basel
Switzerland
www.mdpi.com

Plants Editorial Office
E-mail: plants@mdpi.com
www.mdpi.com/journal/plants

www.ingramcontent.com/pod-product-compliance
Lightning Source LLC
Chambersburg PA
CBHW040835250926
51850CB00007BA/1316

* 9 7 8 3 0 3 6 5 9 2 5 5 8 *